ELECTRON CORRELATION IN MOLECULES

S. WILSON

Department of Theoretical Chemistry
University of Oxford

DOVER PUBLICATIONS, INC.
Mineola, New York

Bibliographical Note

This Dover edition, first published in 2007, is an unabridged and slightly corrected republication of the work originally published by Clarendon Press, Oxford, in 1984 as Volume 11 of the International Series of Monographs on Chemistry. An errata list has been added to the present edition on page xii.

Library of Congress Cataloging-in-Publication Data

Wilson, S. (Stephen), 1950–
Electron correlation in molecules / S. Wilson. — Dover ed.
p. cm.
Originally published: Oxford : Clarendon Press, 1984, in series: International series of monographs on chemistry ; v. 11.
Includes index.
ISBN 0-486-45879-2 (pbk.)
1. Molecular theory. 2. Electron configuration. I. Title.

QD461.W6 2007
541'.22—dc22

2006102935

Manufactured in the United States of America
Dover Publications, Inc., 31 East 2nd Street, Mineola, N.Y. 11501

To Kati

PREFACE

The purpose of this book is to describe methods for the study of electron correlation effects in molecules the accurate description of which has long been recognized to be one of the central problems of theoretical chemistry. Electron correlation effects are known to be important in, for example, the calculation of potential energy curves and surfaces, the study of molecular excitation processes, and in the theory of electron-molecule scattering. Although the energy associated with electron correlation is a small fraction of the total energy of an atom or molecule, it is of the same order of magnitude as most energies of chemical interest. Accurate techniques for the theoretical determination of the effect of electron correlation on molecular properties are therefore vital if solution of the quantum mechanical equations from first principles is to provide a route to quantitative prediction of an accuracy comparable with experiment. There has been significant progress on the problem of correlation in molecules in recent years and the outlook for future developments looks very promising.

Finstock, Oxford S. W.
June 1982

ACKNOWLEDGEMENTS

I am happy to acknowledge the assistance given to me by Professor J. S. Rowlinson, editor of The International Series of Monographs on Chemistry, and by the staff of Oxford University Press.

CONTENTS

NOTE ON ATOMIC UNITS

Atomic units, which are used throughout this book, are the natural units in which to carry out molecular calculations. The basic atomic units are given in Table 1. Some commonly used derived atomic units are given in Table 2.

Table 1

Basic atomic units

Quantity	Atomic unit	Value	SI	cgs
mass	mass of the electron, m_e	9.109 534 (47)	10^{-31} kg	10^{-28} g
length	Bohr radius, a_0	5.291 770 6 (44)	10^{-11} m	10^{-09} cm
angular momentum	$\hbar = h/2\pi$	1.054 588 7 (57)	10^{-34} J s	10^{-27} erg s
charge	electronic charge, e	1.602 189 2 (46)	10^{-19} C	10^{-20} emu
		4.803 242 (14)		10^{-10} esu

Taken from Cohen and Taylor (1973). Uncertainties in the last digits are given in parenthesis.

Table 2

Derived atomic units

Quantity	Value
Energy	1 a.u. = 4.3598×10^{-18} J
Time	1 a.u. = 2.4189×10^{-17} s
Electric dipole moment	1 a.u. = 8.4784×10^{-30} C m
Electric quadrupole moment	1 a.u. = 4.4866×10^{-40} C m^2
Electric octopole moment	1 a.u. = 2.3742×10^{-50} C m^3
Electric field	1 a.u. = 5.1423×10^{11} V m^{-1}
Electric field gradient	1 a.u. = 9.7174×10^{21} V m^{-2}
Polarizability (dipole)	1 a.u. = 1.6488×10^{-41} C^2 m^2 J^{-1}
Hyperpolarizability	1 a.u. = 3.2063×10^{-53} C^3 m^3 J^{-2}
Magnetic moment	1 a.u. = 1.8548×10^{-23} J T^{-1}
Magnetizability	1 a.u. = 7.8910×10^{-29} J T^{-2}
Magnetic vector potential	1 a.u. = 1.2439×10^{-5} m T
Force constant (harmonic)	1 a.u. = 1.5569×10^{3} J m^{-2}
Force constant (cubic)	1 a.u. = 2.9421×10^{13} J m^{-3}
Force constant (quartic)	1 a.u. = 5.5598×10^{23} J m^{-4}
Probability density	1 a.u. = 6.7483×10^{30} m^{-3}

Based on the work of Whiffen (1978), Cohen and Taylor (1973).

Errata

page 113, line 4:

$$\langle \Phi_i | \Lambda | \Phi_j \rangle \text{ should be replaced by } \langle \Phi_i | \Lambda | \Phi_i \rangle$$

page 123, equation (5.2.3):

$$\psi_\ell \left(\Gamma^{(\ell)} \right\rangle \text{ should be replaced by } \psi_\ell \left(\Gamma^{(\ell)} \right) \Big\rangle$$

$$\phi_s \left(\Gamma^{(\ell)} \right) . \Big\rangle \text{ should be replaced by } \phi_s \left(\Gamma^{(\ell)} \right) \Big\rangle$$

page 140, Figure 5.7:
Note that only the upper triangle of this symmetric matrix is shown.

page 175, equation (3.3.4):

$$\langle \Phi | \mathcal{H} | \Phi_I \rangle + \langle \Phi_{II} | \mathcal{H} | \chi \rangle \text{ should be replaced by } \langle \Phi | \mathcal{H} | \Phi \rangle + \langle \Phi | \mathcal{H} | \chi \rangle$$

page 175, equation (6.4.1):

Φ_1 should be replaced by Φ_I

ϕ_{II} should be replaced by Φ_{II}

page 201, Figure 7.1(b):
The final row should read

$$K^{(7)} \vdots \; \vdots \; \vdots \; \vdots \; \vdots \; \vdots 8 \vdots 8 \vdots 9$$

page 224, Figure 7.4:

T_2 should be replaced by Ω_2

V should be replaced by $\mathcal{H}$

Please consult
http://quantumsystems.googlepages.com/electroncorrelationinmolecules
for updates

1

MOLECULAR ELECTRONIC STRUCTURE

1.1 Introduction

The calculation of properties of atoms and molecules from first principles has long been recognized to be of central importance in chemistry and in atomic and molecular physics. Theoretical studies enable the properties of species, which are difficult or indeed impossible to examine experimentally, to be determined by approximate solution of the appropriate quantum mechanical equations. Many compounds are too reactive to be isolated and cannot be studied by means of standard laboratory techniques, such as X-ray crystallography or infra-red spectroscopy. The dipole moment of a charged species cannot be measured experimentally and that of a free radical only with difficulty. However, such molecular properties can often be calculated quite accurately from first principles. (Calculated dipole moments do depend, of course, on the choice of origin.) Radicals and ions have been identified in the interstellar medium on the basis of their calculated properties (Wilson 1980*d*). These species have a very short lifetime in the terrestrial laboratory whereas in space, since collisions are infrequent, their lifetimes are much longer. Some chemical reactions take place in times of the order of a picosecond and experimentally the transition states can only be inferred indirectly by, for example, the effects of substituents on the reaction rate. Theoretical studies can provide valuable information about transition states.

In addition to providing a route to data for a particular molecule or reaction which are not available from experimental studies, theoretical chemistry can also lead to simple and useful chemical concepts which can then be used to rationalize vast quantities of data. For example, orbital theories have led to important models in chemical spectroscopy such as Koopmans' theorem (Koopmans 1933) and in structural chemistry such as lone pairs and bonding pairs (Gillespie and Nyholm 1957). The importance of this interpretive aspect of theoretical chemistry was emphasized by C. A. Coulson (1973) when he remarked "upon the futility of obtaining accurate numbers, whether by computation or by experiment, unless these numbers can provide us with simple and useful chemical concepts; otherwise one might as well be interested in a telephone directory". Theories of electron correlation in molecules must not only afford high accuracy but must also be open to simple qualitative interpretation.

The use of a theoretical rather than an experimental approach to the determination of the properties of molecules can increasingly be justified on economic grounds. Present-day computer programs are usually very flexible and the treatment of a different state or system or different properties of a molecule merely requires a change of the input data. For example, in the field of quantum pharmacology (Richards 1983) properties of molecules which are believed to have potential applications as drugs are calculated, thereby possibly avoiding the painstaking preparation and testing of hundreds of similar molecules before the required pharmacological activity can be found.

The first step in most approaches to the electronic structure of molecules is to decompose the N-electron problem into N one-electron problems. This leads to what are known as independent electron models, or, more simply, orbital models. Orbital models, such as the Hartree–Fock approximation, account for the majority, typically 99.5%, of the total energy of an atom or a molecule. It is unfortunate that the error in such orbital theories is of the same order of magnitude as most energies of chemical interest: binding energies, ionization energies, activation barriers, and the like. This error arises from the correlation of the motions of the individual electrons and this most crucial area is the topic to which this book is addressed.

Of course, the nature of electron correlation effects will depend to a large extent on the particular orbital model with respect to which they are computed. Electron correlation effects are not directly observable; it is only by considering both the results obtained within an orbital model and its correlation corrections that quantities can be obtained which can be compared with data derived from experiment. For this reason Chapter 2 is devoted to independent electron models. In Chapter 3, qualitative and some quantitative aspects of the electron correlation problem in molecules are discussed. Fundamental to modern treatments of electron correlation is the linked diagram theorem and this is addressed in Chapter 4. Powerful methods for the analysis and simplification of molecular correlation calculations can be obtained by employing group theoretical techniques. Group theoretical aspects of the correlation problem are discussed in Chapter 5. In general, molecular calculations can only be rendered tractable by employing the algebraic approximation; that is the orbitals are parameterized by expansion in a finite basis set. The algebraic approximation is discussed in detail in Chapter 6. In Chapter 7, many of the techniques currently being employed in the treatment of electron correlation effects are discussed. These techniques will be analysed in terms of the linked diagram theorem of Chapter 4. and the use of the group theoretical methods described in Chapter 5 will be indicated.

In this first chapter we aim to provide a brief outline of basic molecular physics and quantum chemistry. More detailed discussions can be found elsewhere (e.g. Messiah 1967; Ziman 1969; McWeeny and Sutcliffe 1976; Atkins 1983; Coulson 1961; Dirac 1958; McWeeny 1982; Landau and Lifshitz 1958). In Section 1.2, the Born–Oppenheimer approximation is discussed. This approximation is made in most applications of quantum mechanical methods to molecules. The electronic Schrödinger equation is introduced in Section 1.3. The electronic energy is a function of the nuclear coordinates and gives rise to potential energy curves and surfaces which are the subject of Section 1.4. In Section 1.5, the relationship between quantities which can be calculated and experimental observables is described briefly. The density matrix can provide a useful interpretation of molecular electronic structure and, therefore, this is discussed in Section 1.6. The majority of contemporary quantum mechanical calculations on molecules are performed within the framework of non-relativistic theory. Exact solution of the non-relativistic quantum mechanical equations for molecules leads to the so-called non-relativistic limit and this is the subject of Section 1.7. Finally, relativistic effects are very briefly discussed in Section 1.8.

1.2 The Born–Oppenheimer approximation

The separation of the nuclear and the electronic motion is almost invariably the first step in any application of quantum mechanics to molecules.

For a conservative system, our problem is the solution of the eigenvalue problem

$$\hat{H}\,|\Psi_i\rangle = E_i\,|\Psi_i\rangle \qquad (|\Psi_i\rangle \in \mathscr{h}) \tag{1.2.1}$$

where $\hat{H}$ is the molecular hamiltonian operator, $|\Psi_i\rangle$ is a state function, in Hilbert space $\mathscr{h}$, which completely defines the dynamical state of the system, and E_i is an allowed energy value. Within the framework of non-relativistic quantum mechanics, the total molecular hamiltonian for a system of N electrons and M nuclei takes the form

$$\hat{H} = \hat{T}_{\mathrm{N}} + \hat{T}_{\mathrm{e}} + V_{\mathrm{N}} + V_{\mathrm{e}} + V_{\mathrm{Ne}} \tag{1.2.2}$$

with nuclear kinetic energy

$$\hat{T}_{\mathrm{N}} = -\sum_{\mathrm{A}=1}^{M} \frac{1}{2M_{\mathrm{A}}} \nabla_{\mathrm{A}}^2 \tag{1.2.3}$$

where M_{A} is the mass of nucleus A, electronic kinetic energy

$$\hat{T}_{\mathrm{e}} = -\tfrac{1}{2}\sum_{i=1}^{N} \nabla_i^2 \tag{1.2.4}$$

a nucleus–nucleus repulsion term

$$V_{\mathrm{N}} = \sum_{\mathrm{A}>\mathrm{B}}^{M} \frac{Z_{\mathrm{A}} Z_{\mathrm{B}}}{R_{\mathrm{AB}}} \tag{1.2.5}$$

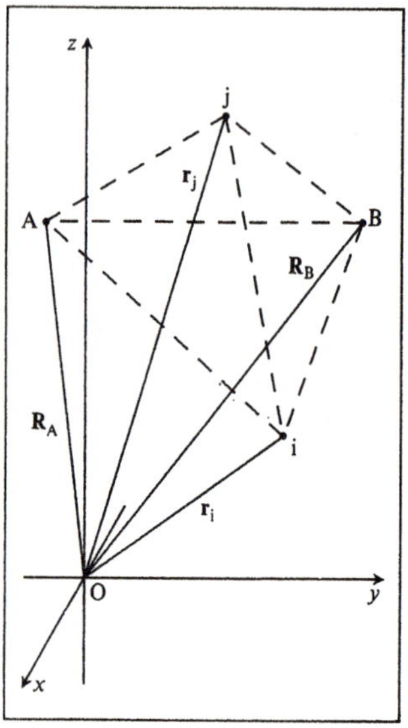

FIG. 1.1. Coordinate system. *A*, *B*, ... denote the nuclei and *i*, *j*, ... denote the electrons.

where Z_A is the charge of nucleus A, the electron–electron repulsion term

$$V_e = \sum_{i>j}^{N} \frac{1}{r_{ij}} \tag{1.2.6}$$

and, finally, an electron–nucleus attraction term

$$V_{Ne} = -\sum_{i=1}^{N} \sum_{A=1}^{M} \frac{Z_A}{r_{iA}}. \tag{1.2.7}$$

Atomic units, which are defined in the Note (p. xi), and the coordinate system defined in Fig. 1.1, have been used throughout.

Equation (1.2.1) is a second-order differential equation in $3(N+M)$ variables. The complexity of the problem can be reduced considerably by invoking the Born–Oppenheimer approximation (Born and Oppenheimer 1927). Physically, it is expected that the disparity of the mass of the electrons and the mass of the nuclei in a molecule will allow the electronic motion to accommodate almost instantaneously any change in the positions of the nuclei. Consequently, the electronic and nuclear motion can be treated separately to a very good approximation. The electrons experience the nuclei as fixed force centres and adiabatically follow any change in the nuclear positions. Conversely, the nuclei experience an average interaction with the electrons.

The nuclear kinetic energy, T_N, is very much smaller than the other terms in eqn (1.2.2). It will be assumed in the present discussion that spin-related effects are small compared with the nuclear kinetic energy. (The reader is referred to the monograph by Kronig (1930) for a discussion of the situation when spin effects are not small.) If, in eqn (1.2.2) $M_A \to \infty$, $\forall_A$, then

$$\hat{H} = \hat{T}_N + \hat{\mathscr{H}} + V_N \to \hat{\mathscr{H}} + V_N \tag{1.2.8}$$

where the electronic hamiltonian operator is given by

$$\hat{\mathscr{H}} = \hat{T}_e + V_e + V_{Ne}. \tag{1.2.9}$$

Since

$$[\hat{\mathscr{H}}, \mathbf{R}] = 0, \tag{1.2.10}$$

the eigenvalues of the electronic hamiltonian may be determined for a particular value of the nuclear position vectors, $\mathbf{R}$. The eigenvalues and eigenfunctions of $\hat{\mathscr{H}}$ are implicit functions of the nuclear geometry; that is

$$(\mathscr{H} + V_N)\,|\Phi_m(\mathbf{R})\rangle = E_m(\mathbf{R})\,|\Phi_m(\mathbf{R})\rangle \tag{1.2.11}$$

where $\mathbf{R}$ has been used to denote the nuclear coordinates. The term arising from repulsions between the nuclei may be treated as a simple additive constant. Equation (1.2.11) is the electronic Schrödinger equation; the problem of describing the electronic motion accurately is the topic of this monograph.

In the remaining part of this section, the problem of describing the nuclear motion in molecules will be briefly addressed. The electronic state functions, $\Phi_m(\mathbf{r}_1\sigma_1, \mathbf{r}_2\sigma_2, \ldots, \mathbf{r}_N\sigma_N; \mathbf{R})$, form a complete orthonormal set and can, therefore, be used to expand the total molecular wave function, for both the electrons and the nuclei, as follows

$$\Psi(\mathbf{r}_1\sigma_1, \ldots, \mathbf{r}_N\sigma_N, \mathbf{R}) = \sum_m \chi_m(\mathbf{R})\Phi_m(\mathbf{r}_1\sigma_1, \ldots, \mathbf{r}_N\sigma_N; \mathbf{R}) \tag{1.2.12}$$

where σ_i denotes the spin function for the ith electron in the system. The Schrödinger equation for the complete system, eqn (1.2.1), may then be written as

$$\sum_m \{(\hat{\mathscr{H}} + \hat{T}_N + V_N)\chi_m\Phi_m - E\chi_m\Phi_m\} = 0 \tag{1.2.13}$$

which may be rewritten as

$$\sum_m \{\chi_m E_m(\mathbf{R})\Phi_m + \hat{T}_N(\chi_m\Phi_m) - E(\chi_m\Phi_m)\} = 0. \tag{1.2.14}$$

Multiplying eqn (1.2.14) from the left by Φ_n^* and integrating over the coordinates of all the electrons leads to

$$(E_n(\mathbf{R})-E)\chi_n(\mathbf{R})+\sum_m \langle\Phi_n \mid \hat{T}_{\mathrm{N}}(\chi_m\Phi_m)\rangle=0 \tag{1.2.15}$$

It can be readily demonstrated that

$$\hat{T}_{\mathrm{N}}(\chi_m\Phi_m)=(\hat{T}_{\mathrm{N}}\chi_m)\Phi_m -\sum_{\mathrm{A}}\frac{1}{2M_{\mathrm{A}}}\{2(\nabla_{\mathrm{A}}\chi_m)(\nabla_{\mathrm{A}}\Phi_m)+(\nabla_{\mathrm{A}}^2\Phi_m)\chi_m\} \tag{1.2.16}$$

and hence

$$\sum_m \langle\Phi_n \mid \hat{T}_{\mathrm{N}}(\chi_m\Phi_m)\rangle=\hat{T}_{\mathrm{N}}\chi_n -\sum_m\left\{\sum_{\mathrm{A}}\frac{1}{M_{\mathrm{A}}}(\langle\Phi_n|\,\nabla_{\mathrm{A}}\,|\Phi_m\rangle\nabla_{\mathrm{A}})+\frac{1}{2M_{\mathrm{A}}}\langle\Phi_n|\,\nabla_{\mathrm{A}}^2\,|\Phi_m\rangle\right\}\chi_m. \tag{1.2.17}$$

Note that the second term on the right-hand side of this equation is zero when $m=n$. We have

$$\sum_m \langle\Phi_n|\,\hat{T}_{\mathrm{N}}\,|\chi_m\Phi_m\rangle=\hat{T}_{\mathrm{N}}\chi_m+\sum_m C_{mn}(\mathbf{R},\nabla)\chi_m(\mathbf{R}) \tag{1.2.18}$$

where

$$C_{mn}=-\sum_{\mathrm{A}}\frac{1}{M_{\mathrm{A}}}(X_{mn}^{(\mathrm{A})}\nabla_{\mathrm{A}}+Y_{mn}^{(\mathrm{A})}) \tag{1.2.19}$$

$$X_{mn}^{(\mathrm{A})}=\langle\Phi_n|\,\nabla_{\mathrm{A}}\,|\Phi_m\rangle \tag{1.2.20}$$

and

$$Y_{mn}^{(\mathrm{A})}=\langle\Phi_n|\,\tfrac{1}{2}\nabla_{\mathrm{A}}^2\,|\Phi_m\rangle. \tag{1.2.21}$$

Substituting eqn (1.2.18), together with eqns (1.2.19–21), into eqn (1.2.15), we obtain

$$(\hat{T}_{\mathrm{N}}+E_{\mathrm{N}n}(\mathbf{R})-E)\chi_n+\sum_{m\neq n} C_{mn}\chi_m=0 \tag{1.2.22}$$

where

$$E_{\mathrm{N}n}(\mathbf{R})=E_n(\mathbf{R})-\sum_{\mathrm{A}}\frac{1}{M_{\mathrm{A}}}Y_{nn}^{(\mathrm{A})}. \tag{1.2.23}$$

In the adiabatic approximation (see, for example, Kolos 1970), it is assumed that the Cs can be neglected giving

$$(\hat{T}_{\mathrm{N}}+E_{\mathrm{N}n}(\mathbf{R}))\chi_n(\mathbf{R})=E_n\chi_n(\mathbf{R}) \tag{1.2.24}$$

or

$$\mathcal{H}_{\mathrm{N}}\chi_n = E_n\chi_n \tag{1.2.25}$$

which is the nuclear eigenvalue equation in which the $E_{\mathrm{N}n}$ is an effective potential for the nuclei. In the Born–Oppenheimer approximation (Born and Oppenheimer 1927; Born and Huang 1954) in addition to taking the Cs to be zero it is assumed that the effective potential for the nuclei is equal to the electronic energy function, that is $E_{\mathrm{N}n}(\mathbf{R}) = E_n(\mathbf{R})$.

It should be noted that the separation of the nuclear and electronic motion described above is only valid as long as there is no significant coupling between different electronic states induced by the nuclear motion. The Born–Oppenheimer approximation is valid only if the energy levels are not degenerate and there are no nearby levels of the same symmetry.

The Born–Oppenheimer approximation is found to be excellent in practice. Only in the most accurate quantum mechanical treatments of molecules containing light atoms are errors arising from this approximation significant (see, for example, the calculations on the hydrogen molecule by Kolos *et al.* 1960, 1963, 1964).

1.3 Molecular electronic structure

The electronic structure of a molecule is defined, within the Born–Oppenheimer approximation, as the solution of the Schrödinger equation for the motion of the electrons in the field of fixed nuclei

$$\mathcal{H}\,|\Phi_m(\mathbf{R})\rangle = E_m(\mathbf{R})\,|\Phi_m(\mathbf{R})\rangle \tag{1.2.11}$$

Although the Born–Oppenheimer approximation simplifies the quantum mechanics of molecules considerably, further approximations are necessary in order to develop theoretical machinery which leads to computationally tractable algorithms. This monograph is concerned with development of methods for performing accurate calculations of the electronic structure of molecules, however, it is important to remember that any theory, no matter how sophisticated, will contain errors, although for useful models these errors will be small.

Molecular calculations are usually performed for one of two reasons:

(i) Testing a new theory or algorithm;
(ii) Complementing, verifying, or interpreting information derived from experiment about a particular molecular system.

Most of the molecular calculations which will be considered in this monograph are concerned with the former aspect; the demonstration of various properties of different approaches to the correlation problem in molecules. However, in this section, we wish to emphasize that some care must be exercised in employing any theory, but particularly a recently developed theory, in studies of molecules performed for the second of the two reasons given above.

In practical applications of the methods of theoretical chemistry there is much advantage in studying a wide range of problems at a uniform level of approximation (see, for example, applications of quantum chemical methods to the study of interstellar molecules (Wilson 1980*d*)). The results of all calculations at one level of approximation constitute what Pople (1973) has termed a 'theoretical model chemistry'. The effectiveness of any model may be evaluated by comparing some of its details with data derived from experiment when it is available. If the results of such a comparison are favourable, the model acquires some predictive credibility and can be used to study molecules for which no experimental data is available.

To qualify as a satisfactory theoretical model chemistry, a method should ideally satisfy a number of conditions and requirements:

(i) It should be 'conventional'. The method should have been applied to a wide range of molecular systems and its strengths and weaknesses should be well documented. Thus when the method is applied to a molecular system, for which experimental information is not available, the likely accuracy of the results can be inferred and empirical corrections to the calculated results can sometimes be made.

(ii) It should provide well-defined results for the energy of any electronic state for any arrangement of fixed nuclei. The method will then afford a set of continuous potential energy curves or surfaces.

(iii) It should be such that the amount of computation is not too large and should not increase too rapidly with the number of electrons in the system. If a method requires a great deal of computer time in order to obtain the desired accuracy, it would be both expensive and time-consuming to test the method by applying it to a wide variety of molecules.

(iv) The method should enable meaningful comparisons of molecules of different sizes to be made. Pople (1973) refers to this property as size-consistency. The application of a technique to two well-separated molecules should yield results which are additive for the energy and for other properties. The widely-used method of configuration mixing, limited to single- and double-excitations with respect to some single-determinantal, or indeed multideterminantal, reference function does not satisfy this requirement.

(v) The calculated electronic energy should ideally be an upper bound or lower bound to the true energy.

(vi) The model should be the simplest possible which will afford the required results.

1.4 Potential energy curves and surfaces

Solution of the electronic Schrödinger equation yields the electronic energy of a molecule for the particular electronic state under consideration for a given fixed position of the nuclei. The electronic energy thus depends parametrically on the nuclear coordinates. These electronic energy values, when regarded as a function of the nuclear configuration, give the potential energy curve or surface on which the nuclei move.

For a diatomic molecule, let us denote the potential energy curve by $U(R)$, where R is the internuclear separation. This function is included in the Schrödinger equation for the nuclear wave function, which takes the form

$$\left(-\frac{1}{2\mu}\nabla^2 + U(R)\right)\Psi_{\text{nuclear}} = E\Psi_{\text{nuclear}} \tag{1.4.1}$$

μ is the reduced mass

$$\mu = M_A M_B/(M_A + M_B) \tag{1.4.2}$$

in this equation. Equation (1.4.1) may be separated by using spherical polar coordinates and putting

$$\Psi_{\text{nuclear}} = \frac{\chi(R)}{R} Y_{l,m}(\theta, \phi) \tag{1.4.3}$$

where $Y_{l,m}(\theta, \phi)$ is a spherical harmonic. This leads to a radial wave equation of the form

$$\left(-\frac{1}{2\mu}\nabla^2 + U(R) + \frac{J(J+1)}{2\mu R^2}\right)\chi(R) = E\chi(R) \tag{1.4.4}$$

where $J = 0, 1, \ldots$ is the rotational quantum number. The nuclear wave function is thus separated into a vibrational function and a rotational function.

The separation of the vibrational and rotational motion in polyatomic molecules requires the construction of the Eckart vectors and Eckart frames (Eckart, 1935). We shall not consider this problem further here but refer the reader to recent reviews (Louck and Galbraith 1976; Sutcliffe 1980).

For a diatomic molecule, if the potential energy curve has been calculated then, in principle, the vibrational and rotational levels may be obtained by solution of the nuclear wave equation. The terms neglected in the separation of the electronic and nuclear parts of the hamiltonian give rise to small diagonal corrections which may be added to the potential energy function. The nuclear Schrödinger equation for diatomic molecules can be solved directly by numerical integration, following the approach first described by Cooley (1961) and Cashion (1963). The major drawback of this direct method for solving the nuclear wave equation is that it is necessary to employ a very large number of points on the curve in order to obtain the required accuracy. Some method of interpolation, therefore, has to be used.

A simpler method is to solve the vibrational equation

$$\left(-\frac{1}{2\mu}\frac{\partial^2}{\partial R^2}+U(R)\right)\chi(R)=E\chi(R) \tag{1.4.5}$$

and treat the term $J(J+1)/2R^2$ as a small perturbation. The vibrational equation can be solved either by the Cooley–Cashion method or by expanding $\chi(R)$ in some set of functions and invoking the variation theorem.

In the vast majority of calculations of vibrational and rotational levels in diatomic molecules, approximate solutions of the electronic Schrödinger equation are obtained in tabular form at a relatively small number of internuclear distances. In order to consider vibrational and rotational effects, or to calculate spectroscopic constants or examine molecular dynamics, it is necessary to interpolate between the calculated values. Power series expansions provide a reasonably general means of achieving this. We note, however, that Padé approximants may also be of some use in this respect (Jordan, Kinsey, and Silbey 1974; Jordan 1975).

There are four factors which can influence the expansion coefficients and hence the calculated vibrational frequencies, rotational constants and equilibrium geometries in a power series expansion for the potential energy function:

(i) The accuracy of the calculated points;
(ii) The order of the polynomial fitted to the calculated points;
(iii) The range, number and distribution of the points to which the power series is fitted;
(iv) The expansion variable involved.

In connection with the first two factors, it has been shown (Meyer and Rosmus 1975; see also Wilson 1978*a*) that there is some advantage in fitting different orders of polynomial to energies obtained by means of an independent electron model, such as the matrix Hartree–Fock method, and also to the correlation corrections. Correlation energies are often not known as accurately as the reference energies and should, therefore, be interpolated by a lower-order polynomial. The third factor is, of course, limited by the total number of nuclear configurations for which electronic structure calculations are performed. It is often useful to test calculations by omitting one or two points from the fitting procedure and observing the effect on the expansion coefficients. With regard to point (iv), there appear to be three choices of expansion variable which can be usefully employed in the power series

$$V(\rho)=A_0\left[1+\sum_{i=1} a_i\rho^i\right] \tag{1.4.6}$$

The coefficients A_0 and a_i are to be determined by a least-squares fitting procedure. The Dunham parameter (Dunham 1932; Sandeman 1940) is

$$\rho_1=\frac{r-r_e}{r_e} \tag{1.4.7}$$

where r is the nuclear separation and r_e is its equilibrium value. Using this parameter a power series is obtained having the radius of convergence

$$0<r<2r_e. \tag{1.4.8}$$

The modified Dunham parameters (Fougere and Nesbet 1966; Simons, Parr, and Finlan 1973; Beckel 1976; Wilson 1978*a*)

$$\rho_2=\frac{r-r_e}{r} \qquad (\tfrac{1}{2}r_e<r<\infty) \tag{1.4.9}$$

and

$$\rho_3=\frac{r-r_e}{r+r_e} \qquad (0<r<\infty) \tag{1.4.10}$$

are often used. The radius of convergence of the respective power series is given in parenthesis. If the expansion coefficients of the three series obtained by employing ρ_1, ρ_2, and ρ_3 are denoted by $A_0, a_1, a_2, \ldots$; $B_0, b_1, b_2, \ldots$; $C_0, c_1, c_2, \ldots$, respectively, then they may be related by the equations

$$A_0=B_0 \tag{1.4.11}$$

$$a_n=b_n+\sum_{i=1}^{n-1}(-1)^i\binom{n+1}{i}b_{n-i}+(-1)^n(n+1) \tag{1.4.12}$$

and

$$A_0=\tfrac{1}{4}C_0 \tag{1.4.13}$$

$$a_n=\left(\frac{1}{2}\right)^n\left\{C_n+\sum_{i=1}^{n-1}(-1)^i\binom{n+1}{i}C_{n-i}+(-1)^n(n+1)\right\} \tag{1.4.14}$$

Of course, in using the expansion parameters ρ_1, ρ_2, and ρ_3, it has been assumed that r_e is known. This is not, in general, the case and therefore, an iterative procedure has to be adopted starting from some initial guess for r_e.

In 1932, Dunham calculated the energy levels of a rotating vibrator using the Wentzel–Kramers–Brillouin method. Using the expansion parameter ρ_1 Dunham obtained the energy level formula

$$F_{v,J}=\sum_{l,j} Y_{l,j}(v+\tfrac{1}{2})^l(J(J+1))^j \tag{1.4.15}$$

Table 1.1

The Dunham coefficients

$$Y_{00} = (B_e/8)(3a_2 - 7a_1^2/4)$$

$$Y_{10} = \omega_e\{1 + (B_e^2/4\omega_e^2)(25a_4 - 95a_1a_3/2 - 67a_2^2/4 + 459a_1^2a_2/8 - 1155a_1^4/64)\}$$

$$Y_{20} = (B_e/2)\{3(a_2 - 5a_1^2/4) + (B_e^2/2\omega_e^2)(245a_6 - 1365a_1a_5/2 - 885a_2a_4/2 - 1085a_3^2/4 + 8535a_1^2a_4/8 + 1707a_2^3/8 + 7335a_1a_2a_3/4$$
$$- 23\,865a_1^3a_3/16 - 62\,013a_1^2a_2^2/32 + 239\,985a_1^4a_2/128 - 209\,055a_1^6/512)\}$$

$$Y_{30} = (B_e^2/2\omega_e)(10a_4 - 35a_1a_3 - 17a_2^2/2 + 225a_1^2a_2/4 - 705a_1^4/32)$$

$$Y_{40} = (5B_e^3/\omega_e^2)(7a_5/2 - 63a_1a_5/4 - 33a_2a_4/4 - 63a_3^2/8 + 543a_1^2a_4/16 + 75a_2^3/16 + 483a_1a_2a_3/8 - 1953a_1^3a_3/32$$
$$- 4989a_1^2a_2^2/64 + 23\,265a_1^4a_2/256 - 23\,151a_1^6/1025)$$

$$Y_{01} = B_e\{1 + (B_e^2/2\omega_e^2)(15 + 14a_1 - 9a_2 + 15a_3 - 23a_1a_2 + 21(a_1^2 + a_1^3)/2)\}$$

$$Y_{11} = (B_e^2/\omega_e)[6(1 + a_1) + (B_e^2/\omega_e^2)\{175 + 285a_1 - 335a_2/2 + 190a_3 - 225a_4/2 + 175a_5 + 2295a_1^2/8 - 459a_1a_2 + 1425a_1a_3/4$$
$$- 795a_1a_4/2 + 1005a_2^2/8 - 715a_2a_3/2 + 1155a_1^3/4 - 9639a_1^2a_2/16 + 5145a_1^2a_3/8 + 4677a_1a_2^2/8 - 14\,259a_1^3a_2/16$$
$$- 31\,185(a_1^4 + a_1^5)/128\}]$$

$$Y_{21} = (6B_e^3/\omega_e^2)\{5 + 10a_1 - 3a_2 + 5a_3 - 13a_1a_2 + 15(a_1^2 + a_1^3)/2\}$$

$$Y_{31} = (20B_e^4/\omega_e^3)\{7 + 21a_1 - 17a_2/2 + 14a_3 - 9a_4/2 + 7a_5 + 225a_1^2/8 - 45a_1a_2 + 105a_1a_3/4 - 51a_1a_4/2 + 51a_2^2/8 - 45a_2a_3/2$$
$$+ 141a_1^3/4 - 945a_1^2a_2/16 + 435a_1^2a_3/8 + 411a_1a_2^2/8 - 1509a_1^3a_2/16 + 3807(a_1^4 + a_1^5)/128\}$$

$$Y_{02} = -(4B_e^3/\omega_e^2)[1 + (B_e^2/2\omega_e^2)\{163 + 199a_1 - 119a_2 + 90a_3 - 45a_4 - 207a_1a_2 + 205a_1a_3/2 - 333a_1^2a_2/2 + 693a_1^2/4$$
$$+ 46a_2^2 + 126(a_1^3 + a_1^4/2)\}]$$

$$Y_{12} = -(12B_e^4/\omega_e^3)(19/2 + 9a_1 + 9a_1^2/2 - 4a_2)$$

$$Y_{22} = -(24B_e^5/\omega_e^4)\{65 + 125a_1 - 61a_2 + 30a_3 - 15a_4 + 495a_1^2/4 - 117a_1a_2 + 26a_2^2 + 95a_1a_3/2 - 207a_1^2a_2/2 + 90(a_1^3 + a_1^4/2)\}$$

$$Y_{03} = 16B_e^5(3 + a_1)/\omega_e^4$$

$$Y_{13} = (12B_e^6/\omega_e^5)(233 + 279a_1 + 189a_1^2 + 63a_1^3 - 88a_1a_2 - 120a_2 + 80a_3/3)$$

$$Y_{04} = -(64B_e^7/\omega_e^6)(13 + 9a_1 - a_2 + 9a_1^2/4)$$

where the first subscript under Y refers to the power of the vibrational quantum number and the second that of the rotational quantum number. The Ys are usually referred to as Dunham coefficients. The first fifteen Dunham coefficient are displayed in Table 1.1. The Dunham coefficients are not exactly equal to the spectral constants obtained by means of perturbation theory (see, for example, Herzberg 1950). The connection between the Dunham coefficients and the band spectrum constants is as follows:

$$\begin{array}{lll} Y_{10} \sim \omega_e & Y_{20} \sim -\omega_e x & Y_{30} \sim \omega_e y \\ Y_{01} \sim B_e & Y_{11} \sim -\alpha_e & Y_{21} \sim \gamma_e \\ Y_{02} \sim D_e & Y_{12} \sim \beta_e & Y_{40} \sim \omega_e z \\ Y_{03} \sim F_e & Y_{13} \sim H_e & \end{array} \tag{1.4.16}$$

1.5 Experimental observables

The ultimate test of the theoretical apparatus and computational methods described in this monograph must be made through comparisons with experimentally observed properties of molecules. Often, however, a theoretical technique is assessed not by comparison with experiment but by comparing with results obtained by means of other theoretical approaches. This is because, although the ultimate aim of theory is to reproduce and rationalize experimental data, one can often learn more about a particular model by comparison with theoretically well-defined quantities than by comparison with experiment. For example, it is not possible to determine the Hartree–Fock limit of the energy or any other expectation value from experiment, neither can the correlation energy be measured. When testing a new basis set for self-consistent-field calculations or a new method for performing calculations of electron correlation energies, it is most valuable to compare the resulting energies with the Hartree–Fock energies or with the results of a full configuration interaction calculation in a limited basis. Full configuration mixing within a limited basis set is an exact solution of the Schrödinger equation within that basis set and thus the extent to which a particular model can reproduce that result is an indication of its accuracy, even though in the comparison the basis set truncation error is extremely large.

When the results of a theoretical study are compared with experiment this must be done with care. Comparisons are very often made with quantities which are derived from experimental data rather than with the experimental observations themselves. For example, after calculating a set of points on the potential energy curve for a diatomic molecule, a force constant and various anharmonicity constants may be obtained by fitting a polynomial and the resulting constants compared with the corresponding constants derived from experimental data. It should be remembered, however, that in the determination of the spectroscopic constants from experiment, the spectroscopist will often assume that various higher-order anharmonicity constants are zero (see, for example, Mills 1974).

In comparing calculated dipole moments with experimental values it should not be forgotten that the experimental value is often an average over vibrational states. In order to compare with experiment the theoretician should, therefore, first calculate the dipole moment as a function of the nuclear geometry, using the calculated energy values to solve the nuclear Schrödinger equation to obtain the vibrational levels, and then take a Boltzmann distribution of vibrational states at the temperature relevant to the experiment to obtain a vibrationally averaged dipole moment. In molecular beam resonance experiments, values of the dipole moment of molecules can be determined so accurately that the variation of the dipole moment with vibrational quantum number can be detected. In comparing theoretical results with such experiments one does not assume a Boltzmann distribution but makes a comparison for each vibrational level.

The recent monograph by Čársky and Urban (1980) provides a useful overview of the accuracy with which experimental observables may be predicted and confirmed by theoretical techniques.

1.6 Density matrices

The wave function for an N-electron system depends on the space and spin coordinates of all N of the electrons in the system. Interpretation of the wave function can, therefore, often be very difficult. Much of the information contained in the wave function is, however, highly redundant. Density matrices provide an invaluable tool for the interpretation of the electronic structure of molecules and also form the basis of important simplifications in quantum chemical calculations (e.g. Löwdin 1955*a*,*b*,*c*; McWeeny 1960; McWeeny and Sutcliffe 1976; Davidson 1976).

Let $\Psi(x_1, x_2, \ldots, x_N)$ be the electronic wave function for an atom or molecule, where $x_i = r_i\sigma_i$ denotes both the space and spin coordinates of the ith electron. The Nth order density matrix is then defined by

$$\Gamma^{(N)}(x'_1, x'_2, \ldots, x'_N; x_1, x_2, \ldots, x_N) = \Psi(x_1, x_2, \ldots, x_N)\Psi^*(x'_1, x'_2, \ldots, x'_N) \tag{1.6.1}$$

with the convention that if an arbitrary operator, $\hat{A}$, is applied to $\Gamma^{(N)}$ it acts only on the x_i and not the x'_i and subsequently the x'_i are replaced by x_i. Hence, the expectation value of the operator $\hat{A}$ is given by

$$\langle\Psi|\,\hat{A}\,|\Psi\rangle = \int \hat{A}\Gamma^{(N)}(x'_1, x'_2, \ldots, x'_N; x_1, x_2, \ldots, x_N)\,dx_1\,dx_2\ldots dx_N. \tag{1.6.2}$$

If A is a sum of p-electron operators

$$\hat{A}=\frac{1}{p!}\sum_{i_1i_2\ldots i_p}\hat{a}_{i_1i_2\ldots i_p} \tag{1.6.3}$$

and it is symmetrical in the electronic coordinates, each term will make the same contribution as the leading term since the wave function is antisymmetric. Thus

$$\langle\Psi|\,\hat{A}\,|\Psi\rangle=\int a_{12\ldots p}\Gamma^{(p)}(x_1', x_2', \ldots, x_p'; x_1, x_2, \ldots, x_p)\,dx_1\,dx_2\ldots dx_p \tag{1.6.4}$$

where $\Gamma^{(p)}$ is the pth order reduced density matrix which is defined by

$$\Gamma^{(p)}(x_1', x_2', \ldots, x_p'; x_1, x_2, \ldots, x_p)$$
$$=\binom{N}{p}\int\Gamma^{(N)}(x_1', x_2', \ldots, x_N'; x_1, x_2, \ldots, x_N)\,dx_{p+1}\,dx_{p+2}\ldots dx_N. \tag{1.6.5}$$

Most quantities of interest in atomic and molecular calculations involve at most two-electron operators and can, therefore, be expressed in terms of the first-order density matrix

$$\Gamma^{(1)}(x_1'; x_1)=N\int\Gamma^{(N)}(x_1', x_2', \ldots, x_N'; x_1, x_2, \ldots, x_N)\,dx_2\ldots dx_N \tag{1.6.6}$$

and the second-order density matrix

$$\Gamma^{(2)}(x_1', x_2'; x_1, x_2)=\binom{N}{2}\int\Gamma^{(N)}(x_1', x_2', \ldots, x_N'; x_1, x_2, \ldots, x_N)\,dx_3\ldots dx_N \tag{1.6.7}$$

The first-order density matrix can be derived from the second-order density matrix by integrating over the coordinates of one of the two electrons, that is

$$\Gamma^{(1)}(x_1'; x_2)=\frac{2}{N-1}\int\Gamma^{(2)}(x_1', x_2'; x_1, x_2)\,dx_2 \tag{1.6.8}$$

In general, the $(p-1)$th order density matrix can be derived from the pth order density matrix in a similar fashion, as in

$$\Gamma^{(p-1)}(x_1', x_2', \ldots, x_{p-1}'; x_1, x_2, \ldots, x_{p-1})$$
$$=\frac{p}{N+1-p}\int\Gamma^{(p)}(x_1', x_2', \ldots, x_p'; x_1, x_2, \ldots, x_p)\,dx_p. \tag{1.6.9}$$

The density matrices are usually normalized according to the condition

$$\int \Gamma^{(p)}(x_1', x_2', \ldots, x_p'; x_1, x_2, \ldots, x_p)\, dx_1\, dx_2 \ldots dx_p = \binom{N}{p} \qquad (1.6.10)$$

It should be noted that the density matrices are hermitian

$$\Gamma^{(p)}(x_1', x_2', \ldots, x_p'; x_1, x_2, \ldots, x_p) = \Gamma^{(p)*}(x_1', x_2', \ldots, x_p'; x_1, x_2, \ldots, x_p) \qquad (1.6.11)$$

and antisymmetric in each pair of indices

$$\Gamma^{(p)}(\ldots x_i' \ldots x_j' \ldots; \ldots x_i \ldots x_j \ldots) = -\Gamma^{(p)}(\ldots x_j' \ldots x_i' \ldots; \ldots x_i \ldots x_j \ldots). \qquad (1.6.12)$$

The first-order density matrix may be interpreted as the probability of finding an electron with spin σ_1 in the volume element dr_1 at r_1 multiplied by the number of electrons in the system. The second-order density matrix represents the probability of finding electrons with spins σ_1 and σ_2, respectively, in dr_1 at r_1 and dr_2 at r_2 multiplied by the number of electron pairs in the system. Similar interpretations of the higher-order density matrices may be given.

If wave functions for different electronic states of a molecule are available then transition matrices and reduced transition matrices may be defined as follows

$$\Gamma_{IJ}^{(N)} = \Psi_J(x_1, x_2, \ldots, x_N)\Psi_I^*(x_1', x_2', \ldots, x_N') \qquad (1.6.13)$$

and

$$\Gamma_{IJ}^{(p)} = \binom{N}{p} \int \Gamma_{IJ}^{(N)}(x_1 x_2 \ldots x_N; x_1, x_2, \ldots, x_N)\, dx_{p+1}\, dx_{p+2} \ldots dx_N \qquad (1.6.14)$$

The density matrices, reduced density matrices, transition matrices and reduced transition matrices which have been discussed above, all depend on both the space and spin coordinates of the electrons. It is sometimes convenient to integrate over all spin coordinates and thus obtain spin-free density matrices. Thus, for example, we have for the one-electron density matrix the following spin-free form

$$\Gamma^{(1)}(r_1'; r_1) = \int \Gamma^{(1)}(r_1'\sigma_1'; r_1\sigma_1)\, d\sigma_1 \qquad (1.6.15)$$

Given that the total energy of an atom or molecule can be written in terms of the first-order and the second-order density matrices leads to the idea that it is, in fact, not necessary to calculate the wave function at all. It suggests that the energy should be calculated directly in terms of the density matrices. However, it is clear that $\Gamma^{(1)}$ and $\Gamma^{(2)}$ must be implicitly derived from an N-electron fully antisymmetric wave function. Thus if the variation theorem is invoked to determine the optimum energy with respect to the variation of the density matrices, $\Gamma^{(1)}$ and $\Gamma^{(2)}$ must be varied subject to the constraint that they correspond to physically realizable wave functions. The variation of the one- and two-electron density matrices must be such that they preserve the N-representability (Coleman 1963, 1977; Davidson 1976). In practice, this has proved to be more difficult than the explicit calculation of the wave function from which the density matrices can then be constructed. The direct calculation of density matrices in quantum chemical calculations is, therefore, not widely used.

The density functional theory is an aspect of density matrix theory which is of some interest in the study of large molecules. This theory, which is widely used in solid state physics, is a development of the Thomas–Fermi statistical model for the atom (see, for example, March 1981). A cornerstone of the density functional theory is the theorem of Hohenberg and Kohn (1964) which states that the energy of a system in its ground state is a universal functional of the electron density. However, this universal function is, at present unknown. (Indeed, there is still some controversy in this area (see, for example, Kryachko 1980; Levy and Perdew 1982.)

The main value of density matrices in molecular calculations has, in the majority of studies, been interpretive. Density matrices are often constructed during the analysis of a molecular wave function. In calculations using multiconfigurational wave functions, the density function is particularly useful in view of the difficulty in interpreting the wave function.

1.7 The non-relativistic limit

The exact solution of the molecular electronic Schrödinger equation yields the energy of the system in the so-called non-relativistic limit. Except for the case of very small molecules containing one or two electrons this limit is never reached in practice. However, the non-relativistic limit is the clearly defined goal of the vast majority of contemporary calculations.

For atoms from the first two rows of the periodic table, and molecules containing them, relativistic effects are small. It appears that non-relativistic quantum mechanics can be used to describe chemical processes involving first-row and second-row atoms rather accurately. Relativistic effects are not important in theoretical organic chemistry. Non-relativistic treatments can provide a good first approximation to a more complete relativistic treatment of molecules containing light atoms.

The situation is somewhat different for molecules containing the heavier elements of the later periods: transition metals, the lathanides, and the actinides. For molecules containing these atoms, relativistic effects are certainly quite large. Indeed, the chemistry of the actinides is dominated by relativistic effects (see, for example, Pitzer 1979). However, non-relativistic quantum mechanics may still provide a useful model for molecules containing transition metal atoms, although the results of calculations are necessarily of a more qualitative nature than studies of molecules containing light atoms.

1.8 Relativistic effects

The effects of relativity on the quantum mechanics of molecules will not be considered in any detail in this monograph. (For a review of relativistic atomic structure calculations see, for example, Grant (1970); while for a review of relativistic quantum chemistry see, for example, Pyykkö (1978).) Relativistic effects are usually completely neglected in studies of the electronic structure of molecules containing light atoms. Even for molecules containing heavy atoms, relativistic effects are very frequently not considered and it is argued that the major part of the relativistic energy is associated with the tightly bound inner core electrons and it is claimed that they can be neglected for chemical studies. However, this question requires more careful consideration.

In a hydrogen-like atom with a nuclear charge of 80, for example, the average velocity of the 1s-electrons, v, can be shown to be about fifty-eight per cent of the speed of light, c. At this velocity, the electron mass increases to

$$m = m_e\{1-(v/c)^2\}^{-\frac{1}{2}} \tag{1.8.1}$$

where m_e is the electron rest mass. Thus a relativistic contraction of the order of twenty per cent is obtained. However, in a many-electron atom, this contraction of the inner core orbitals, when relativistic effects are considered, can significantly affect the valence orbitals which are, of course, important in describing the chemistry of the given system. The relativistic contraction of the s- and p-shells leads to a more efficient screening of the nuclear charge, which in turn leads to an expansion of the d- and f-shells in a relativistic self-consistent-field calculation. The contraction of the s- and p-shells and the expansion of the d- and f-shells are mutually reinforcing effects, making them particularly large in the valence shell.

For a single electron, the relativistic quantum mechanical equation of motion is the well known Dirac equation

$$h\psi = \left\{c\boldsymbol{\alpha}\cdot\left(\mathbf{p}+\frac{1}{c}\,\mathbf{A}\right)+\boldsymbol{\beta}mc^2 - V\right\}\psi = E\psi \tag{1.8.2}$$

where **A** is the magnetic vector potential, V is the electric potential, and ψ is a four-component spinor. The Dirac matrices are written

$$\boldsymbol{\alpha}=\begin{pmatrix}\mathbf{0} & \boldsymbol{\sigma}\\ \boldsymbol{\sigma} & \mathbf{0}\end{pmatrix} \tag{1.8.3}$$

and

$$\boldsymbol{\beta}=\begin{pmatrix}1 & 0 & 0 & 0\\ 0 & 1 & 0 & 0\\ 0 & 0 & -1 & 0\\ 0 & 0 & 0 & -1\end{pmatrix} \tag{1.8.4}$$

with the Pauli spin matrices

$$\boldsymbol{\sigma}_x=\begin{pmatrix}0 & 1\\ 1 & 0\end{pmatrix},\quad \boldsymbol{\sigma}_y=\begin{pmatrix}0 & -i\\ i & 0\end{pmatrix},\quad \boldsymbol{\sigma}_z=\begin{pmatrix}1 & 0\\ 0 & -1\end{pmatrix}. \tag{1.8.5}$$

The Dirac–Fock hamiltonian for a many-electron atom or molecule may be written

$$\hat{\mathcal{H}}=\sum_i h_i+\sum_{i>j}\frac{1}{r_{ij}} \tag{1.8.6}$$

where h is given by eqn (1.8.2) with $\mathbf{A}=0$. This hamiltonian treats one-electron effects relativistically but the electron–electron interaction effects non-relativistically. The single particle state functions are usually written in the form

$$\psi=\begin{pmatrix}g\\ f\end{pmatrix} \tag{1.8.7}$$

where g and f are the 'large' and 'small' components, respectively, and satisfy the normalization condition

$$\int_0^{\infty}(g^2+f^2)\,\mathrm{d}\tau=1. \tag{1.8.8}$$

Higher-order corrections to the Dirac–Fock hamiltonian account for the effects of relativity on the electron–electron Coulomb interaction. For example, Gaunt (1929) proposed a term describing the magnetic interaction of the spin current of one electron with the orbital current of another; Breit (1929, 1930, 1932; see also Bethe and Fermi 1932) included a retardation term. In the non-relativistic limit, the sum of the Dirac–Fock hamiltonian and the Breit interaction reduces to the Breit–Pauli form if terms of order $(v/c)^4$ are included (Itoh 1965). In the non-relativistic limit the hamiltonian takes the form

$$\hat{\mathcal{H}}_{\text{n.r.}}=\sum_{i=1}^{15}\hat{\mathcal{H}}_i \tag{1.8.9}$$

where each of the components, $\hat{\mathcal{H}}_i$, are identified in Table 1.2.

Table 1.2
The Breit–Pauli hamiltonian

Component	Expression	Effect
$\hat{\mathcal{H}}_1$	$\sum_i \frac{1}{2m}\mathbf{p}_i^2$	kinetic energy
$\hat{\mathcal{H}}_2$	$-\sum_i \frac{1}{8m^3c^2}\mathbf{p}_i^4$	mass velocity
$\hat{\mathcal{H}}_3$	$+\sum_i \frac{e}{mc}\mathbf{p}_i \cdot \mathbf{A}(\mathbf{r}_i)$	external magnetic field
$\hat{\mathcal{H}}_4$	$+\sum_i \frac{e^2}{2mc^2}\mathbf{A}^2(\mathbf{r}_i)$	external magnetic field
$\hat{\mathcal{H}}_5$	$-\sum_i eV(\mathbf{r}_i)$	external electric field
$\hat{\mathcal{H}}_6$	$+\sum_i \frac{e}{mc}\mathbf{s}_i \cdot \mathbf{H}(\mathbf{r}_i)$	spin–external magnetic field
$\hat{\mathcal{H}}_7$	$+\sum_i \frac{e}{2m^2c^2}\mathbf{s}_i \cdot (\mathbf{E}(\mathbf{r}_i) \times \mathbf{p}_i)$	spin–orbit coupling
$\hat{\mathcal{H}}_8$	$+\sum_i \frac{\pi e\hbar^2}{2m^2c^2}\rho(\mathbf{r}_i)$	Darwin
$\hat{\mathcal{H}}_9$	$+\sum_{i>j} \frac{e^2}{r_{ij}}$	Coulomb
$\hat{\mathcal{H}}_{10}$	$-\sum_{i>j} \frac{e^2}{2m^2c^2}\mathbf{p}_i \cdot \left(\frac{\mathbf{RR}}{R^3}+\frac{1}{R}\right) \cdot \mathbf{p}_j$	orbit–orbit coupling
$\hat{\mathcal{H}}_{11}$	$+\sum_{i\neq j} \frac{e^2}{m^2c^2}\frac{1}{R^3}\mathbf{s}_i \cdot (\mathbf{R} \times \mathbf{p}_j)$	spin–other-orbit coupling
$\hat{\mathcal{H}}_{12}$	$-\sum_{i\neq j} \frac{e^2}{2m^2c^2}\frac{1}{R^3}\mathbf{s}_i \cdot (\mathbf{R} \times \mathbf{p}_j)$	spin–other-orbit coupling
$\hat{\mathcal{H}}_{13}$	$-\sum_{i<j} \frac{e^2}{m^2c^2}\mathbf{s}_i \cdot \left(\frac{3\mathbf{RR}}{R^5}-\frac{1}{R^3}\right) \cdot \mathbf{s}_j$	spin–spin coupling
$\hat{\mathcal{H}}_{14}$	$-\sum_{i<j} \frac{8\pi e^2}{3m^2c^2}\delta(\mathbf{R})\mathbf{s}_i \cdot \mathbf{s}_j$	spin–spin coupling
$\hat{\mathcal{H}}_{15}$	$-\sum_{i<j} \frac{\pi e^2\hbar^2}{mc^2}\delta(\mathbf{R})$	Darwin

$\mathbf{R} = \mathbf{r}_i - \mathbf{r}_j$ $\quad \mathbf{E}(r) = -\nabla V(r)$ $\quad \mathbf{H}(r) = \nabla \times \mathbf{A}(r)$ $\quad \rho(\mathbf{r}) = Ze\,\delta(\mathbf{r})$
The Coulomb gauge $\nabla \cdot \mathbf{A}(\mathbf{r}) = 0$ is employed.

2

INDEPENDENT ELECTRON MODELS

2.1 Orbital theories

Independent electron models, or orbital theories, are fundamental to most tractable schemes employed in the study of molecular electronic structure. For atoms and diatomic molecules containing a small number of electrons it is possible to use trial wave functions which include the interelectronic distance explicitly (Hylleraas 1929; Sims and Hagstrom 1971*a*,*b*). For systems containing more than a few electrons and for polyatomic molecules this approach is not computationally feasible and independent electron models are invariably employed. In such models the many-electron wave function is written as a product of one-electron functions or orbitals. The orbitals may be chosen to be atomic orbitals, molecular orbitals, or they may be selected in other ways suggested by physical and chemical intuition.

Since correlation effects are essentially the corrections to independent electron models, it is clear that the importance of electron correlation as a whole and the relative importance of various components of the correlation energy will be dependent on the choice of orbital model with respect to which electron correlation is considered. This chapter is, therefore, devoted to independent electron models. In Section 2, the simplest of orbital models, the bare-nucleus model and its derivatives, will be discussed. In the third section, a brief discussion of the Hartree model is presented whilst the much more widely used Hartree–Fock approach is considered in Section 4. Although they are not strictly independent electron models, we shall consider multi-configurational Hartree–Fock approaches in Section 5 of this chapter since these methods are often employed as a reference with respect to which correlation effects are examined. Valence bond orbital theories are analysed in Section 6. Prior to considering specific orbital models, the nature of the orbital approximation in general will be discussed.

Let $\{\chi_i\}$ denote some set of one-electron functions and let Φ be the primitive spatial function for a system containing N electrons

$$\Phi = \Phi(\mathbf{r}_1, \mathbf{r}_2, \ldots, \mathbf{r}_N) \tag{2.1.1}$$

where $\mathbf{r}_i$ denotes the spatial coordinates of the ith electron. Now if the set

$\{\chi_i\}$ is a complete set in a specified interval, Φ may be expanded as follows

$$\Phi = \sum_{i_N} C_{i_N}^{N-1}(\mathbf{r}_1, \mathbf{r}_2, \ldots, \mathbf{r}_{N-1})\chi_{i_N}(\mathbf{r}_N) \tag{2.1.2}$$

where $C_{i_N}^{N-1}$ is a coefficient. Furthermore, repeating this procedure leads to

$$\Phi = \sum_{i_{N-1}i_N} C_{i_{N-1}i_N}^{N-2}(\mathbf{r}_1, \mathbf{r}_2, \ldots, \mathbf{r}_{N-2})\chi_{i_{N-1}}(\mathbf{r}_{N-1})\chi_{i_N}(\mathbf{r}_N) \tag{2.1.3}$$

and hence the primitive spatial function may be written rigorously as an expansion in terms of orbital product functions, or Nth rank tensors (see, for example, Löwdin 1960)

$$\Phi = \sum_{i_1, i_2, \ldots, i_N} C_{i_1 i_2 \ldots i_N}^{0} \chi_{i_1}(\mathbf{r}_1)\chi_{i_2}(\mathbf{r}_2)\ldots\chi_{i_N}(\mathbf{r}_N). \tag{2.1.4}$$

Equation (2.1.4) may be rewritten, after introducing a normalization factor, $\mathcal{N}$, in the form

$$\Phi = \mathcal{N}(\Phi_0 + \eta) \tag{2.1.5}$$

where Φ_0 is the orbital product function

$$\Phi = \chi_{j_1}(\mathbf{r}_1)\chi_{j_2}(\mathbf{r}_2)\ldots\chi_{j_N}(\mathbf{r}_N) \tag{2.1.6}$$

and η is the sum of the remaining terms in eqn (2.1.4). Φ_0 is a good approximation to the exact spatial wave function if

$$\frac{C_{i_1 i_2 \ldots i_N}^{0}}{C_{j_1 j_2 \ldots j_N}^{0}} \ll 1, \qquad \forall i \neq j \tag{2.1.7}$$

η may then be neglected as a first approximation. This is the basis of independent electron models or orbital theories.

If, in the above discussion, the set $\{\chi_i\}$ is replaced by some set $\{\chi_i'\}$ whose members are linearly independent combinations of the $\{\chi_i\}$, then the resulting Φ' is identical to Φ. This result is valid even when the one-electron basis is truncated to a finite size. Thus, there is a limit to the accuracy of a wave function within a given basis set which can be attained irrespective of whether we use atomic orbitals, molecular orbitals, or other physically motivated functions if a sufficient number of terms are included in the expansion (2.1.4). This result justifies the observation (see, for example, Longuet-Higgins 1948) that multistructure valence bond calculations and molecular orbital configuration interaction calculations yield results which tend to the same limit as the number of structures and the number of configurations is increased.

Although arbitrary to some extent, the choice of orbital functions may significantly affect the convergence properties of the expansion (2.1.4) for the primitive spatial wave function. For two-electron systems the form of the expansion (2.1.4) can be greatly simplified by using natural orbitals (Löwdin 1955*a,b,c*; Shull and Löwdin 1959; Coleman 1963; Davidson 1972, 1976). For a two-electron system, the expansion (2.1.4) takes the form

$$\Phi(\mathbf{r}_1, \mathbf{r}_2) = \sum_{ij} a_{ij}\chi_i(\mathbf{r}_1)\chi_j(\mathbf{r}_2) \tag{2.1.8}$$

with $a_{ij} = a_{ji}$. The matrix $\mathbf{a}$ may be reduced to diagonal form by a real orthogonal transformation, $\mathbf{U}$, i.e.

$$\mathbf{U}^T\mathbf{a}\mathbf{U} = \mathbf{\Lambda}. \tag{2.1.9}$$

Equation (2.1.8) can thus be written as

$$\Phi(\mathbf{r}_1, \mathbf{r}_2) = \boldsymbol{\chi}(\mathbf{r}_1)\mathbf{a}\boldsymbol{\chi}^T(\mathbf{r}_2) \tag{2.1.10a}$$

$$= \boldsymbol{\chi}(\mathbf{r}_1)\mathbf{U}\mathbf{\Lambda}\mathbf{U}^T\boldsymbol{\chi}^T(\mathbf{r}_2) \tag{2.1.10b}$$

$$= \boldsymbol{\phi}(\mathbf{r}_1)\mathbf{\Lambda}\boldsymbol{\phi}^T(\mathbf{r}_2) \tag{2.1.10c}$$

which may be re-expressed in the form

$$\Phi(\mathbf{r}_1, \mathbf{r}_2) = \sum_{i} \Lambda_{ii}\phi_i(\mathbf{r}_1)\phi_i(\mathbf{r}_2) \tag{2.1.11}$$

where

$$\boldsymbol{\phi} = \boldsymbol{\chi}\mathbf{U} \tag{2.1.12}$$

are the natural orbitals. The expansion (2.1.11) provides the most rapidly convergent series for the wave function. The diagonal form of the wave function leads to a diagonal form for the first-order density matrix. Diagonalization of the first-order density matrix for a general many-electron system provides a definition of the natural orbitals, or, if the density matrix includes spin, the natural spin orbitals. Unfortunately, for systems containing more than two electrons, the natural orbitals do not provide any finite truncation of the wave function expansion which have optimum convergence properties. Furthermore, since the determination of the natural orbitals requires a knowledge of the first-order density matrix, they can only be determined by an iterative procedure which can often be computationally prohibitive.

The total molecular electronic hamiltonian may be written as

$$\mathcal{H} = \sum_{p=1}^{N} h_p + \sum_{p>q}^{N} g_{pq} \tag{2.1.13}$$

where the one-electron term is

$$h_p = \tfrac{1}{2}\nabla_p^2 - \sum_{A=1}^{M} \frac{Z_A}{r_{pA}} \tag{2.1.14}$$

and the two-electron term is

$$g_{pq} = \frac{1}{r_{pq}}. \tag{2.1.15}$$

Physically, the independent electron models correspond to the use of an effective N-electron hamiltonian of the type

$$\hat{\mathcal{H}}_{\text{eff}} = \sum_{p} h_p + u_p \tag{2.1.16}$$

where u_p is a one-electron potential. The N-electron hamiltonian is written as a sum of N one-electron hamiltonians. u_p will in general attempt to provide a description of the interactions between the electrons and also the screening of the nuclear charge. Clearly, the total hamiltonian may be written as

$$\hat{\mathcal{H}} = \hat{\mathcal{H}}_{\text{eff}} + \left\{ \sum_{p>q} g_{pq} - \sum_{p} u_p \right\} \tag{2.1.17}$$

and the terms in parenthesis will be small if the orbital theory is a good approximation. The terms in parenthesis in eqn (2.1.17) are sometimes called the 'fluctuation potential' (see, for example, Sinanoğlu and Brueckner 1970 and references therein.)

2.2 Bare-nucleus models

The very simplest orbital model of molecular electronic structure can be obtained by putting

$$u_p = 0 \qquad \forall p \tag{2.2.1}$$

in eqn (2.1.16). Electron–electron interactions are then completely ignored and each electron moves in the field of the unscreened or bare nuclei. The bare-nucleus model has a number of unique features which make its use in the study of correlation effects attractive (see, for example, Sinanoğlu 1961):

(a) It is both conceptually and computationally a very simple model.
(b) Unlike the widely used Hartree–Fock orbitals, the bare-nucleus orbitals are defined in a non-iterative manner. Within the algebraic approximation (see Chapter 6), the orbitals of the bare-nucleus model are determined by the diagonalization of a single symmetric matrix of order n, where n is the number of basis functions.
(c) The solution of the bare-nucleus problem within a given basis set requires only one-electron integrals which are very much easier to evaluate than the two-electron integrals which arise in, for example, Hartree–Fock calculations. The design and adjustment of basis sets

employed in bare-nucleus calculations can, therefore, be achieved easily and efficiently.

(d) For atoms and diatomic molecules the exact solutions of the bare-nucleus problem are known. Comparison of calculations made within the algebraic approximation with these exact solutions provides a measure of the quality of the basis set.

(e) There is no difficulty in treating excited states in the bare-nucleus model. In particular, excited states with the same symmetry as a lower state can be handled easily.

(f) The one-electron eigenvalue problem which has to be solved in the bare-nucleus model does not depend on the electronic state or on the number of electrons in the system. It is, therefore, a useful first approximation for the study of excitation and ionization processes. Again, this should be contrasted with the situation in, for example, the Hartree–Fock model in which the potential function u_p depends on the number of electrons in the system and the particular state of interest. Furthermore, the orbitals are optimized for one state and often form a poor representation of an excited state or ionized state.

(g) For atoms the bare-nucleus model can be used to develop a perturbation series in terms of Z^{-1} where Z is the nuclear charge (Chisholm and Dalgarno 1966; Horak 1965). The energy coefficients in the perturbation expansion are then independent of the atom being studied.

The most serious disadvantage of the bare-nucleus model is that it does not lead to an accuracy for energies or other expectation values comparable with that afforded by the Hartree–Fock model. This is illustrated in Table 2.1. However, the bare-nucleus approach may still be a very useful

Table 2.1

Comparison of the use of the bare-nucleus model as a reference for a perturbational treatment of the correlation energy of the helium atom with the use of the Hartree model and the Hartree–Fock model[a]

Energy component	Energy		
	Bare-nucleus	Hartree	Hartree–Fock
E_0	−4.000 00	−1.835 92	−1.835 92
E_1	−2.750 00	−2.861 67	−2.861 67
E_2	−2.907 66	−2.909 84	−2.898 92
E_3	−2.903 32	−2.902 67	−2.902 69
E_4	−2.903 53	−2.903 96	−2.903 54
E_5	−2.903 66	−2.903 70	−2.903 70

[a] Based on the work of Riley and Dalgarno (1971); E_i denotes the energy calculated through ith order in the perturbation r_{12}^{-1}.

zero-order model for atomic and molecular studies. For example, a number of Rayleigh–Schrödinger perturbation theory calculations exist which suggest that the series based on a bare-nucleus zero-order function may in fact converge more rapidly than that based on the use of a Hartree–Fock reference function (e.g. Riley and Dalgarno 1971; Musher and Schulman 1968). Goodisman (1968) found that the simple bare-nucleus model leads to a poorly convergent Rayleigh–Schrödinger perturbation expansion for the helium dimer and for some excited states of the hydrogen molecule. However, satisfactory results were obtained by following Dalgarno and Stewart (1961) and introducing a screening factor, ζ_A, for each nucleus. The one-electron potential u_p is then defined by

$$u_p = \sum_{A=1}^{M} \frac{Z_A - \zeta_A}{r_{pA}} \tag{2.2.2}$$

Electron–electron interactions are still completely neglected but the parameter $\mathscr{S}_A$ allows for the screening of the nuclear charges. In this screened bare-nucleus model the problem of establishing a value for $\mathscr{S}_A$ arises. Goodisman (1973) demonstrated that $\mathscr{S}_A$ could be determined by minimizing the first-order energy.

2.3 The Hartree model

The simplest model which attempts to take account of electron–electron interactions in a molecule is due to Hartree (1927, 1957). The electronic wave function is written in the form

$$\Phi = \phi_1(1)\phi_2(2) \ldots \phi_N(N) \tag{2.3.1}$$

and (Hartree 1957, p. 18) "each one of these functions . . . should be determined as a solution of Schrödinger's equation for one electron in the field of the nucleus and of the total average charge distribution of the electrons in the *other* wavefunctions. In such a treatment, the field of the average electron distribution derived from the wavefunctions . . . must be the same as the field used in evaluating these wave functions." Thus the optimum orbitals in eqn (2.3.1) are determined in an iterative fashion. This is the self-consistent-field idea. The function (2.3.1) is not widely used in molecular electronic structure studies, although it has been employed as a zero-order model in perturbation calculations (Riley and Dalgarno 1971). The principal drawback of the function (2.3.1) is that it does not satisfy the Pauli principle. However, when eqn (2.3.1) is anti-symmetrized to ensure that it obeys the Pauli principle, it forms the basis of one of the most widely used orbital theories, namely the Hartree–Fock model.

2.4 The Hartree–Fock model

The most widely used independent electron model of atomic and molecular electronic structure is undoubtedly the Hartree–Fock model (Hartree 1948 and references therein, Fock 1930, 1932). This model is a quantitative formulation of the molecular orbital theory developed by Hund (1928, 1931) and Mulliken (1928) soon after the advent of modern quantum mechanics as a natural extension of the orbital theory of atomic structure to molecules. Within the Hartree–Fock approximation, the electronic wave function is represented by a single determinant of spin orbitals in which the orbitals are optimized to yield the best single determinantal function according to the variation principle.

For molecules, the Hartree–Fock method has enjoyed considerable attention in the formulation in which the orbitals are parameterized by expansion in a finite basis set. This self-consistent-field technique or matrix Hartree–Fock method was developed in the pioneering work of Hall and Lennard-Jones (1950), Hall (1951), and of Roothaan (1951, 1960). The method is capable of recovering a large fraction of the total energy of a system; for example, the Hartree–Fock model yields 96% of the total energy of the hydrogen molecule at its equilibrium nuclear geometry, 98.5% for He_2, 98.9% for LiH, 99.04% for H_2O, 99.5% for N_2, and 99.5% for NH_3. The matrix Hartree–Fock method is practical for the *ab initio* treatment of moderately large systems. It also leads to expectation values of some one-electron properties which are of an acceptable accuracy (see, for example, Green 1974). The Hartree–Fock model does, however, have a number of deficiencies. The most serious of these is the fact that the error in the calculated total energy, although a small percentage of the total energy, is of the same order of magnitude as the largest energies of chemical interest. The binding energy of the lithium hydride molecule, for example, is approximately 1.1% of the total energy while for the nitrogen molecule it is 0.2% of the total. Furthermore, the dissociation of molecules is incorrectly described by the Hartree–Fock model, unless the species into which the molecule dissociates are closed-shell systems. Even the simplest of molecules, the hydrogen molecule, dissociates incorrectly within the Hartree–Fock approximation. Finally, it should be mentioned that the spin distribution near nuclei is often poorly described in the Hartree–Fock model.

In closed-shell Hartree–Fock theory the molecular wave function is written in the form

$$\Phi = (N!)^{-\frac{1}{2}} |\psi_1(\mathbf{x}_1)\bar{\psi}_1(\mathbf{x}_2) \ldots \psi_n(\mathbf{x}_{N-1})\bar{\psi}_n(\mathbf{x}_N)| \tag{2.4.1}$$

where the spin-orbitals are

$$\psi_i = \phi_i\alpha \qquad \bar{\psi}_i = \phi_i\beta \tag{2.4.2}$$

and the spatial orbitals ϕ_i are taken to be orthonormal. The closed-shell Hartree–Fock function is an eigenfunction of $\hat{S}^2$ with $S=0$. If integration over spin coordinates is carried out explicitly, the energy of an atom or molecule takes the form

$$E=2\sum_i \langle\phi_i|\, h\, |\phi_i\rangle+\sum_{i>j}\{2\langle\phi_i\phi_j|\, g\, |\phi_i\phi_j\rangle-\langle\phi_i\phi_j|\, g\, |\phi_j\phi_i\rangle\} \quad (2.4.3)$$

It is required to make E stationary with respect to variation of the orbitals, subject to the orthonormality constraints $\langle\phi_i \mid \phi_j\rangle=\delta_{ij}$. It is usual in molecular applications of the theory to invoke the algebraic approximation (see Chapter 6) in which the orbitals are expanded in some finite basis set. If there are m basis functions, then the occupied orbitals may be expanded as follows

$$\phi_i=\sum_k \chi_k T_{ki} \quad (2.4.4a)$$

$$\boldsymbol{\phi}=\boldsymbol{\chi}\mathbf{T} \quad (2.4.4b)$$

in which $\mathbf{T}$ is an $m\times n$ matrix whose columns contain the expansion coefficients for each of the molecular orbitals. The total energy may then be expressed as (McWeeny 1975, McWeeny and Sutcliffe 1976; McWeeny and Pickup 1980)

$$E=Tr\mathbf{R}(\mathbf{h}+\tfrac{1}{2}\mathbf{G}) \quad (2.4.5)$$

where

$$\mathbf{R}=\mathbf{T}\mathbf{T}^\dagger. \quad (2.4.6)$$

The necessary and sufficient condition for the orthonormality of the orbitals is

$$\mathbf{RSR}=\mathbf{R} \quad (2.4.7)$$

where $\mathbf{S}$ is the overlap matrix, i.e. $S_{ij}=\langle\chi_i \mid \chi_j\rangle$. The electron–electron interaction matrix is defined by

$$\mathbf{G}=\mathbf{J}(2\mathbf{R})-\mathbf{K}(\mathbf{R}) \quad (2.4.8)$$

where the Coulomb matrix has elements

$$\mathbf{J}(\mathbf{R})_{pq}=\sum_{rs} R_{rs}\langle\chi_p\chi_s|\, g\, |\chi_q\chi_r\rangle \quad (2.4.9)$$

and the exchange matrix is

$$\mathbf{K}(\mathbf{R})_{pq}=\sum_{rs} R_{rs}\langle\chi_p\chi_s|\, g\, |\chi_r\chi_q\rangle. \quad (2.4.10)$$

Now the first-order variation of the total energy when $R\rightarrow R+\delta R$ can be shown to be

$$\delta E = 2Tr\mathbf{h}^F\,\delta\mathbf{R} \tag{2.4.11}$$

where $\mathbf{h}^F$ is the Hartree–Fock hamiltonian given by

$$\mathbf{h}^F = \mathbf{h} + \mathbf{G}. \tag{2.4.12}$$

The requirement that E be stationary subject to the orthonormality conditions then leads to the matrix Hartree–Fock equations

$$\mathbf{h}^F\mathbf{T} = \mathbf{S}\mathbf{T}\boldsymbol{\epsilon} \tag{2.4.13}$$

in which $\boldsymbol{\epsilon}$ is an $n \times n$ matrix of Lagrangian multipliers. Without loss of generality, the matrix $\boldsymbol{\epsilon}$ may be taken to be diagonal, so that

$$\mathbf{h}^F\mathbf{T}(k) = \epsilon_k\mathbf{S}\mathbf{T}(k) \qquad \mathbf{T}(k) = k\text{th column of } \mathbf{T} \tag{2.4.14}$$

which is the pseudo-eigenvalue problem determining the canonical molecular orbitals.

The simplest form of Hartree–Fock theory which can be applied to open-shell systems corresponds to a single determinantal wave function of the form

$$\Phi = (N!)^{-\frac{1}{2}}\,|\psi_1\psi_2 \ldots \psi_N| \tag{2.4.15}$$

where the ψ_i are spin-orbitals. This is the unrestricted Hartree–Fock theory. However, as noted by, for example, Hurley (1976a): "Despite its formal simplicity the unrestricted Hartree–Fock theory does not, in itself, provide an appropriate version of the molecular orbital method . . .". The unrestricted Hartree–Fock method, and indeed the so-called generalized unrestricted Hartree–Fock methods (see Seeger and Pople 1977), do not, in general, provide wave functions which are eigenfunctions of the spin operator $\hat{S}^2$. The unrestricted Hartree–Fock determinant does not correspond to a pure spin state. This is a very serious deficiency of the method which can lead to disturbing results even in the case of spin-independent quantities. For example, potential energy curves and surfaces resulting from unrestricted Hartree–Fock calculations may often display unphysical characteristics. Furthermore, the effective operator which arises in the pseudo-eigenvalue equation of unrestricted Hartree–Fock theory does not, in general, display the full symmetry of the molecular system being investigated and the orbitals do not, therefore, have simple symmetry properties. Although the spatial symmetry and spin symmetry problems associated with the unrestricted Hartree–Fock function can be formally resolved by the use of projection operators, projected unrestricted Hartree–Fock wave functions are both complicated and difficult to interpret.

A much more useful form of the Hartree–Fock theory which can be applied to open-shell systems is the restricted Hartree–Fock theory.

Consider the single determinantal wave function

$$\Phi = (N!)^{-\frac{1}{2}} |\psi_1\bar{\psi}_1 \ldots \psi_p\bar{\psi}_p\psi_{p+1}\psi_{p+2} \ldots \psi_{p+q}| \tag{2.4.16a}$$

where $N = 2p + q$ in which the first p orbitals are doubly occupied and the remaining q orbitals are singly occupied with parallel spins. This wave function is clearly an eigenfunction of the spin operator $\hat{S}^2$ with $S = (\frac{1}{2}N - p)$ and $M = S$. It is suitable for describing doublet and triplet states. The orbitals are taken to be orthonormal. It may be shown that the energy expression corresponding to eqn (2.4.16a) may be cast in the form (see, for example, McWeeny and Sutcliffe 1976)

$$E = \nu_1 Tr\{\mathbf{R}_1(\mathbf{h} + \tfrac{1}{2}\mathbf{G}_1)\} + \nu_2 Tr\{\mathbf{R}_2(\mathbf{h} + \tfrac{1}{2}\mathbf{G}_2)\} \tag{2.4.17}$$

in which $\nu_1(=2)$ and $\nu_2(=1)$ are the closed-shell and open-shell occupation numbers, respectively. $\mathbf{R}_1$ and $\mathbf{R}_2$ are the matrices defining the shell densities and

$$\mathbf{G}_1 = \nu_1\{\mathbf{J}(\mathbf{R}_1) - \tfrac{1}{2}\mathbf{K}(\mathbf{R}_1)\} + \nu_2\{\mathbf{J}(\mathbf{R}_2) - \tfrac{1}{2}\mathbf{K}(\mathbf{R}_2)\} \tag{2.4.18a}$$

$$\mathbf{G}_2 = \nu_2\{\mathbf{J}(\mathbf{R}_1) - \mathbf{K}(\mathbf{R}_1)\} + \nu_2\{\mathbf{J}(\mathbf{R}_2) - \tfrac{1}{2}\mathbf{K}(\mathbf{R}_2)\} \tag{2.4.18b}$$

The restricted Hartree–Fock theory may be extended to allow for any number of closed-shells and open-shells provided that the energy functional can be arranged in the form

$$E = \sum_p \nu_p Tr\mathbf{R}_p(\mathbf{h} + \tfrac{1}{2}\mathbf{G}_p) \tag{2.4.19}$$

where ν_p, $\mathbf{R}_p$, and $\mathbf{G}_p$ refer to the pth shell.

In the closed-shell Hartree–Fock theory, and indeed in the unrestricted Hartree–Fock theory, the invariance of the total wave function under unitary mixing of the occupied orbitals was used to eliminate the off-diagonal Lagrangian multipliers in the equations defining the optimum orbitals and thus obtain a pseudo-eigenvalue equation. This cannot be done in the restricted Hartree–Fock theory. However, it was first shown by Roothaan (1960) that a pseudo-eigenvalue equation can in fact be obtained by defining certain coupling operators. There is now a considerable literature on the formulation of eigenvalue equations which embody the conditions for a stationary expectation value of the energy, (see, for example, McWeeny and Pickup (1980) and references therein), but the equations proposed do not always contain both the necessary and sufficient conditions for the optimum orbitals (see, for example, Albat and Gruen 1973). McWeeny (1974, 1975) has shown that for the general energy expression (2.4.19) the optimal orbitals are given by the solution of a single eigenvalue problem in which the effective hamiltonian is given by

$$\mathbf{h}_{\text{eff}} = \sum_{P} a_p \mathbf{R}_{PZ} \mathbf{h}_P \mathbf{R}_{PZ} + \sum_{P<Q} b_{pq} \mathbf{R}_{PQ} \mathbf{h}_{PQ} \mathbf{R}_{PQ} + \mathbf{R}_Z \mathbf{d}_Z \mathbf{R}_Z + \sum_{P} \mathbf{R}_P \mathbf{d}_P \mathbf{R}_P \tag{2.4.20}$$

in which

$$\mathbf{h}_P = \mathbf{h} + \mathbf{G}_P \tag{2.4.21a}$$

$$\mathbf{R}_{PQ} = \mathbf{R}_P + \mathbf{R}_Q \tag{2.4.21b}$$

$$\mathbf{h}_{PQ} = \nu_P \mathbf{h}_P - \nu_Q \mathbf{h}_Q \tag{2.4.21c}$$

and Z labels the subspace of the unoccupied orbitals. The a_p and b_{pq} are 'level shifters' and 'damping factors', respectively, which can be chosen to improve the convergence properties of the iterative procedure employed in the solution of the pseudo-eigen value problem (see, for example, Hillier and Saunders 1970; Guest and Saunders 1974). $\mathbf{d}_Z$ and $\mathbf{d}_P$ are arbitrary matrices which may be defined to select orbitals with a particular canonical significance, since the energy functional is invariant under a unitary mixing of the orbitals within a shell (see, for example, Hirao 1974).

The canonical molecular orbitals are usually delocalized over the whole of the molecule and, therefore, do not lend themselves to the qualitative discussion of bonding, particularly in polyatomic molecules, which is often empirically understood in terms of localized electron pairs or groups of electrons. However, the invariance of the molecular orbitals with respect to a unitary mixing of the orbitals within a shell allows one to determine localized molecular orbitals. Localized orbitals were first derived for the methane molecule by Coulson (1937). Localized equivalent orbitals were studied by Lennard-Jones (1949). Edmiston and Ruedenberg (1963) devised what is perhaps the most satisfactory computational scheme for the determination of localized molecular orbitals by requiring that the sum of the Coulomb repulsion energies of electrons associated with the same spatial orbital be a maximum. The Edmiston–Reudenberg localization criterion is not, however, unique and Gilbert (1964) has provided a comprehensive discussion of alternative localization criteria.

2.5 Multiconfigurational Hartree–Fock models

The Hartree–Fock model provides the best single determinantal wave function for a given molecular system. One method for extending the description of the electronic structure of an atom or molecule beyond the Hartree–Fock model is to employ a multideterminantal function (see, for example, Hartree, Hartree, and Swirles 1939) of the form

$$\Psi = \sum_{i} C_i \Phi_i \tag{2.5.1}$$

where Φ_i corresponds to a single product of orbitals. In the multiconfiguration self-consistent-field method the variation principle is used to optimize both the coefficients C_i and the expansion coefficients for the orbitals used in the construction of the configuration functions, Φ_i. Obviously, if a sufficiently large number of terms are included in eqn (2.5.1) then the multi-configuration self-consistent-field method can be regarded as a method for attacking the electron correlation problem directly. However, the complexity of such calculations is such that it is difficult to handle large numbers of configurations. Multi-configuration self-consistent-field wave functions are usually employed as a reference with respect to which correlation effects are calculated.

Although in using the function (2.5.1), the independent electron model has been abandoned, this functional form for the wave function can always lead to a correct description of molecular dissociation (see, for example, Wahl and Das 1970, 1977). It can also provide a balanced description of a number of electronic states (Docken and Hinze 1972; Ruedenberg, Cheung, and Elbert 1979).

Certainly the most significant problem which has to be addressed in multiconfiguration self-consistent-field calculations is the choice of configurations to be included in the expansion (2.5.1). For singly bonded diatomic molecules, such as LiH, Li_2, F_2 and NaF, a correct description of the dissociation of the molecule may be obtained by means of only two configuration functions Φ_i. This *optimized double configuration theory* (Wahl and Das 1977) has been applied successfully to a number of diatomic molecules. For multiply bonded diatomic molecules, such as N_2 and CO, a correct description of dissociation can be obtained by including all important valence configurations. This is known as the *optimized valence configuration theory* (Wahl and Das 1977). For polyatomic molecules the number of configurations which have to be included to obtain even a qualitative description of the dissociation of the molecule into its component atoms may be impractically large. If, on the other hand, the variation of only one or two nuclear coordinates is of interest, then a smaller number of configurations may provide satisfactory results.

Two other methods for selecting the configurations Φ_i should be mentioned. The *even-replacement MCSCF method,* due to Roothaan, Detrich and Hopper (1979), is computationally tractable, whereas the conceptually very simple *full orbital reaction space MCSCF method* of Ruedenberg and Sundbarg (1976) involves quite considerable computation. (The latter approach has also been discussed by Roos, Taylor and Siegbahn (1980) who termed it the *complete active space MCSCF method.*)

Many different computational schemes have been proposed for the optimization of wave functions of the form (2.5.1), which has to be performed subject to the orthonormality of the orbitals used in the construction of the Φ_i (Veillard 1966; Clementi and Veillard 1966; Das and Wahl 1966, 1967, 1972; Hinze and Roothaan 1967; Levy 1969, 1970, 1973; Grein and Chang 1971; Hinze 1973; Golebiewski, Hinze, and Yurtsever 1979; Grein and Banerjee 1975; Polezzo 1975; Kuprievich and Schramko 1975; Kendrick and Hillier 1976; Banerjee and Grein 1976, 1977; Wahl and Das 1977; Chang and Schwartz 1977; Ruttink and van Lenthe 1977; Ruedenberg, Cheung, and Elbert 1979; Dalgaard and Jorgensen 1978; Dalgaard 1979; Yeager and Jorgensen 1979; Roothaan, Detrich, and Hopper 1979; Lengsfield 1980; Werner and Meyer 1980). In what are perhaps the most promising of these techniques, the coefficients C_i and the orbital expansion coefficients are optimized simultaneously. Orthonormality conditions are maintained during the optimization by means of exponential transformations (see, for example, Werner and Meyer 1980). Quadratically convergent Newton–Raphson type procedures are used to determine the optimum values of the various coefficients (see, for example, Werner and Meyer 1980; Roos 1983).

2.6 Valence bond models

Interest in atomic and molecular collision processes, reactive scattering, and predissociation phenomena has led to revival of interest in valence bond theory in recent years as a method for obtaining potential energy curves and surfaces (see, for example, Balint-Kurti 1975; Gerratt 1974). In the traditional formulation of valence bond theory, the molecular wave function is written as a product of the wave functions for the component atoms. A qualitatively correct picture of dissociative processes is, therefore, usually obtained. The accuracy of the calculation may be improved by including further structures in the wave function. However, in resorting to a multi-structure valence bond approach, just as in multi-configuration Hartree–Fock theories, we have to abandon the simple picture afforded by the independent electron model. A model for the electronic structure of molecules will be considered in this section which may be regarded as a simple generalization of both the molecular orbital method and the valence bond theory. It is a synthesis of these two methods which, whilst removing many of their inadequacies, maintains the simple physical picture of electronic structure afforded by the independent electron model. In particular, we consider molecular wave functions which are constructed as a product of non-orthogonal orbitals (see, for example, Gerratt and Lipscomb 1968; Goddard 1967*a*,*b*, 1970; Gallup 1968, 1973), that is

$$\Psi_{S,M;k} = \sqrt{N!}\,\mathscr{A}(\phi_1\phi_2\ldots\phi_N\Theta^N_{S,M;k}) \tag{2.6.1}$$

where $\mathscr{A}$ is the indempotent antisymmetrizing operator and $\Theta^N_{S,M;k}$ is the total molecular spin function. If the orbitals are restricted to be atomic orbitals then we obtain the simple valence bond theory, whereas if they are restricted to be doubly occupied, i.e. $\phi_{2i-1}=\phi_{2i}$, the simple Hartree–Fock molecular orbital theory is recovered.

In order to discuss the properties of the generalized valence bond function (2.6.1) let us consider the hydrogen molecule. In 1949, Coulson and Fischer pointed out that the molecular orbital theory for H_2 fails at large nuclear separations because it takes insufficient account of electron repulsion. Working within a minimum basis set of a single 1s-function centred on each of the atoms, they suggested that the usual molecular orbitals for H_2

$$\phi_g = 1s_A + 1s_B \tag{2.6.2a}$$

$$\phi_u = 1s_A - 1s_B \tag{2.6.2b}$$

be replaced by orbitals of the form

$$\phi_1 = 1s_A + \lambda 1s_B \tag{2.6.3a}$$

$$\phi_2 = 1s_B + \lambda 1s_A \tag{2.6.3b}$$

in which λ is a variable parameter. The total spatial wave function for the ground state of H_2 is therefore

$$\{1s_A(1)+\lambda 1s_B(1)\}\{1s_B(2)+\lambda 1s_A(2)\}+\{1s_A(2)+\lambda 1s_B(2)\}\{1s_B(1)+\lambda 1s_A(1)\} \tag{2.6.4}$$

A simple rearrangement of eqn (2.6.4) shows it to be equivalent to the wave function of Weinbaum (1933) who added an ionic structure to the Heitler–London function (1927) to obtain

$$1s_A(1)1s_B(2)+1s_A(2)1s_B(1)+\mu\{1s_A(1)1s_A(2)+1s_B(1)1s_B(2)\}. \tag{2.6.5}$$

A further rearrangement of eqn (2.6.4) shows that it is also equivalent to a two configuration function based on molecular orbital theory, i.e.

$$\phi_g(1)\phi_g(2)+\nu\phi_u(1)\phi_u(2). \tag{2.6.6}$$

Coulson and Fischer (1949) were thus able to show that some of the benefits of configuration mixing and of multi-structure valence bond calculations may be obtained by employing more general orbital product functions. The Coulson–Fischer approach affords a simple physical picture of molecular electronic structure in terms of an orbital model.

When the parameter λ in the spatial wave function (2.6.3) is set to zero, the simple valence bond function is recovered; whereas if it is fixed at unity, we obtain the molecular orbital function. Thus, if λ is optimized at each nuclear separation, the Coulson–Fischer functions will approximate the valence bond function at large nuclear separations, the molecular orbital function as the united atom situation is approached and smoothly interpolate between these two limits for intermediate values of the internuclear distance.

In the most general wave function of the type (2.6.1) for the hydrogen molecule, the orbitals ϕ_1 and ϕ_2 are expanded in a large basis set. This wave function, or a function which is equivalent to it, has been considered by a number of authors (see, for example, Wilson and Gerratt 1975 and references therein). The variation of these generalized Coulson–Fisher orbitals with nuclear separation R is shown in Fig. 2.1, where the smooth transition from the valence bond to the molecular orbital model as R increases can be observed. The potential energy curve for the ground state of H_2 obtained by using the generalized Coulson–Fischer function is

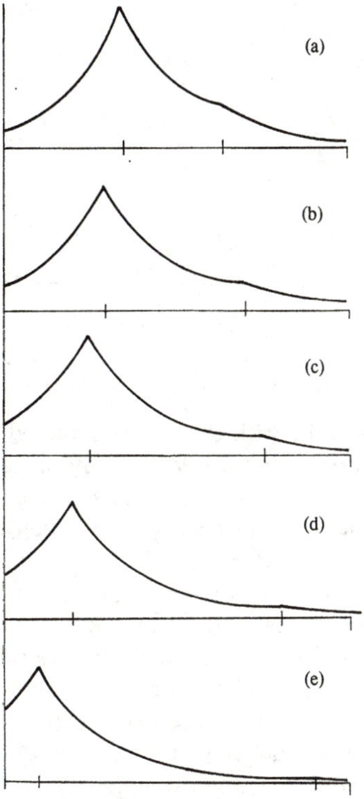

FIG. 2.1. Amplitude of the Coulson–Fischer orbital for the hydrogen molecule at various nuclear separations. (a) $R = 1.4a_0$; (b) $R = 2.0a_0$; (c) $R = 2.5a_0$; (d) $R = 3.0a_0$; (e) $R = 4.0a_0$.

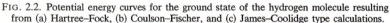

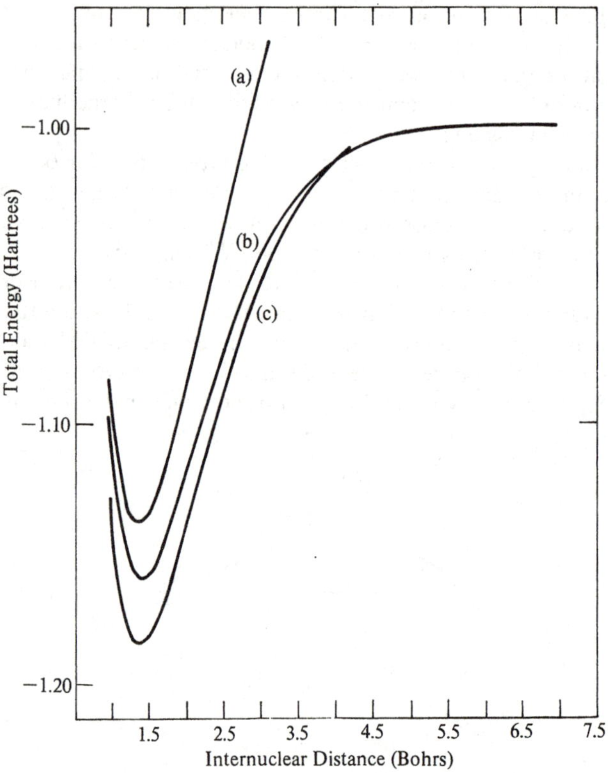

FIG. 2.2. Potential energy curves for the ground state of the hydrogen molecule resulting from (a) Hartree–Fock, (b) Coulson–Fischer, and (c) James–Coolidge type calculations.

given in Fig. 2.2, together with the curve obtained from a Hartree–Fock calculation and the exact curve. Being a more general orbital product function than the Hartree–Fock determinant, the generalized Coulson–Fischer function leads to an energy value below that given by the Hartree–Fock method. Goddard (1967*a*,*b*) has suggested that the difference between the Hartree–Fock energy and the energy resulting from the generalized Coulson–Fischer function be termed 'static correlation energy' and that the difference between the generalized Coulson–Fischer energy and the exact non-relativistic energy be termed the 'dynamic correlation energy'. This will be discussed further in Chapter 3.

The general orbital product wave function (2.6.1) has been investigated by a number of authors (see, for example, Gerratt 1974 and references therein, Kaplan 1975 and references therein). If necessary, the spin function $\Theta^N_{S,M,k}$ is replaced by a linear combination of spin functions to allow a smooth recoupling of the one-electron spin functions as a molecule dissociates (Gerratt 1971). The so-called valence states of traditional valence bond theory are thus avoided (van Vleck and Sherman 1935; Moffitt 1954; Doggett 1969; Gerratt 1974). The wave function (2.6.1) is generally more accurate than the widely used Hartree–Fock function in that in a variation calculation it leads to a lower energy. Furthermore, the function (2.6.1) predicts spin distributions near nuclei well and is useful, therefore, in the interpretation of nuclear magnetic resonance and electron spin resonance experiments. The use of the wave function (2.6.1) does, however, lead to severe computational problems. The non-orthogonality of the orbitals gives rise to expressions for the energy and other expectation values which contain $N!$ terms, where N is the number of electrons in the system. In order to employ the function (2.6.1) in studies of molecules containing many electrons, further restrictions have to be introduced. These restrictions most often take the form of orthogonality conditions which are imposed on the orbitals. Before considering these orthogonality conditions, let us consider the expression for the energy corresponding to function (2.6.1).

The wave function (2.6.1) may be written in the form

$$\Psi_{S,M;k} = \sqrt{N!}\,\mathscr{A}(\Phi\Theta^N_{S,M;k}) \tag{2.6.7}$$

where the N-electron spatial function is

$$\Phi = \prod_{\mu}^{n} \Phi_\mu \tag{2.6.8}$$

with each N_μ-electron function given by

$$\Phi_\mu = \prod_{i=1}^{N_\mu} \phi_{\mu i} \tag{2.6.9}$$

and

$$\sum_{\mu=1}^{n} N_\mu = N. \tag{2.6.10}$$

The energy expectation value corresponding to eqn (2.6.7) may be written as

$$E_{Sk} = (\Delta^S_{kk})^{-1} \sum_{P\in S_N} U^{N,S}_{k,k}(P)\langle P^r\Phi|\,\hat{\mathscr{H}}\,|\Phi\rangle \tag{2.6.11}$$

where P^r permutes the electronic spatial coordinates, Δ^S_{kk} is the normalization integral

$$\Delta^S_{kk} = \sum_{P\in S_N} U^{N,S}_{k,k}(P)\langle P^r\Phi\,|\,\Phi\rangle \tag{2.6.12}$$

and the matrices $\mathbf{U}^{N,S}$ form a representation of the symmetric group (see Chapter 5). For simplicity, we consider only a single spin coupling k. If the electronic hamiltonian is written as a sum of one and two electron parts as in eqn (2.1.13), then it can be shown that (see, for example,

Gerratt 1971)

$$E_{Sk} = \sum_{\mu_p,\nu_q} D(\mu_p\nu_q \mid Skk)\langle\mu_p|\, h\, |\nu_q\rangle$$
$$+\tfrac{1}{2}\sum_{\mu_p\nu_q}{}' \sum_{\sigma_r\tau_s}{}' D(\mu_p\nu_q\sigma_r\tau_s \mid Skk)\langle\mu_p\nu_q|\, g\, |\sigma_r\tau_s\rangle \tag{2.6.13}$$

where

$$D(\mu_p\nu_q \mid Skk) = (\Delta^S_{kk})^{-1} \sum_{P\in S_{N-1}} U^{N,S}_{k,k}(P)\langle P^r\Phi^{N-1}_{(\nu_q)} \mid \Phi^{N-1}_{(\mu_p)}\rangle \tag{2.6.14}$$

and

$$D(\mu_p\nu_q\sigma_r\tau_s \mid Skk) = (\Delta^S_{kk})^{-1} \sum_{P\in S_{N-2}} U^{N,S}_{k,k}(P)\langle P^r\Phi^{N-2}_{(\sigma_r\tau_s)} \mid \Phi^{N-2}_{(\mu_p\nu_q)}\rangle. \tag{2.6.15}$$

$\Phi^{N-2}_{(\sigma_1\sigma_2)}$ denotes the orbital product function Φ with the orbitals which are given in parenthesis removed. Efficient schemes for the evaluation of the D-coefficients have been devised (Gerratt 1971; Pyper and Gerratt 1977).

Now let us examine the effect of the orthogonality restrictions

$$\langle\phi_{\mu i} \mid \phi_{\nu j}\rangle \equiv \langle\mu_i \mid \nu_j\rangle = 0, \qquad \mu \neq \nu \qquad \forall i, j, \tag{2.6.16}$$

that is, the orbitals are divided into groups, and no orthogonality restrictions are imposed between orbitals which are in the same group but orbitals in different groups are required to be orthogonal. For the normalization integral (2.6.12), permutations which interchange electrons between different groups lead to terms which vanish and thus

$$\Delta^S_{kk} = \prod_{\mu}^{n} (d^{Sk}_\mu)^{-1} \tag{2.6.17}$$

where

$$d^{Sk}_\mu = \left(\sum_{P\in S_{N_\mu}} U^{N,S}_{k,k}(P)\langle P^r\Phi^{N_\mu}_\mu \mid \Phi^{N_\mu}_\mu\rangle\right)^{-1} \tag{2.6.18}$$

The $D(\mu_p\nu_q \mid Skk)$, defined in eqn (2.6.14), will clearly vanish unless $\mu = \nu$, that is

$$D(\mu_p\mu_q| \, Skk) = d^{Sk}_\mu \sum_{P\in S_{N_\mu-1}} U^{N,S}_{k,k}(P)\langle P^r\Phi^{N_\mu-1}_{\mu(q)} \mid \Phi^{N_\mu-1}_{\mu(p)}\rangle$$
$$= D_\mu(pq \mid Skk) \tag{2.6.19}$$

where $\Phi^{N_\mu-i}_{\mu(\,)}$ denotes $\Phi^{N_\mu}_\mu$ with the orbitals given in parenthesis removed. $D(\mu_p\nu_q\sigma_r\tau_s \mid Skk)$ will vanish unless

(a) $\mu = \nu = \sigma = \tau$: the intra-group two-electron energy or group 'self-energy' with D-coefficient

$$D(\mu_p\mu_q\mu_r\mu_s \mid Skk) = d_\mu^{Sk} \sum_{P \in S_{N_\mu - 1}} U_{k,k}^{N,S}(P) \langle P^r \Phi_{\mu(rs)}^{N_\mu - 2} \mid \Phi_{\mu(pq)}^{N_\mu - 2} \rangle$$
$$= D_\mu(pqrs \mid Skk) \tag{2.6.20}$$

(b) $\mu = \sigma$, $\nu = \tau$: the inter-group Coulomb energy with D-coefficient

$$D(\mu_p\nu_q\mu_r\nu_s \mid Skk) = d_\mu^{Sk} d_\nu^{Sk} \sum_{\substack{P = P_\mu P_\nu \\ P_\mu \in S_{N_\mu} \\ P_\nu \in S_{N_\nu}}} U_{k,k}^{N,S}(P_\mu) U_{k,k}^{N,S}(P_\nu)$$
$$\cdot \langle P_\mu^r \Phi_{\mu(r)}^{N_\mu - 1} \mid \Phi_{\mu(p)}^{N_\mu - 1} \rangle \langle P_\nu^r \Phi_{\nu(s)}^{N_\mu - 1} \mid \Phi_{\nu(q)}^{N_\mu - 1} \rangle$$
$$= D_\mu(pr \mid Skk) D_\nu(qs \mid Skk) \tag{2.6.21}$$

(c) $\mu = \tau$, $\nu = \sigma$: the inter-group exchange energy with D-coefficient

$$D(\mu_p\nu_q\nu_r\mu_s \mid Skk) = d_\mu^{Sk} d_\nu^{Sk} \sum_{\substack{P = P_\mu P_\nu P_{\mu_p \nu_q} \\ P_\mu \in S_{N_\mu - 1} \\ P_\nu \in S_{N_\nu - 1}}} U_{k,k}^{N,S}(P)$$
$$\cdot \langle P^r \Phi_{\mu(s)}^{N_\mu - 1} \Phi_{\nu(r)}^{N_\nu - 1} \mid \Phi_{\nu(q)}^{N_\nu - 1} \Phi_{\mu(p)}^{N_\mu - 1} \rangle \tag{2.6.22}$$

The total electronic energy may be written in the form

$$E_{Sk} = E_{Sk}^{(\mathrm{intra})} + E_{SK}^{(\mathrm{inter})} \tag{2.6.23}$$

with

$$E_{SK}^{(\mathrm{intra})} = \sum_{\mu}^{n} \mathbb{H}_\mu^{Sk} + \mathbb{J}_\mu^{SK} \tag{2.6.24}$$

and

$$E_{Sk}^{(\mathrm{inter})} = \tfrac{1}{2} {\sum_{\mu,\nu}^{n}}' \mathbb{J}_{\mu\nu}^{SK} + \mathbb{K}_{\mu\nu}^{SK}. \tag{2.6.25}$$

The one-electron energy is

$$\mathbb{H}_\mu^{Sk} = {\sum_{p,q}^{N_\mu}}' D_\mu(pq \mid Skk) \langle \mu_p | \, h \, | \mu_q \rangle \tag{2.6.26}$$

and the intra-group two-electron energy is

$$\mathbb{J}_\mu^{Sk} = \tfrac{1}{2} {\sum_{p,q}^{N_\mu}}' \sum_{r,s}^{N_\mu} D_\mu(pqrs \mid Skk) \langle \mu_p\mu_q | \, g \, | \mu_r\mu_s \rangle. \tag{2.6.27}$$

The inter-group Coulomb energy takes the form

$$\mathbb{J}_{\mu,\nu}^{Sk} = \sum_{p,r}^{N_\mu} \sum_{q,s}^{N_\nu} D_\mu(pr \mid Skk) D_\nu(qs \mid Skk) \langle \mu_p\nu_q | \, g \, | \mu_r\nu_s \rangle \tag{2.6.28}$$

and the inter-group exchange energy is

$$\mathbb{K}^{Skk}_{\mu,\nu} = \sum_{p,s}^{N_\mu} \sum_{q,r}^{N_\nu} D(\mu_p \nu_q \nu_r \mu_s \mid Skk) \langle \mu_p \nu_q | \, g \, | \nu_r \mu_s \rangle. \tag{2.6.29}$$

If the total spin function for each group of electrons is a singlet, then eqn (2.6.29) can be simplified further since

$$D(\mu_p \nu_q \nu_r \mu_s \mid Skk) = -\tfrac{1}{2} D(\mu_p \nu_r \mid Skk) D(\nu_q \mu_s \mid Skk). \tag{2.6.30}$$

By imposing the orthogonality conditions (2.6.16), we can control the non-orthogonality problem which plagues wave functions of the form (2.6.1). For an N-electron system divided into n groups of $N_1, N_2, \ldots, N_n$ electrons, there are

$$\tfrac{1}{2} n(n-1)(N_1! \, N_2! \ldots N_n!) \tag{2.6.31}$$

permutations which have to be considered compared with $N!$ if no orthogonality restrictions are introduced. Most applications of this approach have divided molecules into pairs of electrons and each pair is described by non-orthogonal orbitals which are orthogonal at all other orbitals used in the construction of the wave function (Slater 1965; Kotani *et al.* 1963; Wilson and Gerratt 1975; Goddard *et al.* 1973). This approach has been termed the generalized valence bond method by Goddard *et al.* (1973). The number of permutations which give rise to non-vanishing contributions to the energy expectation value in this pair function model is

$$2^{n-1} n(n-1) \tag{2.6.32}$$

for a system of n pairs and this does not increase dramatically with the number of electrons being considered, as illustrated in Table 2.2.

Table 2.2

Comparison of the number of terms which arise in the energy expression in the pair function model and in the model obtained by relaxing orbital orthogonality restrictions

Number of electrons	$2^{n-1}n(n-1)$	$N!$
4	4	24
6	24	720
8	96	40 320
10	320	3 628 800

For the pair function model, the expression for the energy expectation value takes a particularly simple form if the Serber basis (Serber 1934*a,b*; Pauncz 1979) for the spin functions is employed. In this basis, the one-electron spin functions are first coupled together in pairs and then the pair spin functions are coupled to give a total spin function. (Further details may be found in Chapter 5.) In the Serber basis

$$U^{N,S}_{k,l}(P_{2i-1,2i}) = \pm\delta_{k,l} \tag{2.6.33}$$

and

$$d^{Sk}_{\mu} = (1 + U^{S,N}_{k,k}(P_{\mu_1\mu_2})\langle\mu_1 \mid \mu_2\rangle\langle\mu_2 \mid \mu_1\rangle)^{-1}. \tag{2.6.34}$$

For the components of the energy we obtain

$$\mathbb{H}^{Sk}_{\mu} = d^{Sk}_{\mu}\{\langle\mu_1| h |\mu_1\rangle + \langle\mu_2| h |\mu_2\rangle + 2U^{N,S}_{k,k}(P_{\mu_1\mu_2})\langle\mu_1 \mid \mu_2\rangle\langle\mu_2| h |\mu_1\rangle\} \tag{2.6.35}$$

$$\mathbb{J}^{Sk}_{\mu} = d^{Sk}_{\mu}\{\langle\mu_1\mu_2| g |\mu_1\mu_2\rangle + U^{N,S}_{k,k}(P_{\mu_1\mu_2})\langle\mu_1\mu_2| g |\mu_2\mu_1\rangle] \tag{2.6.36}$$

$$\begin{aligned}
\mathbb{J}^{Sk}_{\mu,\nu} = d^{Sk}_{\mu} d^{Sk}_{\nu}\{&\langle\mu_1\nu_1| g |\mu_1\nu_1\rangle + \langle\mu_1\nu_2| g |\mu_1\nu_2\rangle \\
&+ \langle\mu_2\nu_1| g |\mu_2\nu_1\rangle + \langle\mu_2\nu_2| g |\mu_2\nu_2\rangle \\
&+ 2U^{N,S}_{k,k}(P_{\mu_1\mu_2})\langle\mu_1 \mid \mu_2\rangle(\langle\mu_1\nu_1| g |\mu_2\nu_1\rangle + \langle\mu_1\nu_2| g |\mu_2\nu_2\rangle) \\
&+ 2U^{N,S}_{k,k}(P_{\nu_1\nu_2})\langle\nu_1 \mid \nu_2\rangle(\langle\mu_1\nu_1| g |\mu_1\nu_2\rangle + \langle\mu_2\nu_1| g |\mu_2\nu_2\rangle) \\
&+ 2U^{N,S}_{k,k}(P_{\mu_1\mu_2})U^{N,S}_{k,k}(P_{\nu_1\nu_2})\langle\mu_1 \mid \mu_2\rangle\langle\nu_1 \mid \nu_2\rangle \\
&\times(\langle\mu_1\nu_1| g |\mu_2\nu_2\rangle + \langle\mu_1\nu_2| g |\mu_2\nu_2\rangle)\}
\end{aligned} \tag{2.6.37}$$

$$\begin{aligned}
\mathbb{K}^{Sk}_{\mu,\nu} = d^{Sk}_{\mu} d^{Sk}_{\nu} U^{N,S}_{k,k}(P_{\mu_2\nu_1})\{&\langle\mu_1\nu_1| g |\nu_1\mu_1\rangle + \langle\mu_1\nu_2| g |\nu_2\mu_1\rangle \\
&+ \langle\mu_2\nu_1| g |\nu_1\mu_2\rangle + \langle\mu_2\nu_2| g |\nu_2\mu_2\rangle \\
&+ 2U^{N,S}_{k,k}(P_{\mu_1\mu_2})\langle\mu_1 \mid \mu_2\rangle(\langle\mu_1\nu_1| g |\nu_1\mu_2\rangle + \langle\mu_1\nu_2| g |\nu_2\mu_2\rangle) \\
&+ 2U^{N,S}_{k,k}(P_{\nu_1\nu_2})\langle\nu_1 \mid \nu_2\rangle(\langle\mu_1\nu_1| g |\nu_2\mu_1\rangle + \langle\mu_2\nu_1| g |\nu_2\mu_2\rangle) \\
&+ 2U^{N,S}_{k,k}(P_{\mu_1\mu_2})U^{N,S}_{k,k}(P_{\nu_1\nu_2})\langle\mu_1 \mid \mu_2\rangle\langle\nu_1 \mid \nu_2\rangle \\
&\times(\langle\mu_1\nu_1| g |\nu_2\mu_2\rangle + \langle\mu_1\nu_2| g |\nu_1\mu_2\rangle)\}.
\end{aligned} \tag{2.6.38}$$

For a system containing an odd number of electrons the following additional terms arise for the non-paired (μth) electron

$$\mathbb{H}^{Sk}_{\mu} = \langle\mu| h |\mu\rangle \tag{2.6.39}$$

$$\mathbb{J}^{Sk}_{\mu} = 0. \tag{2.6.40}$$

For the inter-pair terms there are two cases:

(a) for the single orbital μ and the pair function (ν_1, ν_2) the inter-pair Coulomb term is

$$\mathbb{J}^{Sk}_{\mu,\nu} = d^{Sk}_{\nu}[\langle\mu\nu_1| g |\mu\nu_1\rangle + \langle\mu\nu_2| g |\mu\nu_2\rangle + 2U^{N,S}_{k,k}(P_{\nu_1\nu_2})\langle\nu_1 \mid \nu_2\rangle\langle\mu\nu_1| g |\mu\nu_2\rangle \tag{2.6.41}$$

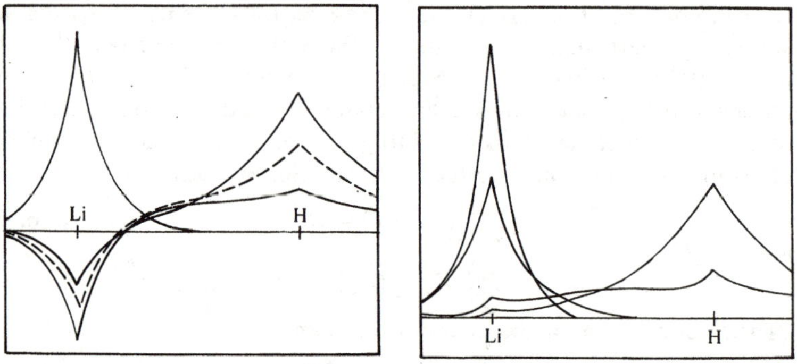

FIG. 2.3. Comparison of (*left*) pair function orbitals (——) and valence molecular orbital (– – – –) with (*right*) orbitals obtained by completely relaxing orbital orthogonality restrictions for the lithium hydride molecule.

and the exchange term is

$$\mathbb{K}^{Sk}_{\mu,\nu} = d^{Sk}_{\nu} U^{N,S}_{k,k}(P_{\mu\nu_2})\{\langle \mu\nu_1 | g | \nu_1\mu \rangle + \langle \mu\nu_2 | g | \nu_2\mu \rangle + 2U^{N,S}_{k,k}(P_{\nu_1\nu_2})\langle \nu_1 | \nu_2 \rangle \langle \mu\nu_1 | g | \nu_2\mu \rangle\} \quad (2.6.42)$$

(b) for two single orbitals, μ and ν, the Coulomb term is

$$\mathbb{J}^{Sk}_{\mu,\nu} = \langle \mu\nu | g | \mu\nu \rangle \quad (2.6.43)$$

and the exchange term is

$$\mathbb{K}^{SK}_{\mu,\nu} = U^{S,N}_{k,k}(P_{\mu\nu})\langle \mu\nu | g | \nu\mu \rangle. \quad (2.6.44)$$

Additional terms will arise if a linear combination of spin couplings is employed (see Gerratt 1971).

In Fig. 2.3 the orbitals obtained from a pair function calculation for the lithium hydride molecule are compared with those obtained by the molecular orbital method and those obtained by relaxing the restriction that the pair functions must satisfy the orthogonality conditions (2.6.16) (Palke and Goddard 1969; Wilson and Gerratt 1974). The orbitals of the core pair function were found to be virtually identical. The pair function valence orbitals, like the molecular orbital, were found to have a node. This occurs, however, in a region where the core orbital dominates the electron density and the orthogonality restrictions (2.6.16) have not significantly altered the shape of the orbitals in the chemically important valence region.

3

ELECTRON CORRELATION

3.1 Electron correlation effects

Electron correlation effects may be loosely termed the corrections to independent electron models, or orbital theories, of molecular electronic structure within the framework of non-relativistic quantum mechanics. Of course, the effect of electron correlation cannot be directly observed since there is no way of 'turning on' the interactions between electrons in a molecule. Because correlation energy is a theoretical concept, there have been a number of differing definitions which have led to confusion in the comparison of different methods. Slater (1953) complained that "the term correlation energy has meant so many things to different people that (Slater) feels it perhaps better not to use the word at all".

Wigner and Seitz (1933) first used the term 'electron correlation' in a study of the electronic structure and cohesive energy of metals. The correlation energy of a system is usually defined as that part of the energy which is ignored when a single determinantal description is employed. However, there is a statistical correlation of the motion of the electrons which is accounted for by the single determinantal wave function. This statistical correlation is known as Fermi correlation and it prevents electrons with the same spin occupying the same region of space. Fermi correlation arises from the antisymmetry of the total wave function with respect to permutation of the electron coordinates. It is regarded as part of the overall correlation of electrons in space whereas the component of the energy which arises from the antisymmetrization of the wave function is not regarded as a part of the electron correlation energy.

Electron correlation effects can be analysed in terms of the concepts of probability theory (see, for example, Kutzelnigg *et al.* 1968). In probability theory two statistical variables x and y, say, with distribution functions $X(x)$ and $Y(y)$ are said to be independent if and only if the total distribution function $\mathrm{F}(x, y)$ is given by the product

$$\mathrm{F}(x, y) = X(x)Y(y) \tag{3.1.1}$$

otherwise the variables are said to be correlated. The one-electron density function $\Gamma^{(1)}(x_1'; x_1)$ defined in Section 1.6, is the probability that an electron will be found at the point $\mathbf{r}_1$. The probability that the electron labelled **1** will be at point $\mathbf{r}_1$ is $\Gamma^{(1)}(\mathbf{r}_1)/N$. The two-electron density function $\Gamma^{(2)}(\mathbf{r}_1, \mathbf{r}_2)$ is the probability that electrons will be found at points $\mathbf{r}_1$ and $\mathbf{r}_2$. The probability that the electron labelled **1** will be found at point $\mathbf{r}_1$ and electron **2** will be found at point $\mathbf{r}_2$ is $2\Gamma^{(2)}(r_1, r_2)/(N(N-1))$. The true two-electron density function is actually smaller than the value corresponding to uncorrelated electrons

$$\Gamma^{(2)}_{\text{uncorrelated}}(\mathbf{r}_1, \mathbf{r}_2) = \frac{N-1}{2N}\Gamma^{(1)}(\mathbf{r}_1)\Gamma^{(1)}(\mathbf{r}_2) \tag{3.1.2}$$

in regions of space where $(\mathbf{r}_1 - \mathbf{r}_2)$ is small and larger in other regions of space, that is, the electrons are mutually repelled.

There are three factors which contribute to the correlation of the motion of the electrons in a molecule in space. Firstly, there is the Coulomb repulsion between the electrons. $1/r_{\text{ij}}$, the Coulomb interaction between electrons i and j, becomes infinite when r_{ij} is zero and, therefore, it is energetically extremely unfavourable for two electrons to be in close proximity. The two-electron density function $\Gamma^{(2)}$ must vanish or become very small when the two electrons are in the same region of space. This type of correlation of the electrons gives rise to what is often termed the Coulomb hole. This is the only type of electron correlation in systems such as the He atom or the H_2 molecule. It can be shown that the Coulomb hole has a cusp (Bingel 1966).

A second source of correlation effects in molecules is associated with the Pauli principle; that is the requirement that the total wave function be antisymmetric with respect to the permutation of the electronic coordinates. The one-electron density function may be written as a sum of a density function associated with α spins and a density function associated with β spins; that is

$$\Gamma^{(1)}(\mathbf{r}) = \Gamma^{(1)}_{\alpha}(\mathbf{r}) + \Gamma^{(1)}_{\beta}(\mathbf{r}). \tag{3.1.3}$$

For the two-electron density function, it can be easily shown that

$$\Gamma^{(2)}(\mathbf{r}_1, \mathbf{r}_2) = \Gamma^{(2)}_{\alpha,\alpha}(\mathbf{r}_1, \mathbf{r}_2) + \Gamma^{(2)}_{\alpha,\beta}(\mathbf{r}_1, \mathbf{r}_2) + \Gamma^{(2)}_{\beta,\alpha}(\mathbf{r}_1, \mathbf{r}_2) + \Gamma^{(2)}_{\beta,\beta}(\mathbf{r}_1, \mathbf{r}_2). \tag{3.1.4}$$

For a single determinantal wave function it can further be shown that

$$\Gamma^{(2)}_{\alpha,\alpha}(\mathbf{r}_1, \mathbf{r}_2) = 0 \tag{3.1.5}$$

$$\Gamma^{(2)}_{\beta,\beta}(\mathbf{r}_1, \mathbf{r}_2) = 0 \tag{3.1.6}$$

$$\Gamma^{(2)}_{\alpha,\beta}(\mathbf{r}_1, \mathbf{r}_2) = \Gamma^{(1)}_{\alpha}(\mathbf{r}_1)\Gamma^{(1)}_{\beta}(\mathbf{r}_2) \tag{3.1.7}$$

and

$$\Gamma^{(2)}_{\beta,\alpha}(\mathbf{r}_1, \mathbf{r}_2) = \Gamma^{(1)}_{\beta}(\mathbf{r}_1)\Gamma^{(1)}_{\alpha}(\mathbf{r}_2). \tag{3.1.8}$$

There is no correlation between electrons with different spins in the description afforded by a single determinant but the probability of finding two electrons with the same spin at a point $\mathbf{r}$ is zero. This is termed the

Fermi correlation of the electrons and gives rise to the Fermi hole. The Fermi hole is found to be deeper than the Coulomb hole (see, for example, Kutzelnigg 1973) but it has no cusp. Correlation of electrons with the same spins is a consequence of the antisymmetry of the wave function. If the wave function is antisymmetric then

$$\Gamma^{(2)}(\mathbf{r}_1, \mathbf{r}_2; \mathbf{r}_1', \mathbf{r}_2) = -\Gamma^{(2)}(\mathbf{r}_1, \mathbf{r}_2; \mathbf{r}_2', \mathbf{r}_1') \tag{3.1.9}$$

and the two electron density functions must be zero when $\mathbf{r}_1 = \mathbf{r}_2$.

The third factor which influences the correlation of the motion of the electrons in a molecule is the symmetry of the system; correlation effects will depend on both the spin and spatial symmetry properties. The effect of the spin symmetry is, of course, closely connected with the Pauli principle discussed above. Correlation effects are also influenced by the spatial symmetry; that is, the point symmetry group according to which the particular configuration of the nuclei transforms and the particular irreducible representation of that point group to which the electronic wave function is required to belong.

3.2 Analysis of correlation effects

In this section, the qualitative analysis of correlation effects in molecules is considered. The ground state of the hydrogen molecule provides a simple system for which some qualitative aspects of the correlation problem, which are applicable to systems containing many more electrons, can be examined.

The ground state of the hydrogen molecule may be described by two spin-orbitals with different spins and thus the correlation effects are purely electrostatic. Kolos and Roothann (1960) reported both a Hartree–Fock and an extended James–Coolidge calculation for the hydrogen molecule with energies which may be denoted by $E_{\mathrm{HF}}(R)$ and $E_{\mathrm{JC}}(R)$. The total correlation energy is then given by

$$E_{\text{correlation}}(R) = E_{\mathrm{JC}}(R) - E_{\mathrm{HF}}(R). \tag{3.2.1}$$

The quantitative analysis of the correlation energy will be delayed until Section 3.5. Here the functional form of the wave function is of more concern.

The Hartree–Fock ansatz provides an increasingly poor description of the hydrogen molecule as the nuclei are drawn apart. This difficulty may be removed if, following Coulson and Fischer (1949), a valence bond function of the type described in Section (2.6) is employed. For the hydrogen molecule, this function consists of a product of two non-orthogonal orbitals, one for each electron

$$\Psi = \sqrt{(2)}\mathscr{A}(\phi_1\phi_2\Theta) \tag{3.2.2}$$

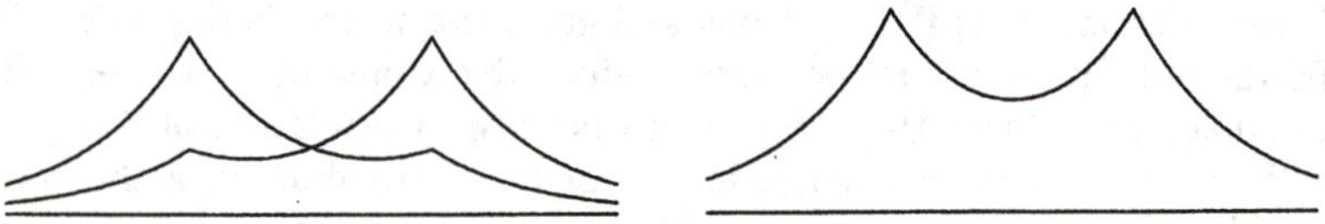

FIG. 3.1. Comparison of orbitals for the hydrogen molecule in the extended valence bond model and the molecular orbital model.

This provides a good description of the molecule for any nuclear geometry. In Fig. 3.1, the orbitals for the hydrogen molecule in the generalized valence bond model are compared with those in the molecular orbital model at the equilibrium nuclear geometry. It can be seen that the success of the former in describing molecular dissociation is attributable to the incorporation of longitudinal or 'left–right' correlation effects. The longitudinal correlation of electrons is the tendency of electron **1** to be located in the region of nucleus **A** when electron **2** is in the region of nucleus **B**. At large nuclear separations, electron **1** (**2**) is totally associated with nucleus **A** (**B**) and longitudinal correlation effects are extremely important. As shown in Fig. 2.1, the amplitude of orbital ϕ_1 (ϕ_2) on nucleus **B** (**A**) increases as the nuclei are brought closer together. The orbitals ϕ_1 and ϕ_2 have $C_{\infty v}$ symmetry and are expanded in terms of basis functions with σ symmetry.

After longitudinal correlation, the next most important source of correlation effects may be termed angular correlation. The importance of angular correlation effects was recognized by Hirschfelder and Linnett (1950) who demonstrated that the Heitler–London function for H_2 could be usefully improved by employing a superposition of the following configurations

$$\begin{aligned}\Psi_0 &= a(r_1)b(r_2)+b(r_1)a(r_2)\\ \Psi_1 &= \{a(r_1)b(r_2)+b(r_1)a(r_2)\}(x_1x_2+y_1y_2)\\ \Psi_2 &= a(r_1)b(r_2)z_{a_1}z_{a_2}+b(r_1)a(r_2)z_{b_1}z_{b_2}\\ \Psi_3 &= a(r_1)a(r_2)+b(r_1)b(r_2)\end{aligned} \tag{3.2.3}$$

with

$$a(r)=\sqrt{\left(\frac{\zeta^3}{\pi}\right)}\mathrm{e}^{-\zeta r_a}; \qquad b(r)=\sqrt{\left(\frac{\zeta^3}{\pi}\right)}\mathrm{e}^{-\zeta r_b}. \tag{3.2.4}$$

Ψ_0 is a covalent structure, Ψ_3 is an ionic structure, Ψ_2 is capable of describing polarization effects, and Ψ_1 is able to describe angular correlation effects since it contains $\cos(\phi_1-\phi_2)$. This wave function, which has been found to be fairly accurate for all values of the nuclear separation, has also been investigated by Shull and Ebbing (1957); Kim and

Hirschfelder (1957); Linnett and Ricra (1969); Urban and Lavicky (1972); and Wilson and Gerratt (1975). Shull (1959) has concluded from an extensive natural orbital analysis that the addition of configurations describing angular correlation effects provides the greatest improvement of the extended Coulson-Fischer function.

A generalized Hirschfelder–Linnett wave function for the hydrogen molecule may be written as (Wilson and Gerratt 1975)

$$\Psi = \sqrt{(2)}\mathscr{A}[\{C_1(\sigma_1\sigma_2) + C_2(\pi_{x_1}\pi_{x_2} + \pi_{y_1}\pi_{y_2})\}\Theta] \tag{3.2.5}$$

where the symmetry type of each orbital is given. This wave function may be written in the alternative form

$$\Psi = \sqrt{(2)}\mathscr{A}[\{\omega(^1\Sigma_g^+)\phi_1\phi_2\}\Theta] \tag{3.2.6}$$

where $\omega(^1\Sigma_g^+)$ is the Wigner projection operator which ensures that the wave function has $^1\Sigma_g^+$ symmetry. This projection operator has the form

$$\omega(^1\Sigma_g^+) = \frac{1}{2\pi}\int_0^{2\pi} C_\phi + \sigma_v(\phi) - S_\phi + C_2(\phi)\,d\phi \tag{3.2.7}$$

where C_ϕ is a rotation about the internuclear axis by an angle ϕ, $C_2(\phi)$ is a rotation about the C_2 axis intersecting the xz plane at an angle ϕ, $\sigma_v(\phi)$ reflects through the plane intersecting the xz plane with an angle ϕ, and $S_\phi = \sigma_h C_\phi$, where σ_h reflects through the plane perpendicular to the internuclear axis. For orbitals of π symmetry, we have

$$\omega(^1\Sigma_g^+)\pi_{x_1}\pi_{x_2} = \pi_{x_1}\pi_{x_2} + \pi_{y_1}\pi_{y_2} \tag{3.2.8}$$

and for orbitals with δ symmetry

$$\omega(^1\Sigma_g^+)\delta_{x^2-y^2,1}\delta_{x^2-y^2,2} = \delta_{x^2-y^2,1}\delta_{x^2-y^2,2} + \delta_{xy,1}\delta_{xy,2}. \tag{3.2.9}$$

The function (3.2.8) describes the 180° angular correlation of the electrons; that is the tendency for the lines joining the electron position coordinate to the internuclear axis to be inclined at 180° when viewed along the internuclear axis. The function (3.2.9) describes the 90° angular correlation of the electrons. In Table 3.1, the expansion coefficients for the extended Hirschfelder–Linnett function are displayed for various nuclear separations. It can be seen that the angular correlation effects become increasingly important as the nuclear separation is reduced. For values of the nuclear separation between 1.1 and 6.0 bohr the orbitals of π symmetry in the function (3.2.5) are found to have approximately $D_{\infty h}$ symmetry. To see why this is so, consider the π configuration

$$\Phi = \pi_{x,1}\pi_{x,2} \tag{3.2.10}$$

Table 3.1
Orbital expansion coefficients for the extended Hirschfelder–Linnett wave function, eqn (3.2.5)

Basis function[a]	$R=1.40a_0$	$R=2.00a_0$	$R=3.00a_0$	$R=4.00a_0$	$R=8.00a_0$
$1s_a$ (see below)	0.680 75	0.813 23	0.954 38	0.995 58	0.999 97
$2s_a$ (1.2200)	0.186 64	0.106 58	0.011 79	−0.004 75	0.000 02
$2p\sigma_a$ (2.0500)	−0.007 00	0.011 67	0.009 50	0.003 42	0.000 01
$3d\sigma_a$ (2.3500)	−0.012 92	−0.005 02	0.000 45	0.000 69	0.000 00
$1s_b$ (see below)	0.135 57	0.099 88	0.060 18	0.024 51	0.000 36
$2s_b$ (1.2200)	0.060 41	0.040 37	0.022 92	0.013 20	0.000 31
$2p\sigma_b$ (2.0500)	0.048 85	0.020 68	0.007 64	0.002 18	−0.000 05
$3d\sigma_b$ (2.3500)	0.014 46	0.005 51	0.001 66	0.000 47	−0.000 02
C_1	0.775 89	0.817 06	0.902 74	0.967 78	0.999 93
$2p\pi_{xa}$ (0.8660)	−0.520 67	−0.550 40	−0.602 47	−0.644 36	0.964 16
$3d\pi_{xa}$ (2.3500)	−0.100 28	−0.129 48	−0.183 83	−0.208 41	0.178 56
$2p\pi_{xb}$ (0.8660)	−0.520 69	−0.550 46	−0.602 59	−0.644 62	0.181 50
$3d\pi_{xb}$ (2.3500)	−0.100 29	−0.129 61	−0.183 48	−0.207 84	0.073 56
C_2	−0.047 72	−0.042 31	−0.026 21	−0.012 36	−0.000 89
$\zeta(1s)$	1.4000	1.1869	1.0315	1.0000	1.0000

[a] Orbital exponent given in parentheses.

and put

$$\begin{aligned}\pi_{x,1} &= \pi_{x,a} + \lambda\pi_{x,b}\\ \pi_{x,2} &= \pi_{x,b} + \lambda\pi_{x,a}\end{aligned} \tag{3.2.11}$$

where $\pi_{x,a}$ is a function on centre A and $\pi_{x,b}$ is a function on centre B. When the parameter λ is set to zero

$$\Phi = \pi_{x,a}\pi_{x,b} \tag{3.2.12}$$

whilst when $\lambda = 1$

$$\Phi = \pi_{x,a}\pi_{x,b} + \pi_{x,b}\pi_{x,a} + \pi_{x,a}\pi_{x,a} + \pi_{x,b}\pi_{x,b}. \tag{3.2.13}$$

It is for this later function that the orbitals have $D_{\infty h}$ symmetry. $\lambda = 1$ allows for the angular correlation of the electron near one of the nuclei; $\lambda = 0$ allows for angular correlation for those dispositions in which the two electrons are near different nuclei. As discussed by Linnett and Ricra (1969), the function (3.2.13) is expected to be much more important than (3.2.12).

In the extended Coulson–Fischer function (3.2.1), the orbitals are required to have $C_{\infty v}$ symmetry and are related by a σ_h operation. Since

$$D_{\infty h} = C_{\infty v} + \sigma_h C_{\infty v} \tag{3.2.14}$$

the total wave function has $D_{\infty h}$ symmetry, as indeed it must. However,

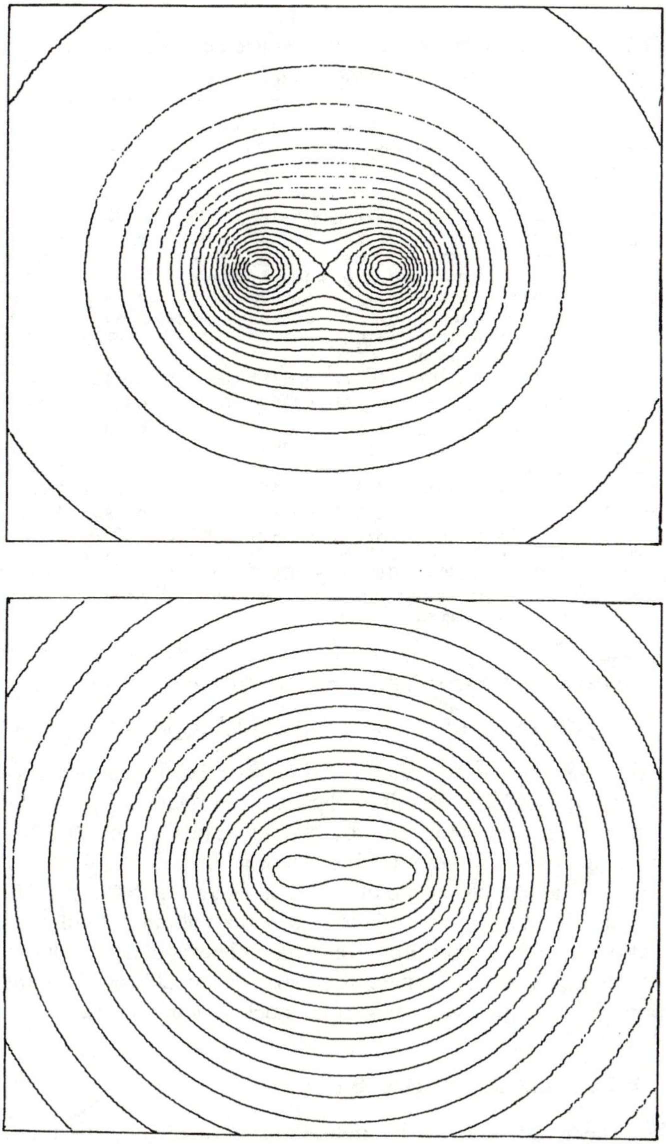

FIG. 3.2. Orbitals describing axial correlation in the hydrogen molecule.

Table 3.2

Orbital expansion coefficients for hydrogen molecule wave function, eqn (3.2.15), describing axial correlation effects†

Orbital	Basis function	Nuclear separation (bohr) $R=1.00$	$R=1.20$	$R=1.40$	$R=1.60$	$R=1.80$	$R=2.00$	$R=4.00$	$R=6.00$
	C_1	−0.751 23	−0.770 20	0.791 90	−0.820 25	0.845 74	0.878 39	0.957 23	0.996 94
	1s(a)	0.531 14	0.581 09	0.632 65	0.666 47	0.711 15	0.744 82	0.992 33	0.999 36
	2s(a)	0.253 26	0.257 64	0.239 45	0.227 80	0.196 33	0.175 24	0.000 73	0.000 43
	2p(a)	−0.055 60	−0.039 30	−0.027 01	−0.023 61	−0.018 98	−0.018 18	−0.003 96	−0.000 90
ϕ_1	3d(a)	−0.023 15	−0.021 04	−0.018 90	−0.019 07	−0.017 06	−0.016 36	−0.002 09	−0.000 35
	1s(b)	0.145 42	0.109 57	0.087 37	0.067 63	0.063 95	0.059 96	0.014 33	0.002 01
	2s(b)	0.153 60	0.131 57	0.118 08	0.118 02	0.107 23	0.100 08	0.018 12	0.002 43
	2p(b)	0.064 98	0.044 69	0.026 81	0.009 32	0.001 37	−0.007 29	−0.005 40	−0.001 18
	3d(b)	0.017 43	0.014 67	0.009 77	0.003 38	0.000 55	−0.004 23	−0.001 74	−0.000 33
	C_2	0.123 69	0.119 52	0.112 61	0.126 30	0.118 02	0.124 89	0.026 83	0.003 92
	1s(a)	−0.556 41	−0.579 62	0.576 43	−0.539 45	0.487 13	0.393 23	0.664 10	−0.317 20
ϕ_3	2s(a)	0.636 14	0.698 39	−0.753 33	0.778 26	−0.793 88	−0.767 69	−0.437 77	0.048 55
	2p(a)	−0.274 22	−0.272 21	0.284 63	−0.306 32	0.317 71	0.321 58	0.540 33	−0.599 77
	3d(a)	−0.087 99	−0.090 53	0.104 79	−0.124 34	0.132 41	0.150 34	0.207 65	−0.221 80
	1s(a)	0.161 88	0.159 22	0.148 93	0.183 91	0.179 71	0.198 92	0.428 71	−0.634 56
ϕ_4	2s(a)	0.399 79	0.402 24	0.411 89	0.384 04	0.391 23	0.377 96	0.205 09	0.011 35
	2p(a)	−0.096 81	−0.094 07	−0.098 44	−0.090 70	−0.102 96	−0.104 52	−0.205 16	0.297 43
	3d(a)	−0.034 67	−0.030 43	−0.025 34	−0.020 07	−0.021 15	−0.019 62	−0.084 10	0.109 92

† Orbital exponents are given in Table 3.1; $\phi_2 = \sigma_h \phi_1$.

the orbitals comprising the wave function could also be required to have $D_{\infty h}$ symmetry. A two-configuration wave function of the form

$$\Psi = \sqrt{(2)}\mathscr{A}([C_1\{\phi_1(C_{\infty v})\phi_2(C_{\infty v})\} + C_2\{\phi_3(D_{\infty h})\phi_4(D_{\infty h})\}]\Theta) \quad (3.2.15)$$

could then be constructed (Wilson and Gerratt 1975). The second configuration in eqn (3.2.15) describes the axial correlation of the electrons, that is, the tendency for one of the electrons to be far from the internuclear axis when the other is close ot it. In Fig. 3.2, the form of the orbitals ϕ_3 and ϕ_4 at the equilibrium nuclear separation, 1.4 bohr, is illustrated. For nuclear separations of less than about 1.1 bohr, the axial correlation is in fact more important than the angular correlation. This can be seen in Table 3.2 where the expansion coefficients for the wave function (3.2.15) are given for various nuclear separations.

3.3 Definition of the correlation energy

One of the most widely used definitions of the correlation energy in atomic and molecular systems is due to Löwdin (1959):

> "The correlation energy for a certain state with respect to a specified Hamiltonian is the difference between the exact eigenvalue of the Hamiltonian and its expectation value in the Hartree–Fock approximation for the state under consideration."

Often it is convenient to replace the Hartree–Fock approximation in this definition by some other well-defined approximation of the type discussed in Chapter 2. For example, even in the case of the hydrogen molecule it is perhaps more appropriate to discuss correlation effects with respect to an extended valence bond reference function especially if large internuclear separations are to be considered (see, for example, Goddard 1967*a*,*b*; Wilson and Gerratt 1975). Goddard (1967*a*,*b*) has suggested that correlation energy calculated with respect to the Hartree–Fock model be subdivided into static correlation energy, that is, the energy lowering achieved when orbital orthogonality restrictions are relaxed, and dynamic correlation energy. Sinanoğlu (1964), on the other hand, has divided correlation effects into dynamical and non-dynamical.

In order to obtain an expression for the correlation energy, let us consider a system described by the hamiltonian $\hat{\mathcal{H}}$. Let Ψ_i denote the exact wave function for the particular state of interest for this hamiltonian operator and let Φ_i denote a reference function used to describe the system in the ith state to a first approximation. Furthermore, let

$$\Psi_i = \Phi_i + \chi_i \tag{3.3.1}$$

and

$$\langle \Phi_i \mid \Phi_i \rangle = 1 \qquad \langle \Phi_i \mid \Psi_i \rangle = 1 \tag{3.3.2}$$

which implies that

$$\langle \Phi_i \mid \chi_i \rangle = 0. \tag{3.3.3}$$

Multiplying the exact Schrödinger equation, $\hat{\mathcal{H}}\Psi_i = E_i\Psi_i$, for the system from the left, by the reference function Φ_i, gives immediately

$$\begin{aligned} E_i &= \langle \Phi_i | \hat{\mathcal{H}} | \Phi_i \rangle + \langle \Phi_i | \hat{\mathcal{H}} | \chi_i \rangle \\ &= E_{\text{reference}} + E_{\text{correlation}}. \end{aligned} \tag{3.3.4}$$

$E_{\text{reference}}$ is the expectation of the total hamiltonian for the reference wave function; this will be the Hartree–Fock energy in the definition given by Löwdin. $E_{\text{correlation}}$ is the correlation energy for the system with respect to the chosen reference model.

In actual calculations, neither the exact wave function within the chosen reference model nor, of course, the exact wave function itself are known. In the calculation of the expectation of the hamiltonian for the reference wave function it is usual, at least in molecular calculations, to parameterize the orbitals in terms of a finite basis set. This, inevitably, introduces an error associated with basis set truncation effects (see Chapter 6). In the calculation of the correlation correction to the wave function, Φ_i, it is necessary not only to employ a finite basis set but also to use only a subset of the terms which are required to obtain a complete

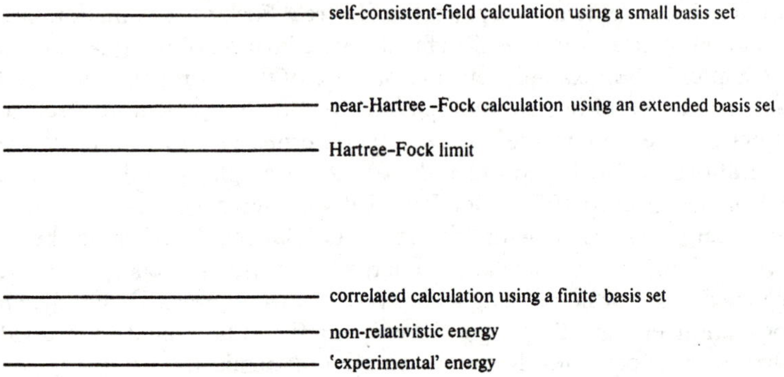

FIG. 3.3. Relation between calculations, the exact model, and the non-relativistic limit.

description of electron correlation effects within that basis set. In Fig. 3.3, the relation between the results of actual calculations, the exact reference model results and exact correlated results is illustrated schematically.

In the determination of energies of chemical interest, such as the calculation of binding energies, ionization energies, or excitation energies, it should be remembered that these energies are often quite small compared with the total energy of the molecule, or molecules, involved. In Fig. 3.4, the possible changes in the total energy which can take place during the process A → B are indicated. Small energy differences can be

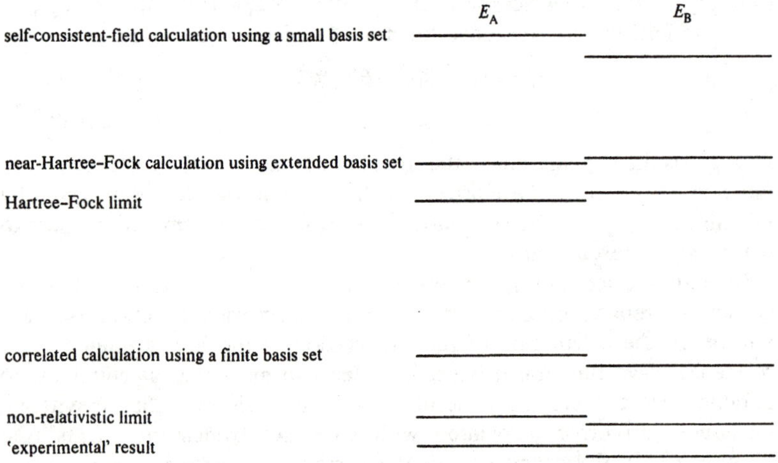

FIG. 3.4. Schematic representation of the energy changes which may take place during the reaction A → B.

significantly affected by both the truncation of the basis set and the truncation of the expansion for the correlation wave function.

3.4 Magnitude of the correlation energy

The correlation energy cannot be determined from experiment. It is determined from the experimental total energy of a molecule and the energy of the molecule in some independent electron model, usually the Hartree–Fock model. An estimate of the energy of the molecule in the non-relativistic limit is obtained by subtracting the relativistic energy $E_{\text{relativistic}}$ from the experimental energy $E_{\text{experimental}}$; thus

$$E_{\text{non-relativistic}} = E_{\text{experimental}} - E_{\text{relativistic}}. \tag{3.4.1}$$

An empirical estimate of the correlation energy is then given by the difference between the non-relativistic energy and the energy given by an orbital model. Thus, with respect to the Hartree–Fock model, the correlation energy is given by

$$E_{\text{correlation}} = E_{\text{Hartree–Fock}} - E_{\text{non-relativistic}}. \tag{3.4.2}$$

This is the 'experimental' correlation energy with which calculated correlation energies are usually compared.

There are, however, a number of uncertainties which arise in the above procedure. First of all, there is a degree of uncertainty associated with the extrapolation of the matrix Hartree–Fock energies, that is, energies obtained within a finite basis set, to the basis set limit. This may lead to error of a few per cent in the 'experimental' correlation energy. Secondly, current understanding of relativistic effects for many electron systems is, at present, insufficient for reliable estimates of the relativistic energies to be made, except for atoms of the first row of the periodic table. Indeed, correlation energies and relativistic energies are not strictly additive as eqns (3.4.1) and (3.4.2) suggest. The errors present in most contemporary calculations of electron correlation energies are somewhat larger than the errors resulting from the uncertainties in relativistic energies and comparisons with 'experimental' correlation energies, as defined above, provide the most useful means of assessing the accuracy of a theoretical approach.

The Hartree–Fock ground state energies for some atoms of the first row and the second row of the periodic table are displayed in Table 3.3, together with the corresponding relativistic energies and correlation energies. For the first-row atoms, the correlation energy is much larger than the energy associated with relativistic effects, whereas for argon, in the second row, the relativistic energy is more than twice as large as the

Table 3.3
Hartree–Fock energy, correlation energy, relativistic energy, and lamb shift for some atomic systems[a]

System	Hartree–Fock energy	Correlation energy	Relativistic energy	Lamb shift
$C(^3P)$	−37.688 61	−0.1551	−0.0138	+0.0014
$N(^4S)$	−54.400 91	−0.1861	−0.0273	+0.0025
$O(^3P)$	−74.809 37	−0.2539	−0.0494	+0.0040
$Ar(^1S)$	−526.817 34	−0.732	−1.7609	+0.0552

[a] All energies are in atomic units. This table is based on the work of Veillard and Clementi (1968).

correlation energy. The energy associated with the Lamb shift, a quantum electrodynamical correction, is also given in Table 3.3.

Electron correlation energies are particularly large in negative ions (Simons 1977; Massey 1976, 1979). Indeed, if correlation energies are calculated with respect to a Hartree–Fock reference function, they are often vital if a description of a negative ion is to be obtained which is stable when compared with the neutral system plus an electron (Ahlrichs 1975; England 1980; Cooper and Wilson 1982*d*). The instability of some negative ions within the Hartree–Fock approximation suggests that it may be preferable to use the Hartree–Fock energy of the neutral species in the definition of the correlation energy (Cook 1980).

Changes in the magnitude of the correlation energy with the geometry of the nuclei or between different electronic states are of particular interest in quantum chemistry. The correlation energy is typically of the same order of magnitude as the dissociation energy of a molecule. In

Table 3.4
Empirical correlation energy for the ground state of the nitrogen molecule

Experimentally determined ground state energy of the nitrogen atom, E_N	−54.6122 a.u.
Dissociation energy, D_0	9.760 eV
Fundamental vibrational frequency, ω_e	2358.07 cm^{-1}
Experimental ground state energy of the nitrogen molecule, $E_{N_2} = 2E_N + D_0 + \frac{1}{2}\hbar\omega_e$	−109.5884 a.u.
Relativistic energy of the nitrogen atom, E_N^{rel}	−0.027 32 a.u.
Empirical relativistic energy for the nitrogen molecule, $E_{N_2}^{rel} = 2E_N^{rel}$	−0.054 64 a.u.
Non-relativistic energy for the nitrogen molecule, $E_2^{n\text{-}rel} = E_{N_2} - E_{N_2}^{rel}$	−109.5338 a.u.
Hartree–Fock energy for the nitrogen molecule at its equilibrium nuclear geometry, $E_{N_2}^{HF}$	−108.9956 a.u.
Empirical correlation energy for the N_2 molecule $E_{N_2}^{corr} = E_{N_2}^{HF} - E_{N_2}^{n\text{-}rel}$	−0.5382 a.u.
Compare with	
Correlation energy for two isolated nitrogen atoms	−0.3722 a.u.
Thus	
Change in correlation energy on forming N—N bond	−0.1660 a.u. (=−4.517 eV)

Table 3.4 the Hartree–Fock energies and the correlation energies for the ground state of the nitrogen atom and of the nitrogen molecule are compared. The experimental binding energy of the nitrogen molecule is 0.33% of the energy of two isolated nitrogen atoms. The correlation energy changes by 44.6% when the two nitrogen atoms are bound.

The Hartree–Fock model does not provide a qualitatively correct description of molecular dissociation in many cases, including the ground state of the nitrogen molecule discussed above. If the correlation energy is to be examined as a function of nuclear geometry, it is preferable to determine correlation effects with respect to a model which provides a qualitatively correct description of the system for all geometries. The valence bond models, described in Section 2.6, are most useful in this respect. In Fig. 3.5, the correlation energy of the hydrogen molecule is shown as a function of internuclear separation with respect to a Hartree–Fock reference function and with respect to the extended valence bond function defined in eqn (2.6.1).

3.5 Analysis of the correlation energy

A detailed analysis of the correlation energy of a molecular system can often provide valuable insight into the origin of correlation effects. There is, of course, no unique method of analysis for electron correlation

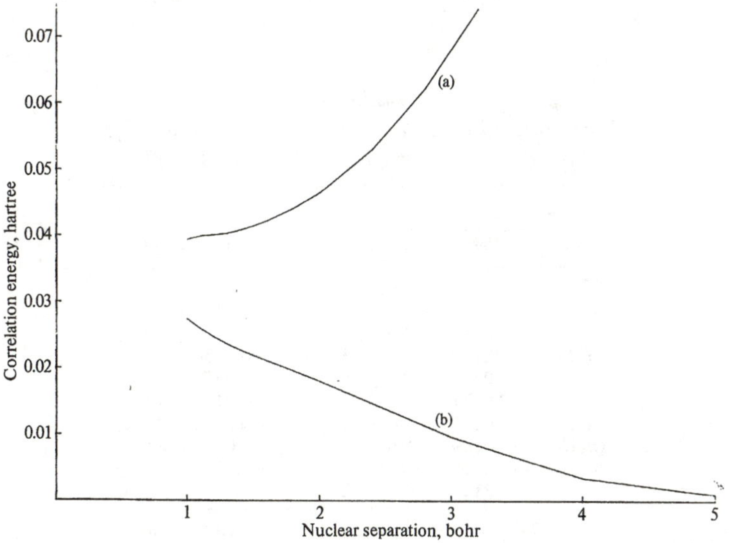

FIG. 3.5. Correlation energy of the hydrogen molecule as a function of nuclear separation (a) with respect to the Hartree–Fock function, (b) with respect to an extended valence bond function.

energies and the insight which is obtained is dependent, to a large extent, on the chosen analytical technique.

For diatomic molecules the correlation energy can be analysed in terms of the concepts introduced in Section 3.2; namely, longitudinal (or left–right) correlation, axial correlation, and angular correlation. These terms can be employed in a more qualitative fashion to discuss the components of the correlation energy in polyatomic molecules.

The most widely used method for analysing correlation energy is the electron pair theory. Simple geminal theories lead to a decomposition of the correlation energy $N/2$ pairs for an N-electron system. More generally, diagrammatic perturbation theory, coupled pair many-electron theory, and the coupled electron pair approximation, for example, lead to the analysis of the correlation energy in terms of $\frac{1}{2}N(N-1)$ electron pairs. In second-order perturbation theory the correlation energy can be written exactly as a sum of pair energies. By employing the scaling procedure, described in Section 7.6, the total correlation energy through any order of the perturbation series can be analysed in terms of pair energies. For example, in Table 3.5, an analysis of the valence shell correlation energy of the water molecule is compared with a similar analysis for the H_2S molecule. This Table serves to emphasize the similarity in the nature of the correlation effects in the valence shells of these two molecules as one would expect on simple chemical grounds.

For molecules at their equilibrium nuclear configuration the virial

Table 3.5

Pair correlation energies for the H_2O and H_2S molecules in their ground electronic states[a]

H_2O		H_2S	
Pair	Pair energy	Pair	Pair energy
$2a_1^2$	−0.0105 (4.2%)	$4a_1^2$	−0.0092 (5.3%)
$2a_1 1b_1$	−0.0226 (9.1%)	$4a_1 2b_1$	−0.0158 (9.1%)
$1b_1^2$	−0.0218 (8.8%)	$2b_1^2$	−0.0179 (10.3%)
$3a_1 2a_1$	−0.0197 (8.0%)	$5a_1 4a_1$	−0.0134 (7.8%)
$3a_1 1b_1$	−0.0370 (15.0%)	$5a_1 2b_1$	−0.0251 (14.5%)
$3a_1^2$	−0.0211 (8.5%)	$5a_1^2$	−0.0173 (10.0%)
$2a_1 1b_2$	−0.0205 (8.3%)	$4a_1 2b_2$	−0.0122 (7.0%)
$1b_1 1b_2$	−0.0349 (14.1%)	$2b_1 2b_2$	−0.0205 (11.8%)
$3a_1 1b_2$	−0.0375 (15.2%)	$5a_1 2b_2$	−0.0252 (14.5%)
$1b_2^2$	−0.212 (8.6%)	$2b_2^2$	−0.0170 (9.8%)

[a] These pair correlation energies were obtained bv using the scaling method, described in Section 7.6, in third-order diagrammatic many-body perturbation theory calculations. The geometries and basis sets are given by Guest and Wilson (1980).

theorem is obeyed. Thus

$$E_{\text{total}} = -T_{\text{total}}. \tag{3.5.1}$$

Since the virial theorem is also obeyed by the Hartree–Fock energy

$$E_{\text{HF}} = -T_{\text{HF}} \tag{3.5.2}$$

and, therefore, for the correlation energy we have

$$E_{\text{corr}} = -T_{\text{corr}}. \tag{3.5.3}$$

Since the kinetic energy often has a simpler form than the total energy, the relation (3.5.3) can prove most useful in the analysis of correlation effects. For example, in strongly orthogonal geminal models, of the type discussed in Section 7.3, the correlation energy is a sum of both intra-geminal and inter-geminal terms. However, the kinetic energy, T_{corr}, may be written entirely in terms of intra-geminal components

$$T_{\text{corr}} = \sum_{\mu=1}^{\mu=\frac{1}{2}N} T_{\text{corr},\,\mu}. \tag{3.5.4}$$

By using the virial (3.5.1), the total energy for a molecule, within a strongly orthogonal geminal model, may be written as a sum of the energies of the component parts with no interaction terms. Allen and Shull (1961) have suggested that this approach provides a quantum mechanical basis for the concept of bond energy which is widely used in chemical discussions.

3.6 Correlation effects for atomic and molecular properties

Some properties of atoms and molecules depend directly on the electronic energy and are, therefore, affected by the correlation energy and the accuracy with which it is calculated. For example, in molecules the equilibrium nuclear geometry is directly determined by the energy. The equilibrium geometry determined by employing a limited basis set with little or no account of electron correlation is often found to lead to quite accurate bond lengths. This degree of agreement with experiment is more often than not fortuitous. As the basis set is extended the matrix Hartree–Fock method usually leads to bond lengths which are somewhat shorter than the experimentally determined value. Correlation effects extend the bond length, bringing it into closer agreement with the experimental value. Correlation energy is also important in the determination of the shape of potential curves and surfaces. In Table 3.6, the values of the fundamental frequency of vibration ω_e, for various diatomic hydrides, are shown for the matrix Hartree–Fock model and for calculations which take account of electron correlation. In this Table we give the coupled electron pair approximation results of Meyer and Rosmus (1975). The correlation energy can obviously affect calculated energy changes, such as barriers to internal rotation or ionization potentials.

Table 3.6
The effects of electron correlation on fundamental frequencies of vibration[a]

	$\omega_e(\mathrm{cm}^{-1})$		
Molecule	Matrix Hartree–Fock	Coupled-electron pair approximation	Experiment
LiH	1428.2	1401.5	1405.7
BeH	2145.0	2064.6	2060.8
BH	2484.6	2352.1	2366.9
CH	3044.1	2841.7	2858.5
NH	3546.5	3269.3	3282.1
OH	4054.5	3743.6	3739.9
HF	4476.2	4169.3	4138.7
NaH	1184.2	1172.3	1172.2
MgH	1583.6	1492.3	1497.0
AlH	1730.8	1691.7	1682.6
SiH	2126.2	2034.7	2041.8
PH	2487.6	2365.9	2380.0
SH	2839.9	2676.4	2697.0
HCl	3141.1	2977.2	2991.1

[a] Taken from the work of Meyer and Rosmus (1975).

Table 3.7
The effects of electron correlation on dipole moments[a]

	Dipole moment (atomic units)		
Molecule	Matrix Hartree–Fock	Coupled-electron pair approximation	Experimental
LiH	−6.028	−5.837	−5.882
BeH	−0.277	−0.170	—
BH	1.785	1.398	1.270
CH	1.602	1.430	1.460
NH	1.649	1.578	—
OH	1.768	1.689	1.668
FH	1.902	1.833	1.827
NaH	−7.077	−6.470	—
MgH	−1.481	−1.236	—
AlH	0.107	−0.098	—
SiH	0.276	0.141	—
PH	0.546	0.481	—
SH	0.875	0.810	0.620
ClH	1.194	1.136	1.093

[a] Taken from the work of Meyer and Rosmus (1975).

A second class of atomic and molecular properties are given as expectation values. In the calculation of such properties, correlation effects which were not very significant in the calculation of the correlation energy may turn out to be of great importance. In the determination of dipole moments (see, for example, Green 1974), single excitations are found to be most important in a configuration mixing expansion, whereas in the calculation of correlation energies the double excitations are found to be most important. In Table 3.7 calculated dipole moments taken from the work of Meyer and Rosmus (1975) are presented for a number of diatomic hydrides. Results obtained by using the matrix Hartree–Fock model and the coupled-electron pair approximation are compared with experimental values.

Finally, we consider second-order properties such as polarizability. This describes the response of the system to an applied electric field. The polarizability is related to the second derivative of the energy with respect to the applied external electric field in the limit that this field is infinitesimally small. This derivative may be evaluated by finite difference methods and the calculated value will then be affected by the accuracy with which correlation energy is calculated in zero-field and in a small applied field. As an example, in Table 3.8, we give some calculated polarizabilities for the F^- ion. The effects of electron correlation are particularly important in the calculation of polarizabilities of negative ions such as F^-. Perturbation theory was used to estimate the effects of electron correlation on the polarisability of F^- presented in Table 3.8.

Table 3.8

The effects of electron correlation on polarizability calculated by diagrammatic perturbation theory through second-order and third-order[a]

	Basis set A	Basis set B	Basis set C
α_{scf}	9.23	10.57	10.62
$\alpha_{scf}+\alpha_2$	12.58	16.84	16.54
$\alpha_{scf}+\alpha_2+\alpha_3$	11.00	12.93	12.56

[a] Based on the work of Wilson and Sadlej (1981); atomic units are used throughout; basis sets consist of electric-field-variant Gaussian-type orbitals in the contractions: basis set A: (13s, 8p, 1d/7s, 4p, 1d), basis set B (14s, 9p, 1d/8s, 5p, 1d), basis set C (14s, 9p, 2d/8s, 5p, 2d).

4

THE LINKED DIAGRAM THEOREM

4.1 N-dependence of expectation values

Chemistry is primarily concerned not with the properties of single molecules but with periodic trends, homologous series, functional groups, and the like. In chemistry the systemization of the properties of a series of molecules is just as important as the determination of all of the properties of one particular species. Theoretically, therefore, it is important that any method which is applied to the problem of molecular electronic structure leads to expressions for expectation values which are directly proportional to the number of electrons in the system being studied. Meaningful comparisons of atoms and molecules of different sizes are then possible. This property has been termed 'size-consistency' (Pople, Binkley, and Seeger 1976; Davidson and Silver 1977; Wilson 1981*c*). Independent electron models, such as the widely used Hartree–Fock approximation, provide a size-consistent theory of atomic and molecular structure. Many of the theoretical techniques which are used to take account of electron correlation effects, however, have a non-linear dependence on the number of electrons being considered. For example, the method of configuration mixing, when limited to single and double excitations with respect to a single determinantal reference function, leads to expressions for correlation energy which depend on the square root of the number of electrons in the system. Methods for treating electron correlation effects which do lead to expressions for the correlation energy and other correlated expectation values directly proportional to the number of electrons in the system are based, directly or indirectly, on the linked diagram theorem. The linked diagram theorem is the topic of this chapter.

Size-consistency, or direct proportionality to N, the number of electrons, is of crucial importance in many chemical studies. Consider a system A to which some technique is applied to give a total energy $E(\mathrm{A})$; and a system B to which we apply the same technique to obtain an energy $E(\mathrm{B})$. If the same technique is applied to the supersystem AB in which A and B are an infinite distance apart, then it is to be expected that the total energy of the supersystem will be given by

$$E(\mathrm{AB}) = E(\mathrm{A}) + E(\mathrm{B}). \tag{4.1.1}$$

The energy of any system should be written as a sum of the energies of its component parts if those parts are an infinite distance apart. Furthermore, this result should hold no matter how the component parts are defined. If expectation values are directly proportional to the number of electrons in a system, then the energy of that system can be expressed as a sum of the energies of the component parts. Size-consistency, and the closely related concept of additive separability (Primas 1965), is clearly important in the calculation of molecular interaction energies, dissociation energies, ionization energies and, indeed, in any process in which the number of interacting electrons changes. The linked diagram theorem is fundamental to the development of size-consistent theories of molecular electronic structure.

The linked diagram expansion can most easily be approached via perturbational techniques. In Section 4.2, we discuss the perturbation theory of Lennard-Jones, Brillouin, and Wigner. This is formally perhaps the most straightforward and simple perturbation expansion but unfortunately contains spurious terms which have an unphysical dependence on the number of electrons in the system. Rayleigh–Schrödinger perturbation theory is somewhat more complicated than the Lennard-Jones–Brillouin–Wigner expansion from which it can be derived by expansion of certain energy denominators. Furthermore, the Rayleigh–Schrödinger perturbation theory also contains terms which have an unphysical dependence on the number of electrons in the molecule. However, it was first shown by Brueckner that these spurious terms mutually cancel in each order of the perturbation series. Rayleigh–Schrödinger perturbation theory is described in Section 4.3. The particle–hole formalism which will be required in order to explicitly carry out the cancellation of unphysical terms is introduced in Section 4.4; and the many-body perturbation theory, which is directly proportional to the number of electrons being considered, is described in Section 4.5. The linked diagram theorem is discussed in Section 4.6. The use of diagram techniques is elaborated in Section 4.7 and this enables the diagrammatic perturbation theory, that is, the diagrammatic representation of the many-body perturbation theory, to be given in Section 4.8. In Section 4.9, the diagrammatic perturbation theory of open-shell systems is considered. Quasi-degeneracy in diagrammatic perturbation theory is discussed in Section 4.10. Finally, in Section 4.11, the use of valence bond functions of the type described in Section 2.6 in diagrammatic perturbation theory studies is considered.

4.2 Lennard-Jones–Brillouin–Wigner perturbation theory

The partitioning technique, due to Feshbach (1962) and Löwdin (1962*a,b* and references therein), provides a straightforward and general introduction to perturbation expansions. The Lennard-Jones–Brillouin–Wigner perturbation theory (Lennard-Jones 1930; Brillouin 1932; Wigner 1935) is formally the simplest perturbative expansion. It is instructive to examine the Lennard-Jones–Brillouin–Wigner expansion in some detail before considering more physically acceptable approaches.

Let $\hat{P}$ denote the projector onto some zero-order model wave function, $|\Phi\rangle$, and let $\hat{Q}$ be its complement

$$\hat{Q} = \hat{I} - \hat{P}. \tag{4.2.1}$$

The electronic Schrödinger equation

$$\hat{\mathcal{H}}\,|\Psi\rangle = \mathcal{E}\,|\Psi\rangle \tag{4.2.2}$$

where $\hat{\mathcal{H}}$ is the molecular electronic Hamiltonian, may then be written as a two-by-two block matrix equation

$$\begin{pmatrix} \hat{P}\hat{\mathcal{H}}\hat{P} & \hat{P}\hat{\mathcal{H}}\hat{Q} \\ \hat{Q}\hat{\mathcal{H}}\hat{P} & \hat{Q}\hat{\mathcal{H}}\hat{Q} \end{pmatrix} \begin{pmatrix} \hat{P}\Psi \\ \hat{Q}\Psi \end{pmatrix} = \mathcal{E} \begin{pmatrix} \hat{P}\Psi \\ \hat{Q}\Psi \end{pmatrix} \tag{4.2.3}$$

where $|\Phi\rangle = \hat{P}\,|\Psi\rangle$. $\hat{Q}\,|\Psi\rangle$ can be eliminated to produce the effective Schrödinger equation

$$[\hat{P}\hat{\mathcal{H}}\hat{P} + \hat{P}\hat{\mathcal{H}}\hat{Q}(\mathcal{E} - \hat{Q}\hat{\mathcal{H}}\hat{Q})^{-1}\hat{Q}\hat{\mathcal{H}}\hat{P}]\,|\Phi\rangle = \mathcal{E}\,|\Phi\rangle \tag{4.2.4}$$

or

$$\hat{\mathcal{H}}_{\text{eff}}\Phi = \mathcal{E}\Phi \tag{4.2.5}$$

where

$$\hat{\mathcal{H}}_{\text{eff}} = [\hat{P}\hat{\mathcal{H}}\hat{P} + \hat{P}\hat{\mathcal{H}}\hat{Q}(\mathcal{E} - \hat{Q}\hat{\mathcal{H}}\hat{Q})^{-1}\hat{Q}\hat{\mathcal{H}}\hat{P}]. \tag{4.2.6}$$

This effective hamiltonian has eigenfunctions in the model space but has the exact energy as a eigenvalue.

Various forms of perturbation theory result from different expansions of the inverse in the effective hamiltonian (4.2.6) using the operator identity

$$(\hat{X} - \hat{Y})^{-1} = \sum_{n=0}^{\infty} \hat{X}^{-1}(\hat{Y}\hat{X}^{-1})^n \tag{4.2.7}$$

If $\hat{\mathcal{H}}_0$ denotes some zero-order hamiltonian then the perturbation expansion of Lennard-Jones–Brillouin–Wigner is obtained by putting

$$\hat{X} = \mathcal{E} - \hat{\mathcal{H}}_0 \tag{4.2.8}$$

and

$$\hat{Y} = \hat{\mathcal{H}} - \hat{\mathcal{H}}_0. \tag{4.2.9}$$

Let the total hamiltonian be written as a sum of the zero-order operator and a perturbation operator

$$\hat{\mathcal{H}} = \hat{\mathcal{H}}_0 + \hat{\mathcal{H}}_1 \tag{4.2.10}$$

with

$$\hat{\mathscr{H}}_0 |\Phi_i\rangle = E_i |\Phi_i\rangle \qquad i = 0, 1, 2, \ldots \tag{4.2.11}$$

and

$$\hat{\mathscr{H}} |\Psi_i\rangle = \mathscr{E}_i |\Psi_i\rangle \qquad i = 0, 1, 2, \ldots \tag{4.2.12}$$

where

$$\mathscr{E}_i = E_i + \Delta E_i \tag{4.2.13}$$

and ΔE_i is referred to as the level shift for the ith state.

Confining our attention, for simplicity, to the ground state, $i = 0$, we introduce the projection operator

$$\hat{P}_0 = |\Phi_0\rangle\langle\Phi_0| \tag{4.2.14}$$

and its orthogonal complement

$$\hat{Q}_0 = \hat{I} - |\Phi_0\rangle\langle\Phi_0|. \tag{4.2.15}$$

It is useful to assume, without loss of generality, that the ground state wave function is normalized

$$\langle\Phi_0 \mid \Phi_0\rangle = 1 \tag{4.2.16}$$

and that the intermediate normalization condition

$$\langle\Phi_0 \mid \Psi_0\rangle = 1 \tag{4.2.17}$$

is obeyed. It is also useful to define the wave operator, $\hat{\Omega}$, such that

$$|\Psi_0\rangle = \hat{\Omega} |\Phi_0\rangle. \tag{4.2.18}$$

Thus the action of the wave operator on the ground state model wave function is to produce the exact ground state wave function. Conversely, the action of the projection operator $\hat{P}_0$ on the exact ground state wave function is to produce the ground state model wave function. $\hat{\Omega}$ has the following properties

$$\hat{P}_0\hat{\Omega} = \hat{P}_0 \tag{4.2.19a}$$

$$\hat{\Omega}\hat{P}_0 = \hat{\Omega} \tag{4.2.19b}$$

$$\hat{\Omega}\hat{Q}_0 = 0 \tag{4.2.19c}$$

Using the wave operator, $\hat{\Omega}$, an expression for the level shift, ΔE_0, can be obtained

$$\Delta E_0 = \langle\Phi_0| \hat{\mathscr{H}}_1 |\Psi_0\rangle \tag{4.2.20a}$$

$$\Delta E_0 = \langle\Phi_0| \hat{\mathscr{H}}_1\hat{\Omega} |\Phi_0\rangle \tag{4.2.20b}$$

$$\Delta E_0 = \langle\Phi_0| \hat{\mathscr{V}} |\Phi_0\rangle \tag{4.2.20c}$$

where $\hat{\mathscr{V}}$ is the reaction operator defined by

$$\hat{\mathscr{V}} = \hat{\mathscr{H}}_1\hat{\Omega}. \tag{4.2.21}$$

In the Lennard-Jones–Brillouin–Wigner perturbation theory, the wave operator is written as

$$\hat{\Omega}=\left[\hat{I}+\sum_{n=1}^{\infty}\left(\frac{\hat{Q}_0}{E_0+\Delta E_0-\hat{\mathscr{H}}_0}\hat{\mathscr{H}}_1\right)^n\right]\hat{P}_0 \tag{4.2.22}$$

and the level shift, therefore, takes the form

$$\Delta E_0=\langle\Phi_0|\,\hat{\mathscr{H}}_1+\sum_{n=1}^{\infty}\hat{\mathscr{H}}_1\left(\frac{\hat{Q}_0}{E_0+\Delta E_0-\hat{\mathscr{H}}_0}\hat{\mathscr{H}}_1\right)^n|\Phi_0\rangle. \tag{4.2.23}$$

The expressions (4.2.22) and (4.2.23) for the wave operator and the reaction operator are formally equivalent to the integral equations

$$\hat{\Omega}=\hat{P}_0+\frac{\hat{Q}_0}{E_0+\Delta E_0-\hat{\mathscr{H}}_0}\hat{\mathscr{H}}_1\hat{\Omega} \tag{4.2.24}$$

and

$$\hat{\mathscr{V}}=\hat{\mathscr{H}}_1\hat{P}_0+\hat{\mathscr{H}}_1\frac{\hat{Q}_0}{E_0+\Delta E_0-\hat{\mathscr{H}}_0}\hat{\mathscr{V}} \tag{4.2.25}$$

respectively, from which the corresponding perturbation expansions can be obtained by iteration.

The level shift is most often written as

$$\Delta E_0=\sum_{i=1}^{\infty}E_0^{(i)} \tag{4.2.26}$$

where $E_0^{(i)}$ depends on the ith power of the perturbing operator, $\hat{\mathscr{H}}_1$. The first few terms of the Lennard-Jones–Brillouin–Wigner perturbation expansion may be written explicitly as follows:

$$E_0^{(0)}=\langle\Phi_0|\,\hat{\mathscr{H}}_0\,|\Phi_0\rangle \tag{4.2.27a}$$

$$E_0^{(1)}=\langle\Phi_0|\,\hat{\mathscr{H}}_1\,|\Phi_0\rangle \tag{4.2.27b}$$

$$E_0^{(2)}=\langle\Phi_0|\,\hat{\mathscr{H}}_1\hat{R}\hat{\mathscr{H}}_1\,|\Phi_0\rangle \tag{4.2.27c}$$

$$E_0^{(3)}=\langle\Phi_0|\,\hat{\mathscr{H}}_1(\hat{R}\hat{\mathscr{H}}_1)^2\,|\Phi_0\rangle \tag{4.2.27d}$$

$$E_0^{(4)}=\langle\Phi_0|\,\hat{\mathscr{H}}_1(\hat{R}\hat{\mathscr{H}}_1)^3\,|\Phi_0\rangle \tag{4.2.27e}$$

$$E_0^{(5)}=\langle\Phi_0|\,\hat{\mathscr{H}}_1(\hat{R}\hat{\mathscr{H}}_1)^4\,|\Phi_0\rangle. \tag{4.2.27f}$$

In these expressions $\hat{R}$ is the resolvant

$$\hat{R}=\frac{Q_0}{\mathscr{E}_0-\mathscr{H}_0}=\sum_{k\neq 0}\frac{|\Phi_k\rangle\langle\Phi_k|}{\mathscr{E}_0-E_k}. \tag{4.2.28}$$

It is clear that the general ith order terms take the form

$$E_0^{(i)}=\langle\Phi_0|\,\hat{\mathscr{H}}_1(\hat{R}\hat{\mathscr{H}}_1)^{i-1}\,|\Phi_0\rangle \tag{4.2.29}$$

The Lennard-Jones–Brillouin–Wigner perturbation expansion is a simple geometric series. However, the unknown exact energy, $\mathscr{E}_0$, is contained within the denominators in second-order and higher-order terms. The expansion is, therefore, not a simple power series in the perturbation, $\hat{\mathscr{H}}_1$. The perturbation expansion of Lennard-Jones, Brillouin, and Wigner does not lead to expressions which are directly proportional to the number of electrons in the system being studied. This can be readily demonstrated by considering the application of the second-order theory to the energy of two helium atoms an infinite distance apart. The energy obtained is not equal to twice the second-order energy of a single helium atom.

4.3 Rayleigh–Schrödinger perturbation theory

In Rayleigh–Schrödinger perturbation theory (see, for example, Hirschfelder, Byers Brown, and Epstein 1964) the unknown exact total energies which arise in the denominators in the expansion of Lennard-Jones, Brillouin, and Wigner are avoided. This enables a size-consistent theory of electron correlation effects in molecules to be developed; a theory which leads to expressions which are directly proportional to the number of electrons in the system.

The Rayleigh–Schrödinger perturbation expansion can be derived from the Lennard-Jones–Brillouin–Wigner expansion by replacing the exact energy in the denominators which arise in the latter by the energy corresponding to the model wave function. The Rayleigh–Schrödinger perturbation theory is obtained by putting

$$\hat{X} = E_0 - \hat{\mathscr{H}}_0 \tag{4.3.1}$$

and

$$\hat{Y} = \hat{\mathscr{H}} - \hat{\mathscr{H}}_0 - \mathscr{E}_0 + E_0 \tag{4.3.2}$$

in the expansion (4.2.7) of the inverse operator in the effective hamiltonian (4.2.6). The wave operator $\hat{\Omega}$ is written in an alternative form by making a different division of the hamiltonian to that used in Section 4.2 for the Lennard-Jones–Brillouin–Wigner theory. In particular, the following alternative zero-order hamiltonian is defined as

$$\hat{\mathscr{H}}_0 + \hat{Q}_0 \, \Delta E_0 \hat{Q}_0 \tag{4.3.3}$$

together with the corresponding perturbation

$$\hat{\mathscr{H}}_1 - \hat{Q}_0 \, \Delta E_0 \hat{Q}_0. \tag{4.3.4}$$

This leads to the following expressions for the wave operator

$$\hat{\Omega} = \left[\hat{I} + \sum_{n=1}^{\infty} \left(\frac{\hat{Q}_0}{E_0 - \hat{\mathscr{H}}_0} (\hat{\mathscr{H}}_1 - \Delta E_0) \right)^n \right] P_0 \tag{4.3.5}$$

which may be written in the form

$$\hat{\Omega}=\hat{P}_0+\frac{\hat{Q}_0}{E_0-\hat{\mathcal{H}}_0}(\hat{\mathcal{H}}_1-\Delta E_0)\hat{\Omega}. \tag{4.3.6}$$

Using the expansion for the level shift in powers of the perturbation operator and making a similar expansion for the wave operator

$$\hat{\Omega}=\sum_{i=1}^{\infty}\hat{\Omega}^{(i)} \tag{4.3.7}$$

we can develop an expansion for the wave operator in terms of powers of the perturbation, $\hat{\mathcal{H}}_1$; that is

$$\hat{\Omega}^{(i)}=\frac{\hat{Q}_0}{E_0-\hat{\mathcal{H}}_0}\left\{\hat{\mathcal{H}}_1\hat{\Omega}^{(i-1)}-\sum_{j=1}^{i-1}E_0^{(j)}\hat{\Omega}^{(i-j)}\right\}. \tag{4.3.8}$$

The ith order energy is then given by

$$E_0^{(i)}=\langle\Phi_0|\,\hat{\mathcal{H}}_1\hat{\Omega}^{(i-1)}\,|\Phi_0\rangle. \tag{4.3.9}$$

Explicitly, the first few terms in the Rayleigh–Schrödinger perturbation expansion are

$$E_0^{(0)}=\langle\Phi_0|\,\hat{\mathcal{H}}_0\,|\Phi_0\rangle \tag{4.3.10a}$$

$$E_0^{(1)}=\langle\Phi_0|\,\hat{\mathcal{H}}_1\,|\Phi_0\rangle \tag{4.3.10b}$$

$$E_0^{(2)}=\langle\Phi_0|\,\hat{\mathcal{H}}_1\hat{R}_0\hat{\mathcal{H}}_1\,|\Phi_0\rangle \tag{4.3.10c}$$

$$E_0^{(3)}=\langle\Phi_0|\,\hat{\mathcal{H}}_1\hat{R}_0(\hat{\mathcal{H}}_1-E_0^{(1)})\hat{R}_0\hat{\mathcal{H}}_1\,|\Phi_0\rangle \tag{4.3.10d}$$

$$\begin{aligned}E_0^{(4)}=&\langle\Phi_0|\,\hat{\mathcal{H}}_1\hat{R}_0(\hat{\mathcal{H}}_1-E_0^{(1)})\hat{R}_0(\hat{\mathcal{H}}_1-E_0^{(1)})\hat{R}_0\hat{\mathcal{H}}_1\,|\Phi_0\rangle\\&-E_0^{(2)}\langle\Phi_0|\,\hat{\mathcal{H}}_1\hat{R}_0^2\hat{\mathcal{H}}_1\,|\Phi_0\rangle\end{aligned} \tag{4.3.10e}$$

$$\begin{aligned}E_0^{(5)}=&\langle\Phi_0|\,\hat{\mathcal{H}}_1\hat{R}_0\{(\hat{\mathcal{H}}_1-E_0^{(1)})\hat{R}_0\}^3\hat{\mathcal{H}}_1\,|\Phi_0\rangle\\&-E_0^{(2)}\{\langle\Phi_0|\,\hat{\mathcal{H}}_1\hat{R}_0^2(\hat{\mathcal{H}}_1-E_0^{(1)})\hat{R}_0\hat{\mathcal{H}}_1\,|\Phi_0\rangle\\&+\langle\Phi_0|\,\hat{\mathcal{H}}_1\hat{R}_0(\hat{\mathcal{H}}_1-E_0^{(1)})\hat{R}_0^2\hat{\mathcal{H}}_1\,|\Phi_0\rangle\}\\&-E_0^{(3)}\langle\Phi_0|\,\hat{\mathcal{H}}_1\hat{R}_0^2\hat{\mathcal{H}}_1\,|\Phi_0\rangle\end{aligned} \tag{4.3.10f}$$

in which $\hat{R}_0$ is the reduced resolvant

$$\hat{R}_0=\frac{\hat{Q}_0}{E_0-\hat{\mathcal{H}}_0}=\sum_{k\neq 0}\frac{|\Phi_k\rangle\langle\Phi_k|}{E_0-E_k}. \tag{4.3.11}$$

The Rayleigh–Schrödinger perturbation expansion does lead to expressions for molecular properties which are directly proportional to the number of electrons in the system being investigated. This can be readily demonstrated by considering the application of the second-order theory to the energy of two helium atoms separated by an infinite distance. The energy obtained is equal to twice the second-order energy of a single helium atom—a result which should be contrasted with that obtained for the Lennard-Jones–Brillouin–Wigner perturbation theory discussed in Section 4.2.

4.4 Particle–hole formalism

In the discussion of the Lennard-Jones–Brillouin–Wigner perturbation theory and the Rayleigh–Schrödinger perturbation theory given in the preceeding two sections, the perturbation series have been developed in terms of N-electron state functions, where N is the number of electrons in the atom or molecule being considered. For example, when considering a double-excitation from the reference configuration in which orbitals i and j are replaced by orbitals a and b, the N-electron function $\Phi\begin{pmatrix} ab \\ ij \end{pmatrix}$ has to be constructed and all of the electrons in the system must necessarily be explicitly included in the calculation. However, the number of electrons N can often be quite large and it is, therefore, usually much more profitable to employ the particle–hole formalism as opposed to the particle formalism. In the particle–hole formalism only the particle states, a and b, created above the Fermi level, and the hole states, i and j, created below the Fermi level, have to be considered during an excitation. The relation between the particle formalism and the particle–hole formalism is illustrated in Fig. 4.1. Before discussing the particle–hole formalism

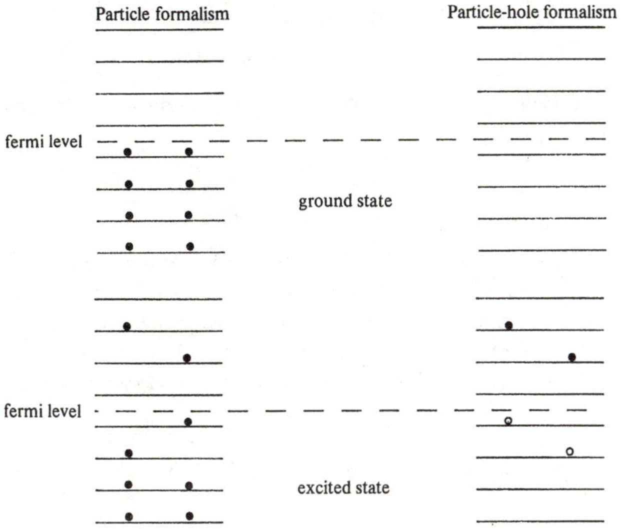

FIG. 4.1. Relation between the particle formalism and the particle–hole formalism.

in detail it is necessary to introduce the second-quantization technique (see, for example, Ziman 1969).

Let $|i_k\rangle$ denote some complete orthonormal set of spin orbitals and let us denote the antisymmetrized product of N of these spin orbitals as follows

$$|\{i_1\, i_2 \ldots i_k \ldots i_N\}\rangle. \tag{4.4.1}$$

The annihilation operator, $\hat{a}$, is defined by

$$\hat{a}_{i_k}\, |\{i_1\, i_2 \ldots i_{k-1}\, i_k\, i_{k+1} \ldots i_N\}\rangle = |\{i_1\, i_2 \ldots i_{k-1}\, i_{k+1} \ldots i_N\}\rangle \tag{4.4.2}$$

and

$$\hat{a}_j\, |\{i_1\, i_2 \ldots i_N\}\rangle = 0 \quad \text{if} \quad j \neq i_k, \qquad \forall k. \tag{4.4.3}$$

When $N=1$

$$\hat{a}_i\, |i\rangle = |0\rangle \tag{4.4.4}$$

where $|0\rangle$ is the true vacuum state which is taken to be normalized. Finally, it should be noted that

$$\hat{a}_i\, |0\rangle = 0. \tag{4.4.5}$$

The corresponding creation operator, $\hat{a}^\dagger$, is the hermitian conjugate of the annihilation operator, and thus

$$\hat{a}_i^\dagger\, |0\rangle = |i\rangle \tag{4.4.6}$$

$$\hat{a}_j^\dagger\, |\{i_1\, i_2 \ldots i_N\}\rangle = |\{j\, i_1\, i_2 \ldots i_N\}\rangle \quad \text{if} \quad j \neq i_k, \qquad \forall k \tag{4.4.7}$$

$$\hat{a}_{i_k}^\dagger\, |\{i_1\, i_2 \ldots i_k \ldots i_N\}\rangle = 0. \tag{4.4.8}$$

From these definitions the following anticommutation relations can be obtained

$$[\hat{a}_i, \hat{a}_j]_+ = [\hat{a}_i^\dagger, \hat{a}_j^\dagger]_+ = 0 \tag{4.4.9}$$

$$[\hat{a}_i^\dagger, \hat{a}_j]_+ = \langle i \mid j\rangle = \delta_{ij} \tag{4.4.10}$$

where $[\hat{x}, \hat{y}]_+ = \hat{x}\hat{y} + \hat{y}\hat{x}$ and δ_{ij} is the Kronecker delta. The anticommutation relations (4.4.9) and (4.4.10) ensure that the system obeys Fermi–Dirac statistics and maintain the antisymmetry of the total wave function. It is clear that an N-electron wave function is given by the action of N distinct creation operators on the true vacuum state

$$\hat{a}_{i_1}^\dagger\, \hat{a}_{i_2}^\dagger \ldots \hat{a}_{i_N}^\dagger\, |0\rangle, \qquad i_1 \neq i_2 \neq \ldots \neq i_N. \tag{4.4.11}$$

The *normal product* of the operators, $\hat{x}_{j_1}\, \hat{x}_{j_2} \ldots \hat{x}_{j_t}$, where $\hat{x} = \hat{a}$ or $\hat{a}^\dagger$, is defined by

$$n[\hat{x}_{j_1}\, \hat{x}_{j_2} \ldots \hat{x}_{j_t}] = (-1)^p \hat{a}_{i_1}^\dagger\, \hat{a}_{i_2}^\dagger \ldots \hat{a}_{i_k}^\dagger\, \hat{a}_{i_{k+1}}\, \hat{a}_{i_{k+2}} \ldots \hat{a}_{i_t} \tag{4.4.12}$$

where p is the parity of the permutation operator (not uniquely defined)

$$\begin{pmatrix} j_1 & j_2 & \dots & j_t \\ i_1 & i_2 & \dots & i_t \end{pmatrix} \tag{4.4.13}$$

which ensures that all of the creation operators are to the left of the annihilation operators. A *contraction* (or *pairing*) of two creation or annihilation operators is defined by

$$\underline{\hat{x}_1 \hat{x}_2} = \hat{x}_1 \hat{x}_2 - n[\hat{x}_1 \hat{x}_2]. \tag{4.4.14}$$

All contractions vanish excepting the case

$$\underline{\hat{a}_i \hat{a}_j^\dagger} = \langle i \mid j \rangle. \tag{4.4.15}$$

A normal product with contractions is defined by

$$n[\hat{x}_1 \hat{x}_2 \dots \hat{x}_{i_1} \dots \hat{x}_{i_2} \dots \hat{x}_{j_2} \dots \hat{x}_{j_1} \dots \hat{x}_{i_3} \dots \hat{x}_{i_4} \dots \hat{x}_{j_3} \dots \hat{x}_{j_4} \dots \hat{x}_t]$$

$$= (-1)^p \underline{\hat{x}_{i_1} \hat{x}_{j_1}}\, \underline{\hat{x}_{i_2} \hat{x}_{j_2}} \dots \underline{\hat{x}_{i_r} \hat{x}_{j_r}} \cdot n[\hat{x}_{k_1} \dots \hat{x}_{i_1 - 1} \hat{x}_{i_1 + 1} \dots \hat{x}_{k_s}] \tag{4.4.16}$$

where p is the parity of the permutation

$$\begin{pmatrix} 1 & 2 & \dots & 2r & 2r+1 & \dots & t \\ i_1 & j_1 & \dots & j_r & k_1 & \dots & k_s \end{pmatrix}, \qquad 2r + s = t. \tag{4.4.17}$$

Wick's theorem (Wick 1950; March, Young, and Sampanthar 1967; Paldus and Čižek 1975) can now be formulated. Wick's theorem states that

$$\hat{x}_1 \hat{x}_2 \dots \hat{x}_t = n[\hat{x}_1 \hat{x}_2 \dots \hat{x}_t] + \sum n[\hat{x}_1 \underline{\hat{x}_2 \dots \hat{x}_t}] \tag{4.4.18}$$

where the summation runs over normal products with all possible contractions. This allows the vacuum mean value of a product of creation and annihilation operators to be reduced to a sum of all possible fully contracted terms. Terms which are not fully contracted are zero, that is

$$\langle 0 | \hat{x}_1 \hat{x}_2 \dots \hat{x}_t | 0 \rangle = \langle 0 | \sum_{\substack{\text{fully} \\ \text{contracted}}} n[\underline{\hat{x}_1 \hat{x}_2} \dots \underline{\hat{x}_t}] | 0 \rangle$$

$$= \sum (-1)^p \langle i_1 \mid j_1 \rangle \langle i_2 \mid j_2 \rangle \dots \langle i_t \mid j_t \rangle \tag{4.4.19}$$

The power of the second-quantized formalism and the theorems discussed above can be realized when it is recognized that any operator can be expressed in terms of creation and annihilation operators. The one- and two-electron operators

$$\hat{f} = \sum_p \hat{f}_p \tag{4.4.20}$$

and

$$\hat{g} = \sum_{p>q} \hat{g}_{pq} \tag{4.4.21}$$

may be written as

$$\hat{f}=\sum_{i_1i_2}\langle i_1|\hat{f}|i_2\rangle a^{\dagger}_{i_1}a_{i_2} \tag{4.4.22}$$

and

$$\hat{g}=\tfrac{1}{2}\sum_{i_1i_2i_3i_4}\langle i_1i_2|g|i_3i_4\rangle a^{\dagger}_{i_1}a^{\dagger}_{i_2}a_{i_4}a_{i_3} \tag{4.4.23}$$

where

$$\langle i|\hat{f}|j\rangle=\int\langle i\mid \mathbf{r}_p\rangle\hat{f}(\mathbf{r}_p)\langle\mathbf{r}_p\mid j\rangle\,\mathrm{d}\mathbf{r}_p \tag{4.4.24}$$

and

$$\langle ij|\hat{g}|kl\rangle=\int\langle i\mid\mathbf{r}_p\rangle\langle j\mid\mathbf{r}_q\rangle\hat{g}(\mathbf{r}_p,\mathbf{r}_q)\langle\mathbf{r}_q\mid l\rangle\langle\mathbf{r}_p\mid k\rangle\,\mathrm{d}\mathbf{r}_p\,\mathrm{d}\mathbf{r}_q. \tag{4.4.25}$$

Thus the total molecular hamiltonian may be written in second-quantized form as follows

$$\mathscr{H}=\sum_{i_1i_2}\langle i_1|\hat{f}|i_2\rangle\hat{a}^{\dagger}_{i_1}a_{i_2}+\tfrac{1}{2}\sum_{i_1i_2i_3i_4}\langle i_1i_2|\hat{g}|i_3i_4\rangle\hat{a}^{\dagger}_{i_1}\hat{a}^{\dagger}_{i_2}\hat{a}_{i_4}\hat{a}_{i_3}. \tag{4.4.26}$$

Let us now return to the particle–hole formalism. It is most convenient when dealing with the ground states and low-lying excited states to redefine the vacuum state to contain the spin orbitals occupied in the ground state instead of the true vacuum. This new vacuum state

$$|\Psi_0\rangle=|\{i_1\,i_2\ldots i_N\}\rangle \tag{4.4.27}$$

will be referred to as the Fermi vacuum in order to distinguish it from the true, or physical, vacuum. The states which are occupied in $|\Psi_0\rangle$ are termed 'hole states' and will be labelled by the indices $i, j, k, \ldots$ whilst the states which are unoccupied in $|\Psi_0\rangle$ are called 'particle states' and will be labelled by the indices $a, b, \ldots$. By employing this particle–hole formalism, a considerable decrease is achieved in the number of creation and annihilation operators which have to be handled when treating the ground or low-lying excited states.

A set of creation and annihilation operators can be defined with respect to the Fermi vacuum as follows

$$\begin{aligned}\hat{b}_i&=\hat{a}^{\dagger}_i & \hat{b}^{\dagger}_i&=\hat{a}_i\\ \hat{b}_a&=\hat{a}_a & \hat{b}^{\dagger}_a&=\hat{a}^{\dagger}_a\end{aligned} \tag{4.4.28}$$

These operators satisfy the anticommutation relations

$$[\hat{b}_p,\hat{b}_q]_+=[\hat{b}^{\dagger}_p,\hat{b}^{\dagger}_q]_+=0 \tag{4.4.29}$$

$$[\hat{b}^{\dagger}_p,\hat{b}_q]_+=\langle p\mid q\rangle \tag{4.4.30}$$

and also

$$\hat{b}_i \,|\Psi_0\rangle = 0. \tag{4.4.31}$$

Theorems and relations analogous to those given above for the particle formalism can be obtained in the particle–hole formalism. The normal product is written as

$$N[\hat{y}_1\,\hat{y}_2 \ldots \hat{y}_t] \qquad \hat{y} = \hat{b} \quad \text{or} \quad \hat{b}^\dagger \tag{4.4.32}$$

and contractions as

$$\overbracket{y_i\, y_j}. \tag{4.4.33}$$

Wick's theorem can be written as

$$y_1\, y_2 \ldots y_t = N[y_1\, y_2 \ldots y_t] + \sum N[\overbracket{y_1\, y_2 \ldots y_t}]. \tag{4.4.34}$$

Again, the only non-vanishing contractions are of the type

$$\langle\Psi_0|\, y_1\, y_2 \ldots y_t \,|\Psi_0\rangle = \langle\Psi_0| \sum_{\substack{\text{fully}\\ \text{contracted}}} N[\overbracket{y_1\, y_2 \ldots y_t}]\,|\Psi_0\rangle \tag{4.4.35}$$

It is sometimes useful to employ a 'mixed' operator representation in which both the particle formalism and the particle–hole formalism are used. It is then necessary to use the appropriate expressions for the following contractions (Paldus and Čižek 1975)

$$\begin{array}{llll}
\overbracket{\hat{a}_p\hat{b}_q} = 0 & \overbracket{\hat{b}_p\hat{a}_i} = \langle p \,|\, i\rangle & \overbracket{\hat{b}_p\hat{a}_a} = 0 & \overbracket{\hat{a}_p\hat{a}_q} = 0 \\
\overbracket{\hat{a}_p^\dagger\hat{b}_q} = 0 & \overbracket{\hat{b}_p^\dagger\hat{a}_q} = 0 & \overbracket{\hat{a}_i^\dagger\hat{a}_q} = \langle i \,|\, q\rangle & \overbracket{\hat{a}_a^\dagger\hat{a}_q} = 0 \\
\overbracket{\hat{a}_i\hat{b}_q^\dagger} = 0 & \overbracket{\hat{a}_a\hat{b}_q^\dagger} = \langle a \,|\, q\rangle & \overbracket{\hat{b}_p\hat{a}_i^\dagger} = 0 & \overbracket{\hat{b}_p\hat{a}_a^\dagger} = \langle p \,|\, a\rangle \\
\overbracket{\hat{a}_i\hat{a}_q^\dagger} = 0 & \overbracket{\hat{a}_a\hat{a}_q^\dagger} = \langle a \,|\, q\rangle & \overbracket{\hat{a}_i^\dagger\hat{b}_q^\dagger} = \langle i \,|\, q\rangle & \overbracket{a_a^\dagger b_q^\dagger} = 0. \\
\overbracket{\hat{b}_p^\dagger\hat{a}_q^\dagger} = 0 & \overbracket{\hat{a}_p^\dagger\hat{a}_q^\dagger} = 0 & &
\end{array} \tag{4.4.36}$$

where p and q denote hole or particle states.

We conclude this section by introducing the field operators

$$\psi(\mathbf{x}) = \sum_k \phi_k(\mathbf{x})a_k \tag{4.4.37}$$

in which $\phi_k(\mathbf{x}) = \langle \mathbf{x} \,|\, k\rangle$ and

$$\psi^\dagger(\mathbf{x}) = \sum_k \phi_k(\mathbf{x})a_k^\dagger. \tag{4.4.38}$$

The operator $\psi(\mathbf{x})$ annihilates an electron at the point $\mathbf{x}$ whilst the operator $\psi^\dagger(\mathbf{x})$ creates an electron at the point $\mathbf{x}$. In terms of these field operators, the one-electron operator given in eqn (4.4.22), for example, may be written as

$$\hat{f} = \int \psi(\mathbf{r}_p^\dagger)\hat{f}(\mathbf{r}_p)\psi(\mathbf{r}_p)\, d\mathbf{r}_p \qquad (4.4.39)$$

Similar expressions may be obtained for other one-electron and for two-electron operators discussed above.

4.5 Many-body perturbation theory

In each order beyond second-order, terms arise in the Rayleigh–Schrödinger perturbation expansion for the energy which have a non-linear dependence on N. However, Brueckner (1955) showed, for the first few orders of the expansion, that the terms having a non-linear dependence on N mutually cancel in each order. Goldstone (1957) demonstrated, using the methods of time-dependent perturbation theory, that these unphysical terms mutually cancel through all orders of the perturbation series.

In this section, the cancellation of terms which are not directly proportional to N will be demonstrated for the first few orders of the Rayleigh–Schrödinger perturbation expansion. By explicitly cancelling these unphysical terms, the many-body perturbation theory will be obtained. In Section 4.6, the second-quantized particle–hole formalism introduced in Section 4.4 will be used to generalize this cancellation and thus obtain the linked diagram theorem which is fundamental to most modern approaches to the correlation problem in atoms and molecules.

Let us recall that the first few terms in the Rayleigh–Schrödinger perturbation series for the energy may be written as

$$E_0 = \langle\Phi_0|\,\hat{\mathcal{H}}_0\,|\Phi_0\rangle \qquad (4.5.1)$$

$$E_1 = \langle\Phi_0|\,\hat{\mathcal{H}}_1\,|\Phi_0\rangle \qquad (4.5.2)$$

$$E_2 = \sum_{k\neq 0} \frac{\langle\Phi_0|\,\hat{\mathcal{H}}_1\,|\Phi_k\rangle\langle\Phi_k|\,\hat{\mathcal{H}}_1\,|\Phi_0\rangle}{\mathcal{E}_0 - \mathcal{E}_k} \qquad (4.5.3)$$

$$E_3 = \sum_{k,l\neq 0} \frac{\langle\Phi_0|\,\hat{\mathcal{H}}_1\,|\Phi_k\rangle\langle\Phi_k|\,\hat{\mathcal{H}}_1 - E_1\,|\Phi_l\rangle\langle\Phi_l|\,\hat{\mathcal{H}}_1\,|\Phi_0\rangle}{(\mathcal{E}_0 - \mathcal{E}_k)(\mathcal{E}_0 - \mathcal{E}_l)} \qquad (4.5.4)$$

$$E_4 = \sum_{k,l,m\neq 0} \frac{\langle\Phi_0|\,\hat{\mathcal{H}}_1\,|\Phi_k\rangle\langle\Phi_k|\,\hat{\mathcal{H}}_1 - E_1\,|\Phi_l\rangle\langle\Phi_l|\,\hat{\mathcal{H}}_1 - E_1\,|\Phi_m\rangle\langle\Phi_m|\,\hat{\mathcal{H}}_1\,|\Phi_0\rangle}{(\mathcal{E}_0 - \mathcal{E}_k)(\mathcal{E}_0 - \mathcal{E}_l)(\mathcal{E}_0 - \mathcal{E}_m)}$$

$$- E_2 \sum_{k\neq 0} \frac{\langle\Phi_0|\,\hat{\mathcal{H}}_1\,|\Phi_k\rangle\langle\Phi_k|\,\hat{\mathcal{H}}_1\,|\Phi_0\rangle}{(\mathcal{E}_0 - \mathcal{E}_k)^2} \qquad (4.5.5)$$

where we now use the notation

$$\mathscr{E}_k = \langle \Phi_k | \hat{\mathscr{H}}_0 | \Phi_k \rangle \tag{4.5.6}$$

and Φ_k is a single determinantal wave function. The spin orbitals which are occupied in the ground state determinant, Φ_0, will be denoted by $i, j, k, \ldots$; and those which are not occupied will be denoted by $a, b, c, \ldots$. Explicit evaluation of E_0, E_1, and E_2 leads to the following expressions for a closed-shell system

$$E_0 = \sum_i \epsilon_i \tag{4.5.7}$$

$$E_1 = \tfrac{1}{2} \sum_{ij} \langle ij | \hat{O} | ij \rangle \tag{4.5.8}$$

$$E_2 = \tfrac{1}{4} \sum_{ij} \sum_{ab} \frac{\langle ij | \hat{O} | ab \rangle \langle ab | \hat{O} | ij \rangle}{\epsilon_i + \epsilon_j - \epsilon_a - \epsilon_b} \tag{4.5.9}$$

where ϵ_p is an orbital energy, $\hat{O} = (1-(12))/r_{12}$ and (12) interchanges the coordinates of electrons 1 and 2. In obtaining eqn (4.5.9), it has been assumed that Φ_0 is a Hartree–Fock function and thus the only excited states Φ_k in eqn (4.5.3) which have to be considered are the double excitations $(i, j) \rightarrow (a, b)$.

Now let us consider the addition of a pair of electrons to a given closed-shell system in such a manner that the resulting system is also a closed-shell species. Let us suppose that the additional pair of electrons are placed in spin orbitals p and $\bar{p}$ with α and β spin, respectively. The zero-order, first-order, and second-order energies of the $N+2$ electron system, which will be denoted by E'_0, E'_1, and E'_2, take the form

$$E'_0 = E_0 + 2\epsilon_p \tag{4.5.10}$$

$$E'_1 = E_1 + \sum_i \{\langle pi | \hat{O} | pi \rangle + \langle \bar{p}i | \hat{O} | \bar{p}i \rangle\} + \langle p\bar{p} | \hat{O} | p\bar{p} \rangle = E_1 + \xi_1 \tag{4.5.11}$$

$$\begin{aligned} E'_2 = E_2 &+ \tfrac{1}{2} \sum_{i,a,b} \left\{ \frac{\langle pi | \hat{O} | ab \rangle \langle ab | \hat{O} | pi \rangle}{\epsilon_p + \epsilon_i - \epsilon_a - \epsilon_b} \right. \\ &\left. + \frac{\langle \bar{p}i | \hat{O} | ab \rangle \langle ab | \hat{O} | \bar{p}i \rangle}{\epsilon_{\bar{p}} + \epsilon_i - \epsilon_a - \epsilon_b} \right\} \\ &+ \tfrac{1}{2} \sum_{ab} \frac{\langle p\bar{p} | \hat{O} | ab \rangle \langle ab | \hat{O} | p\bar{p} \rangle}{\epsilon_p + \epsilon_{\bar{p}} - \epsilon_a - \epsilon_b} \end{aligned} \tag{4.5.12}$$

The energy of the $N+2$ electron system can be written as the energy of the N electron systems plus additional terms. This argument can be repeated as further pairs of electrons are added to the system and thus E_0, E_1, and E_2 are shown to be directly proportional to N.

The third-order term in the Rayleigh–Schrödinger expansion for the energy is the lowest order in which components have to be cancelled in order to obtain a physically acceptable dependence on N. The third-order energy may be written in the form

$$E_3 = \sum_{\substack{k,l\neq 0\\ k\neq l}} \frac{\langle\Phi_0|\hat{\mathcal{H}}_1|\Phi_k\rangle\langle\Phi_k|\hat{\mathcal{H}}_1|\Phi_l\rangle\langle\Phi_l|\hat{\mathcal{H}}_1|\Phi_0\rangle}{(\mathcal{E}_0-\mathcal{E}_k)(\mathcal{E}_0-\mathcal{E}_l)}$$

$$+\sum_{k\neq 0} \frac{\langle\Phi_0|\hat{\mathcal{H}}_1|\Phi_k\rangle\langle\Phi_k|\hat{\mathcal{H}}_1|\Phi_k\rangle\langle\Phi_k|\hat{\mathcal{H}}_1|\Phi_0\rangle}{(\mathcal{E}_0-\mathcal{E}_k)^2}$$

$$-E_1\sum_{k\neq 0} \frac{\langle\Phi_0|\mathcal{H}_1|\Phi_k\rangle\langle\Phi_k|\mathcal{H}_1|\Phi_0\rangle}{(\mathcal{E}_0-\mathcal{E}_k)^2}. \quad (4.5.13)$$

The last term in eqn (4.5.13) has the form

$$E_1\,\Delta_{11} \quad (4.5.14)$$

where

$$\Delta_{11} = \sum_{k\neq 0} \frac{\langle\Phi_0|\hat{\mathcal{H}}_1|\Phi_k\rangle\langle\Phi_k|\hat{\mathcal{H}}_1|\Phi_0\rangle}{(\mathcal{E}_0-\mathcal{E}_k)^2}. \quad (4.5.15)$$

Δ_{11} differs from E_2 (eqn 4.5.9) only by the presence of an additional denominator factor. Clearly, addition of a pair of electrons to an N electron system leads to

$$\Delta'_{11} = \Delta_{11} + \tfrac{1}{2}\sum_{i,a,b}\left\{\frac{\langle pi|\hat{O}|ab\rangle\langle ab|\hat{O}|pi\rangle}{(\epsilon_p+\epsilon_i-\epsilon_a-\epsilon_b)^2} + \frac{\langle\bar{p}i|\hat{O}|ab\rangle\langle ab|\hat{O}|\bar{p}i\rangle}{(\epsilon_{\bar{p}}+\epsilon_i-\epsilon_a-\epsilon_b)^2}\right\}$$

$$+\tfrac{1}{2}\sum_{a,b}\frac{\langle p\bar{p}|\hat{O}|ab\rangle\langle ab|\hat{O}|p\bar{p}\rangle}{(\epsilon_p+\epsilon_{\bar{p}}-\epsilon_a-\epsilon_b)^2}$$

$$=\Delta_{11}+\delta_{11}$$

and since

$$E'_1\,\Delta'_{11} = (E_1+\xi_1)(\Delta_{11}+\delta_{11}) \quad (4.5.16)$$

$E_1\,\Delta_{11}$ can be seen to depend on N^2. Fortunately, the $E_1\,\Delta_{11}$ term can be shown to cancel exactly with portions of the second term in eqn (4.5.13) and the remaining terms have a linear dependence on the number of electrons in the system. After carrying out this cancellation explicitly the third-order energy expression takes the form (see, for example, Kumar (1962))

$$E_3 = \sum_{a,b} \left\{ \frac{\langle ij|\,\hat{O}\,|ab\rangle}{\epsilon_i+\epsilon_j-\epsilon_a-\epsilon_b} \left(\tfrac{1}{8} \sum_{c,d} \frac{\langle ab|\,\hat{O}\,|cd\rangle\langle cd|\,\hat{O}\,|ij\rangle}{\epsilon_i+\epsilon_j-\epsilon_c-\epsilon_d} + \sum_{k,c} \frac{\langle ic|\,\hat{O}\,|ak\rangle\langle jk|\,\hat{O}\,|bc\rangle}{\epsilon_j+\epsilon_k-\epsilon_b-\epsilon_c} + \tfrac{1}{8} \sum_{k,l} \frac{\langle ij|\,\hat{O}\,|kl\rangle\langle ab|\,\hat{O}\,|kl\rangle}{\epsilon_k+\epsilon_l-\epsilon_a-\epsilon_b} \right) \right\} \tag{4.5.17}$$

All of the determinants Φ_k which contribute to E_3 are doubly excited with respect to the Hartree–Fock reference function, Φ_0. Hence, if a third-order Rayleigh–Schrödinger perturbation theory calculation is made using eqn (4.5.13) and all possible double-excitation terms are included, then a size-consistent result is obtained. However, many of the terms which are evaluated are exactly cancelled by other third-order terms.

In fourth-order of the Rayleigh–Schrödinger perturbation series, terms involving factors of E_1 and E_2 arise which depend non-linearly on N. Explicit algebraic manipulations show that these terms are cancelled by terms in the summation over k, l, and m in eqn (4.5.5). In fourth-order of the perturbation expansion, terms arise which are singly-excited, doubly-excited, triply-excited, and quadruply-excited with respect to the Hartree–Fock reference configuration. The cancellation of unphysical terms only takes place if all of these configurations are included. In particular, if only double excitations are considered, a term of the form $E_2\,\Delta_{11}$ arises which is only cancelled by a term corresponding to quadruple excitations. This $E_2\,\Delta_{11}$ term is the source of the dominant error (besides basis set truncation effects) in limited double-excitation configuration interaction calculations and will be discussed further in Chapter 7.

4.6 The linked diagram theorem

In the preceding section it was demonstrated that for the first few orders of the Rayleigh–Schrödinger perturbation theory two types of terms can be identified: those which are directly proportional to the number of electrons in the system, N, and those which are proportional to some power of N other than unity. The former are physically acceptable whereas the latter are not. However, it is possible to demonstrate that the unphysical terms which are not directly proportional to N mutually cancel, leading to the many-body perturbation theory—a perturbation expansion which is valid for any number of electrons. Now the terms which depend on N non-linearly may be associated with unlinked diagrams, that is diagrams that can be divided into two or more separate parts, whilst the terms having the desired property of being proportional

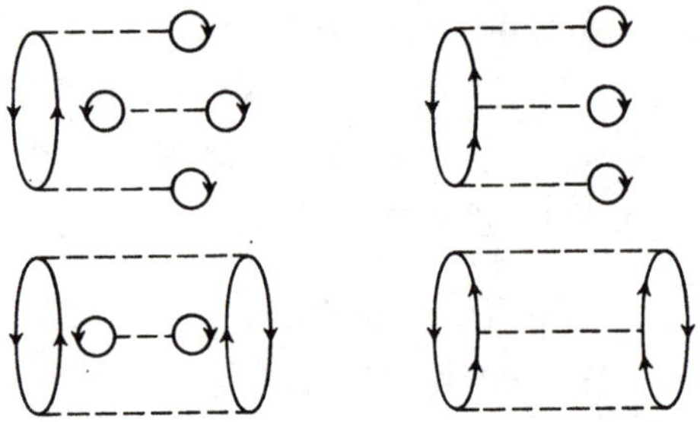

FIG. 4.2. Some linked and unlinked diagrams which correspond to the third-order energy.

to N may be associated with linked diagrams. This is the linked diagram theorem. It provides a pictorial and highly convenient form of the many-body perturbation theory.

In Fig. 4.2 the linked and unlinked diagrams which correspond to the third-order energy discussed in Section 4.5 are displayed. It can readily be seen that the terms which correspond to unlinked diagrams are those which cancel and those which correspond to linked diagrams survive. A detailed review of diagrammatic rules and conventions will be given in Section 4.7. In this section the derivation of the linked diagram expansion will be outlined.

There are a considerable number of derivations of the linked diagram theorem. These derivations can be broadly categorized into two classes—those which use time-dependent perturbation theory and those which employ time-independent methods. Goldstone's original derivation (1957) used time-dependent methods. The treatments given by Hugenholtz (1957), Brandow (1967), Paldus and Čižek (1975), and Hubač and Čársky (1978) are time-independent. Here we shall briefly outline both the time-dependent and the time-independent derivations and refer the reader to the original literature for a more detailed discussion. We shall proceed historically and begin with the time-dependent derivation of the linked diagram theorem.

The diagrammatic many-body perturbation theory may be developed in terms of some set of single particle states, ϕ_p, which are eigenfunctions of a single particle operator

$$\hat{f}\phi_p = \epsilon_p\phi_p \tag{4.6.1}$$

with eigenvalues ϵ_p. In the second-quantized formalism, discussed in Section 4.4, the zero-order hamiltonian has the form

$$\hat{\mathcal{H}}_0 = \int dr_1 \psi^\dagger(r_1)\hat{f}(r_1)\psi(r_1) \tag{4.6.2}$$

and the perturbation operator is

$$\hat{\mathcal{H}}_1 = \tfrac{1}{2}\int dr_1 \int dr_2 \psi^\dagger(r_1)\psi^\dagger(r_2) g(r_1, r_2)\psi(r_2)\psi(r_1) - \int dr_1 \psi^\dagger(r_1) V(r_1)\psi(r_2) \tag{4.6.3}$$

where $\psi^\dagger$ and ψ are the creation and annihilation field operators, g is the two-electron potential, and V is the effective one-electron potential which is added to the bare-nucleus hamiltonian to give the one-electron operator $\hat{f}$. There is, of course, considerable freedom in the choice of the effective potentials V as was discussed more fully in Chapter 2.

Use of the interaction representation in time-dependent perturbation theory (Messiah 1967) and an adiabatic switching $\exp(|\alpha|\, t)$ (Gell-Mann and Low 1951) of the perturbation operator yields the evolution operator

$$\hat{U}^\alpha(t, -\infty) = \hat{I} + \sum_{n=1}^{\infty} \hat{U}_n^\alpha(t, -\infty) \tag{4.6.4}$$

where $\hat{I}$ is the identity operator and $\hat{U}_n^\alpha$ is proportional to the nth power of the perturbation. Thus, we have

$$U_n^\alpha(t, -\infty) = (-i)^n \int_{-\infty}^{t} dt_1 \int_{-\infty}^{t_1} dt_2 \ldots \int_{-\infty}^{t_{n-1}} dt_n \hat{\mathcal{H}}_1(t_1)\hat{\mathcal{H}}_1(t_2)\ldots\hat{\mathcal{H}}_1(t_n),$$

$$\hat{\mathcal{H}}_1(t) = \exp(i\hat{\mathcal{H}}_0 t)\hat{\mathcal{H}}_1 \exp(|\alpha|\, t)\exp(-i\hat{\mathcal{H}}_0 t). \tag{4.6.5}$$

The function

$$|\Psi_0\rangle = \left[\lim_{\alpha\to 0} \frac{\hat{U}^\alpha(t, -\infty)\,|\Phi_0\rangle}{\langle\Phi_0|\, \hat{U}^\alpha(t, -\infty)\,|\Phi_0\rangle}\right]_{t=0} \tag{4.6.6}$$

which obeys the intermediate normalization condition is a well-defined eigenfunction of the total hamiltonian. The total energy is given by

$$E = E_0 + \Delta E_0 \tag{4.6.7}$$

where the level shift is

$$\Delta E_0 = \left[\lim_{\alpha\to 0} \frac{\langle\Phi_0|\, \hat{\mathcal{H}}_1(t)\hat{U}^\alpha(t, -\infty)\,|\Phi_0\rangle}{\langle\Phi_0|\, \hat{U}^\alpha(t, -\infty)\,|\Phi_0\rangle}\right]_{t=0}. \tag{4.6.8}$$

The denominator in this equation may be written as

$$\exp[\langle\Phi_0|\, \hat{U}^\alpha(0, -\infty)\,|\Phi_0\rangle]_{\mathrm{L}} \tag{4.6.9}$$

where the subscript L indicates that only terms associated with linked diagrams are to be included. From eqns (4.6.6) and (4.6.8), we obtain, after taking the limit, the Goldstone expression for the wave function

$$|\Psi_0\rangle = \hat{U}(0, -\infty)\,|\Phi_0\rangle_L \tag{4.6.10}$$

and the level shift

$$\Delta E_0 = \langle\Phi_0|\,\hat{\mathcal{H}}_1(0)\hat{U}(0, -\infty)\,|\Phi_0\rangle_L. \tag{4.6.11}$$

Analysis of the products of field operators in these equations leads to a representation of the wave function and the level shift in terms of diagrams of the type first introduced by Feynman (1948).

It is perhaps more natural for systems which do not involve time-dependent effects to employ a time-independent derivation of the linked diagram theorem. A time-independent derivation of the linked diagram theorem was given by Brandow (1967) starting from the Lennard-Jones–Brillouin–Wigner perturbation theory, which was developed in Section 4.2. The level shift in the Lennard-Jones–Brillouin–Wigner theory is

$$\Delta E_0 = E_0^{(1)} + \langle\Phi_0|\,\hat{\mathcal{H}}_1\hat{R}\hat{\mathcal{H}}_1\,|\Phi_0\rangle + \langle\Phi_0|\,\hat{\mathcal{H}}_1\hat{R}\hat{\mathcal{H}}_1\hat{R}\hat{\mathcal{H}}_1\,|\Phi_0\rangle + \dots \tag{4.6.12}$$

The resolvant in this expansion can be related to that of the Rayleigh–Schrödinger perturbation theory as follows:

$$\hat{R} = \hat{R}_0 + \hat{R}_0(-\Delta E_0)\hat{R}_0 + \hat{R}_0(-\Delta E_0)\hat{R}_0(-\Delta E_0)\hat{R}_0 + \dots \tag{4.6.13}$$

Thus, for example, the second term on the right-hand side of eqn (4.6.12) leads to

$$\langle\Phi_0|\,\hat{\mathcal{H}}_0\hat{R}_0\hat{\mathcal{H}}_0\,|\Phi_0\rangle - \Delta E_0\langle\Phi_0|\,\hat{\mathcal{H}}_1\hat{R}_0^2\hat{\mathcal{H}}_1\,|\Phi_0\rangle + \dots \tag{4.6.14}$$

Brandow showed that the linked diagram theorem can be proved by substituting for ΔE_0 in eqn (4.6.14) and similar terms using (4.6.12), repeatedly.

In order to obtain a completely general form of the linked diagram theorem, two further developments are required—the factorization theorem and the exclusion principle violating term. Since these aspects are most easily discussed in terms of the diagrams which are described in detail in the following sections and since a knowledge of these aspects is not essential at this stage, we discuss the factorization theorem and exclusion principle violating terms in the appendix to this chapter.

4.7 Diagram techniques

The whole theoretical apparatus of the many-body perturbation theory can be set up in entirely algebraic terms and it should be recognized that the use of diagrams is not at all obligatory. However, the diagrams are both more physical and easier to handle than the algebraic expressions and

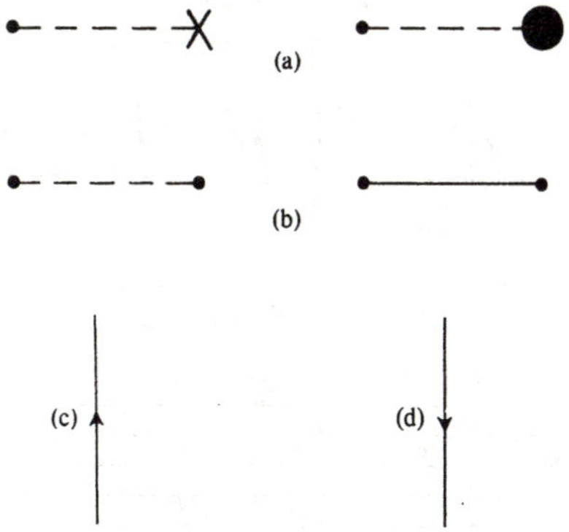

FIG. 4.3. Diagram elements: (a) one-electron operator; (b) two-electron operator; (c) particle line; (d) hole line.

it is well worth the additional effort required to familiarize oneself with the diagrammatic rules and conventions.

Diagrams provide a simple physical picture of the effects which contribute to electron correlation. A diagram is also a device through which a corresponding algebraic expression may be obtained. A number of different diagrammatic conventions are in use. We shall follow the work of Brandow (1967) initially and then discuss its relation with other commonly used conventions.

The basic elements of the Brandow diagrams are shown in Fig. 4.3. Figure 4.3(a) shows the diagrammatic representation of a one-electron operator matrix element. Figure 4.3(b) shows the representation of a two-electron operator matrix element which in the convention proposed by Brandow includes the permutation of the two electrons involved. Upward directed lines, as illustrated in Fig. 4.3(c), represent particles created above the Fermi level when an electron is excited. Downward directed lines, as illustrated in Fig. 4.3(d), are used to represent the holes created below the Fermi level upon excitation of an electron.

A time-dependent physical interpretation of a diagram may be given. Time is taken to increase from the bottom to the top of the diagram. An example of the interpretation of a typical diagram in terms of the particle–hole formalism is displayed in Fig. 4.4.

The algebraic expression corresponding to a given diagram of the Brandow form can be obtained by applying the following simple set of rules:

(a) Each downward directed line is labelled with a unique 'hole' index: $i, j, k, \ldots$.

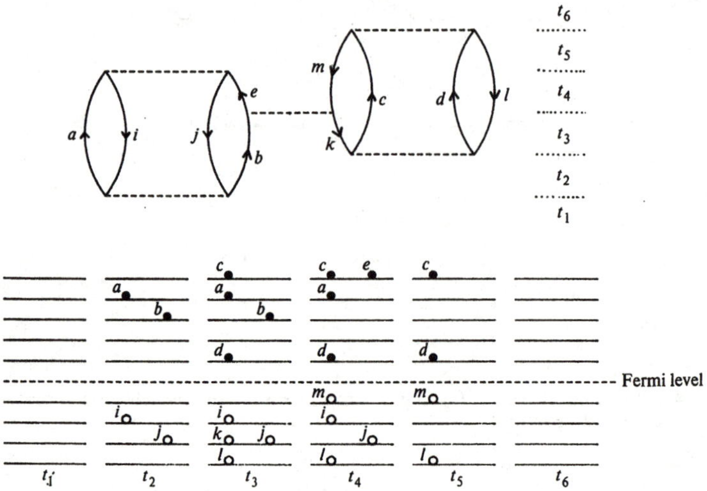

FIG. 4.4. Interpretation of diagrams in terms of the particle–hole formalism.

(b) Each upward directed line is labelled with a unique 'particle' index: $a, b, c, \ldots$.

(c) There is a summation over each unique hole and each unique particle index, covering all possible values of these indices.

(d) The numerator of the summand consists of a product of one- and/or two-electron integrals. The possible types of one-electron integrals which can arise are summarized in Fig. 4.5. The two-electron integrals occur in the antisymmetrized form

$$\mathcal{J}_{pqrs} = \int \mathrm{d}r_1 \int \mathrm{d}r_2 \phi_p^*(r_1)\phi_q^*(r_2) \frac{1}{r_{12}} (P_{12}\phi_r(r_1)\phi_s(r_2)) \qquad (4.7.1)$$

where ϕ is a one-electron function, r_{12} is the interelectronic separation, and P_{12} is the permutation operator which interchanges the coordinates of the two electrons. There is an integral of this type for every interaction line, i.e. horizontal dashed line. The orbital indices p, q, r, s can be read

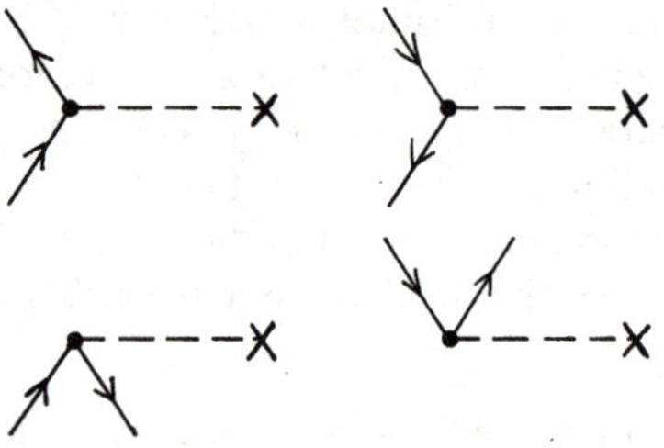

FIG. 4.5. Types of one-electron terms which can arise in diagrams.

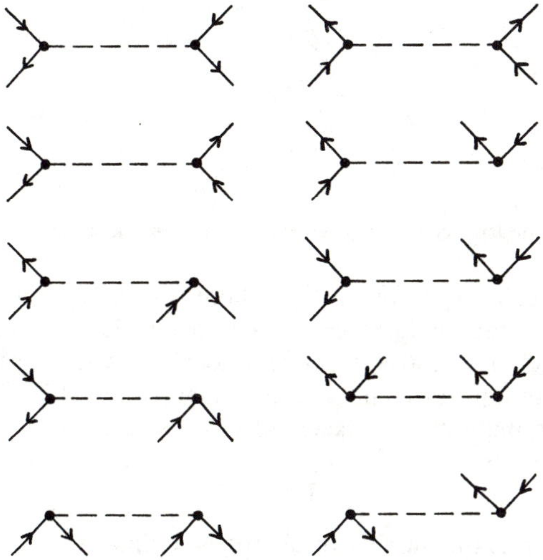

FIG. 4.6. Types of two-electron terms which can arise in diagrams.

from the diagram and correspond to the hole and particle lines entering or leaving the interaction in the order—left–out, right–out, left–in, right–in, respectively. For example, the diagram

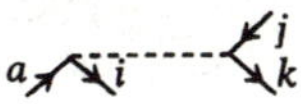

(4.7.2)

and

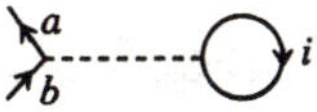

(4.7.3)

correspond to the integrals $\mathscr{I}_{ikaj}$ and $\mathscr{I}_{aibi}$, respectively. The possible types of two-electron interactions which can arise are summarized in Fig. 4.6.
(e) The denominator of the summand consists of a product of factors of the form

$$\sum_i \epsilon_i - \sum_a \epsilon_a \tag{4.7.4}$$

where the first summation is over all hole lines that extend between adjacent interactions and the second summation is over all particle lines that also extend between the two adjacent interactions. ϵ_p denotes the one-electron orbital energy. There are $n-1$ factors of this type corresponding to an nth order diagram such that there is a denominator factor arising between adjacent interactions in the diagram. An example of the correspondence between diagrams and denominators is given in Fig. 4.7.

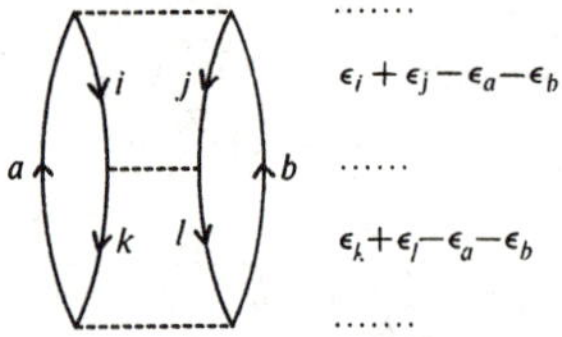

FIG. 4.7. Correspondence between energy diagrams and algebraic expressions.

(f) For each pair of equivalent lines there is a multiplicative factor of $\frac{1}{2}$. An equivalent pair of lines is defined as two lines, beginning at one interaction and ending at another and both going in the same direction. The concept of equivalent lines is illustrated in Fig. 4.7.
(g) There is a multiplicative factor of $(-1)^p$ where

$$p=h+l \qquad (4.7.5)$$

and where (i) h is the number of unique hole lines,
(ii) l is the number of fermion loops.
A fermion loop is determined by following the hole and particle lines in the direction of the arrows to form a continuous loop. The determination of this sign factor is illustrated in Fig. 4.7.

The application of these simple rules will be illustrated by example. Consider the diagram shown in Fig. 4.8(a). The lines in this diagram may be labelled to give Fig. 4.8(b). The numerator in the summand is given by the following product of integrals (reading the diagram from bottom to top)

$$\mathcal{I}_{adli}\mathcal{I}_{lckd}\mathcal{I}_{kbjc}\mathcal{I}_{jiab} \qquad (4.7.6)$$

and the denominator by

$$(\epsilon_i+\epsilon_l-\epsilon_a-\epsilon_d)(\epsilon_i+\epsilon_k-\epsilon_a-\epsilon_c)(\epsilon_i+\epsilon_j-\epsilon_a-\epsilon_b). \qquad (4.7.7)$$

There are no pairs of equivalent lines, 4 unique hole lines and 2 fermion loops. The algebraic expression corresponding to the diagram given in

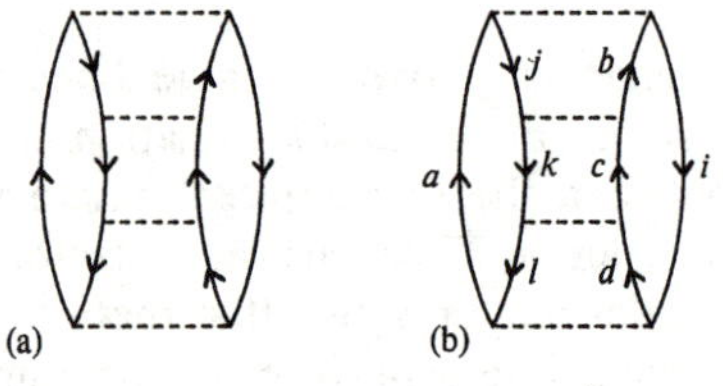

FIG. 4.8

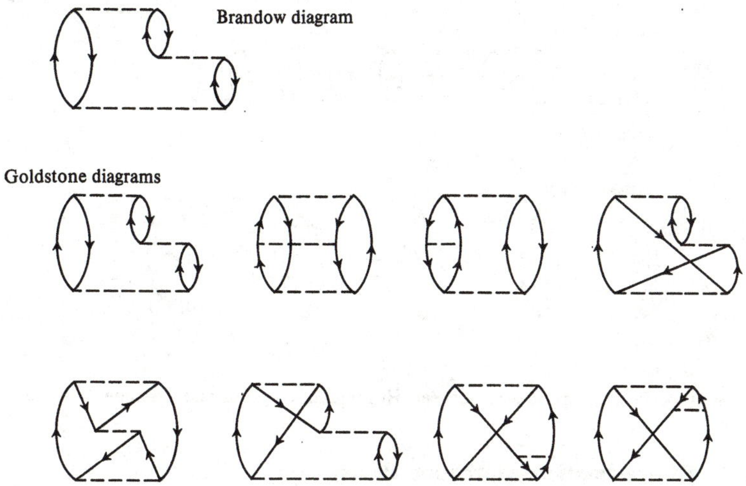

FIG. 4.9. Correspondence between Goldstone diagrams and Brandow diagrams.

Fig. 4.8 is therefore:

$$\sum_{ijkl}\sum_{abcd}\frac{\mathcal{I}_{adli}\mathcal{I}_{lckd}\mathcal{I}_{kbjc}\mathcal{I}_{jiab}}{(\epsilon_i+\epsilon_l-\epsilon_a-\epsilon_d)(\epsilon_i+\epsilon_k-\epsilon_a-\epsilon_c)(\epsilon_i+\epsilon_j-\epsilon_a-\epsilon_b)} \tag{4.7.8}$$

In addition to the Brandow diagrammatic convention discussed above, two other schemes are commonly used in perturbational studies of electron correlation effects in atoms and molecules. The Goldstone (1957) diagrams are similar to those of Brandow, except that the interaction lines do not include permutation of the two electrons involved. Thus, there is a set of Goldstone diagrams which are related to one another by exchange, corresponding to each Brandow diagram. Examples of the correspondence between Goldstone diagrams and Brandow diagrams are given in Fig. 4.9. It can be seen that the Brandow diagrams provide a much more compact representation of the terms in the perturbation series than do the Goldstone diagrams. The diagrams of Hugenholtz (1957) are in one-to-one correspondence with those of Brandow. The Hugenholtz diagrams can be obtained from those of Brandow by replacing the interaction lines in the latter by single dots. The correspondence between Hugenholtz diagrams and Brandow diagrams is illustrated in Fig. 4.10. Finally, it should be noted that some authors (see, for example, Paldus and Čižek 1975) use diagrams which are identical to those discussed above, except that they are turned through 90°.

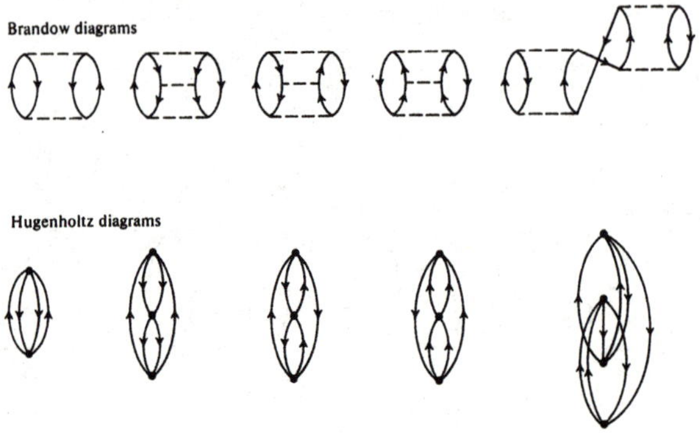

FIG. 4.10. Correspondence between Hugenholtz diagrams and Brandow diagrams.

4.8 Diagrammatic perturbation theory

By using the linked diagram theorem described in Section 4.6 and the diagrammatic conventions presented in Section 4.7, the many-body perturbation theory of Section 4.5 can be developed diagrammatically. The diagrammatic many-body perturbation theory not only forms the basis of a powerful method for the calculation of electron correlation effects in atoms and molecules but also provides an elegant analytical tool for investigations of other approaches to the correlation problem.

For simplicity, attention will be restricted in this section to closed-shell systems which can be described in zero order by a single determinantal wave function. Let us begin by considering the development of the diagrammatic perturbation series with respect to the simplest of the independent electron models discussed in Chapter 2, namely, the bare-nucleus model. In this model, the electron–electron interactions are completely neglected in zero-order and each electron in the system experiences the full unscreened charge of the nuclei. The zero-order hamiltonian in the bare-nucleus model is just the sum of the kinetic energy and electron–nucleus attraction terms for each of the electrons, that is

$$\begin{aligned}\hat{\mathscr{H}}_0 &= -\tfrac{1}{2}\sum_i \nabla_i^2 + \sum_i \sum_A \frac{Z_A}{r_{Ai}} \\ &= \sum_i h_i \end{aligned} \tag{4.8.1}$$

and the perturbation is the sum of the electron–electron repulsion terms

$$\hat{\mathscr{H}}_1 = \sum_{i>j} r_{ij}^{-1} \tag{4.8.2}$$

FIG. 4.11. Diagrammatic representation of the first-order energy.

The zero-order energy is simply the sum of spin orbital energies

$$E_0 = \sum_i \epsilon_i. \tag{4.8.3}$$

In terms of the Rayleigh–Schrödinger perturbation theory the first-order energy is

$$E_1 = \langle \Phi_0 | \hat{\mathcal{H}}_1 | \Phi_0 \rangle \tag{4.8.4}$$

which corresponds to the diagram shown in Fig. 4.11 in the diagrammatic perturbation theory. Using the rules given in Section 4.7, it can be seen that this diagram leads to the expression

$$E_1 = \tfrac{1}{2} \sum_{ij} \mathcal{I}_{ijij} \tag{4.8.5}$$

for the first-order energy, where the indices $i, j, k, \ldots$ are used to denote hole lines and the indices $a, b, c, \ldots$ are used to denote particle lines.

The second-order energy may be written in terms of N-electron functions as follows

$$E_2 = \sum_S \frac{\langle \Phi_0 | \hat{\mathcal{H}}_1 | \Phi_S \rangle \langle \Phi_S | \hat{\mathcal{H}}_1 | \Phi_0 \rangle}{E_0 - \langle \Phi_S | \hat{\mathcal{H}}_0 | \Phi_S \rangle} + \sum_D \frac{\langle \Phi_0 | \hat{\mathcal{H}}_1 | \Phi_D \rangle \langle \Phi_D | \hat{\mathcal{H}}_1 | \Phi_0 \rangle}{E_0 - \langle \Phi_D | \hat{\mathcal{H}}_0 | \Phi_D \rangle} \tag{4.8.6}$$

where Φ_S denotes a determinant which is singly excited with respect to the reference determinant and Φ_D denotes a determinant which is doubly excited with respect to the reference function. The second-order diagrams which arise in the diagrammatic perturbation theory series are displayed in Fig. 4.12. One of these includes only single excitations and leads to the expression

$$E_{2S} = \sum_{ijk} \sum_a \frac{\mathcal{I}_{ijaj} \mathcal{I}_{akik}}{\epsilon_i - \epsilon_a} \tag{4.8.7}$$

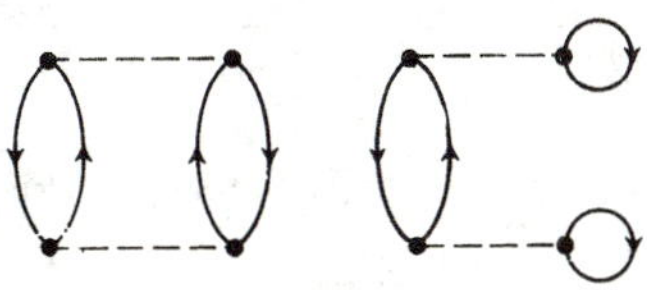

FIG. 4.12. Second-order energy diagrams.

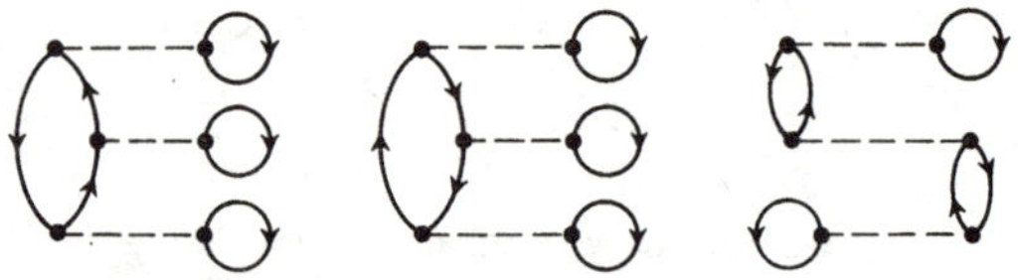

FIG. 4.13. Third-order energy diagrams involving singly excited intermediate states.

on applying the rules of Section 4.7. The other second-order diagram involves only double excitations and leads to the expression

$$E_{2D} = \tfrac{1}{4} \sum_{ij} \sum_{ab} \frac{\mathcal{I}_{ijab} \mathcal{I}_{abij}}{\epsilon_i + \epsilon_j - \epsilon_a - \epsilon_b}. \tag{4.8.8}$$

In third order of the Rayleigh–Schrödinger perturbation series four terms, involving different degrees of excitation with respect to the reference determinant, arise. The first term involves only single excitations

$$E_{3S} = \sum_{S_1} \sum_{S_2} \frac{\langle \Phi_0 | \hat{\mathcal{H}}_1 | \Phi_{S_1} \rangle \langle \Phi_{S_1} | \hat{\mathcal{H}}_1 - E_1 | \Phi_{S_2} \rangle \langle \Phi_{S_2} | \hat{\mathcal{H}}_1 | \Phi_0 \rangle}{(E_0 - \langle \Phi_{S_1} | \hat{\mathcal{H}}_0 | \Phi_{S_1} \rangle)(E_0 - \langle \Phi_{S_2} | \hat{\mathcal{H}}_0 | \Phi_{S_2} \rangle)}. \tag{4.8.9}$$

The diagrams corresponding to this expression are displayed in Fig. 4.13. It can be shown that the algebraic expressions corresponding to these diagrams are, respectively,

$$E_{3S(a)} = \sum_{i} \sum_{ab} \sum_{jkl} \frac{\mathcal{I}_{ajij} \mathcal{I}_{bkak} \mathcal{I}_{ilbl}}{(\epsilon_i - \epsilon_a)(\epsilon_i - \epsilon_b)} \tag{4.8.10a}$$

$$E_{3S(b)} = -\sum_{ij} \sum_{a} \sum_{klm} \frac{\mathcal{I}_{akik} \mathcal{I}_{iljl} \mathcal{I}_{jmam}}{(\epsilon_i - \epsilon_a)(\epsilon_j - \epsilon_a)} \tag{4.8.10b}$$

$$E_{3S(c)} = \sum_{ij} \sum_{ab} \sum_{kl} \frac{\mathcal{I}_{akik} \mathcal{I}_{bija} \mathcal{I}_{jlbl}}{(\epsilon_i - \epsilon_a)(\epsilon_j - \epsilon_b)} \tag{4.8.10c}$$

Two terms arise in third-order Rayleigh–Schrödinger perturbation theory which involve both single-excitations and double-excitations

$$E_{3SD} = \sum_{S} \sum_{D} \frac{\langle \Phi_0 | \hat{\mathcal{H}}_1 | \Phi_S \rangle \langle \Phi_S | \hat{\mathcal{H}}_1 | \Phi_D \rangle \langle \Phi_D | \hat{\mathcal{H}}_1 | \Phi_0 \rangle}{(E_0 - \langle \Phi_S | \hat{\mathcal{H}}_0 | \Phi_S \rangle)(E_0 - \langle \Phi_D | \hat{\mathcal{H}}_0 | \Phi_D \rangle)} \tag{4.8.11a}$$

$$E_{3DS} = \sum_{D} \sum_{S} \frac{\langle \Phi_0 | \hat{\mathcal{H}}_1 | \Phi_D \rangle \langle \Phi_D | \hat{\mathcal{H}}_1 | \Phi_S \rangle \langle \Phi_S | \hat{\mathcal{H}}_1 | \Phi_0 \rangle}{(E_0 - \langle \Phi_D | \hat{\mathcal{H}}_0 | \Phi_D \rangle)(E_0 - \langle \Phi_S | \hat{\mathcal{H}}_0 | \Phi_S \rangle)}. \tag{4.8.11b}$$

The diagrams corresponding to these terms are shown in Fig. 4.14. The

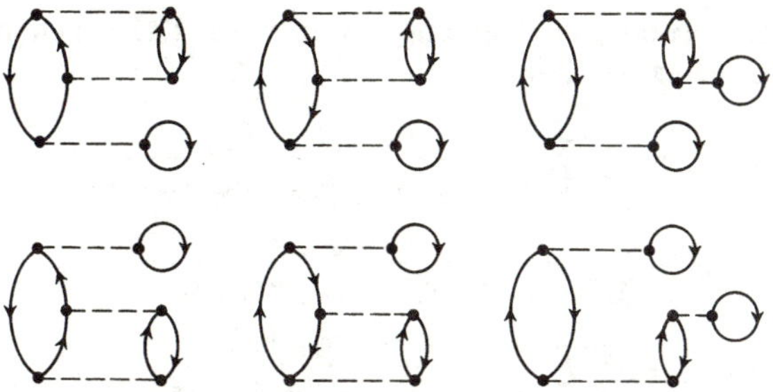

FIG. 4.14. Third-order energy diagrams involving both singly excited and doubly excited intermediate states.

algebraic expressions which result from these diagrams are, respectively,

$$E_{3SD(a)} = \sum_{ij} \sum_{abc} \sum_{k} \frac{\mathcal{I}_{akik}\mathcal{I}_{bcaj}\mathcal{I}_{ijbc}}{(\epsilon_i - \epsilon_a)(\epsilon_i + \epsilon_j - \epsilon_a - \epsilon_b)} \tag{4.8.12a}$$

$$E_{3SD(b)} = -\sum_{ijk} \sum_{ab} \sum_{l} \frac{\mathcal{I}_{alil}\mathcal{I}_{ibjk}\mathcal{I}_{jkab}}{(\epsilon_i - \epsilon_a)(\epsilon_i + \epsilon_j - \epsilon_a - \epsilon_b)} \tag{4.8.12b}$$

$$E_{3SD(c)} = \sum_{ij} \sum_{ab} \sum_{kl} \frac{\mathcal{I}_{akik}\mathcal{I}_{bljl}\mathcal{I}_{ijab}}{(\epsilon_i - \epsilon_a)(\epsilon_i + \epsilon_j - \epsilon_a - \epsilon_b)} \tag{4.8.12c}$$

$$E_{3DS(a)} = \sum_{ij} \sum_{abc} \sum_{k} \frac{\mathcal{I}_{abij}\mathcal{I}_{cjab}\mathcal{I}_{ikck}}{(\epsilon_i + \epsilon_j - \epsilon_a - \epsilon_b)(\epsilon_i - \epsilon_c)} \tag{4.8.12d}$$

$$E_{3DS(b)} = -\sum_{ijk} \sum_{ab} \sum_{l} \frac{\mathcal{I}_{abij}\mathcal{I}_{ijkb}\mathcal{I}_{klal}}{(\epsilon_i + \epsilon_j - \epsilon_a - \epsilon_b)(\epsilon_k - \epsilon_a)} \tag{4.8.12e}$$

$$E_{3DS(c)} = \sum_{ij} \sum_{ab} \sum_{kl} \frac{\mathcal{I}_{abij}\mathcal{I}_{jkbk}\mathcal{I}_{ilal}}{(\epsilon_i + \epsilon_j - \epsilon_a - \epsilon_b)(\epsilon_i - \epsilon_a)}. \tag{4.8.12f}$$

Finally, there is a term in third-order Rayleigh–Schrödinger perturbation theory which involves only double excitations

$$E_{3D} = \sum_{D_1D_2} \frac{\langle\Phi_0|\,\hat{\mathcal{H}}_1\,|\Phi_{D_1}\rangle\langle\Phi_{D_1}|\,\hat{\mathcal{H}}_1 - E_1\,|\Phi_{D_2}\rangle\langle\Phi_{D_2}|\,\hat{\mathcal{H}}_1\,|\Phi_0\rangle}{(E_0 - \langle\Phi_{D_1}|\,\hat{\mathcal{H}}_0\,|\Phi_{D_1}\rangle)(E_0 - \langle\Phi_{D_2}|\,\hat{\mathcal{H}}_0\,|\Phi\rangle)}. \tag{4.8.13}$$

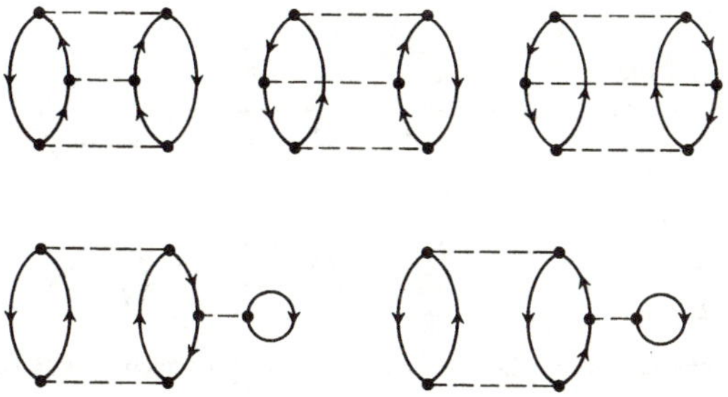

FIG. 4.15. Third-order energy diagrams involving doubly excited intermediate states.

This term gives rise to the diagrams shown in Fig. 4.15 which in turn lead to the algebraic expressions

$$E_{3D(a)} = \frac{1}{8} \sum_{ij} \sum_{abcd} \frac{\mathcal{I}_{ijab} \mathcal{I}_{abcd} \mathcal{I}_{cdij}}{(\epsilon_i + \epsilon_j - \epsilon_a - \epsilon_b)(\epsilon_i + \epsilon_j - \epsilon_c - \epsilon_d)} \tag{4.8.14a}$$

$$E_{3D(b)} = \sum_{ijk} \sum_{abc} \frac{\mathcal{I}_{ijab} \mathcal{I}_{akic} \mathcal{I}_{bcjk}}{(\epsilon_i + \epsilon_j - \epsilon_a - \epsilon_b)(\epsilon_j + \epsilon_k - \epsilon_b - \epsilon_c)} \tag{4.8.14b}$$

$$E_{3D(c)} = \frac{1}{8} \sum_{ijkl} \sum_{ab} \frac{\mathcal{I}_{ijab} \mathcal{I}_{klij} \mathcal{I}_{abkl}}{(\epsilon_i + \epsilon_j - \epsilon_a - \epsilon_b)(\epsilon_k + \epsilon_l - \epsilon_a - \epsilon_b)} \tag{4.8.14c}$$

$$E_{3D(d)} = -\frac{1}{2} \sum_{ijk} \sum_{ab} \sum_{l} \frac{\mathcal{I}_{ijab} \mathcal{I}_{kljl} \mathcal{I}_{abik}}{(\epsilon_i + \epsilon_j - \epsilon_a - \epsilon_b)(\epsilon_i + \epsilon_k - \epsilon_a - \epsilon_b)} \tag{4.8.14d}$$

$$E_{3D(e)} = \frac{1}{2} \sum_{ij} \sum_{abc} \sum_{k} \frac{\mathcal{I}_{ijab} \mathcal{I}_{kbkc} \mathcal{I}_{acij}}{(\epsilon_i + \epsilon_j - \epsilon_a - \epsilon_b)(\epsilon_i + \epsilon_j - \epsilon_a - \epsilon_c)}. \tag{4.8.14e}$$

If, instead of a bare-nucleus zero-order model, correlation effects are examined with respect to the Hartree–Fock independent electron model discussed in Section 2.4, the zero-order hamiltonian operator becomes

$$\hat{\mathcal{H}}_0 = \sum_i (\hat{h}_i + \hat{V}_i^{\mathrm{N}}) \tag{4.8.15}$$

where $\hat{V}_i^{\mathrm{N}}$ is the Hartree–Fock potential. The perturbing operator is

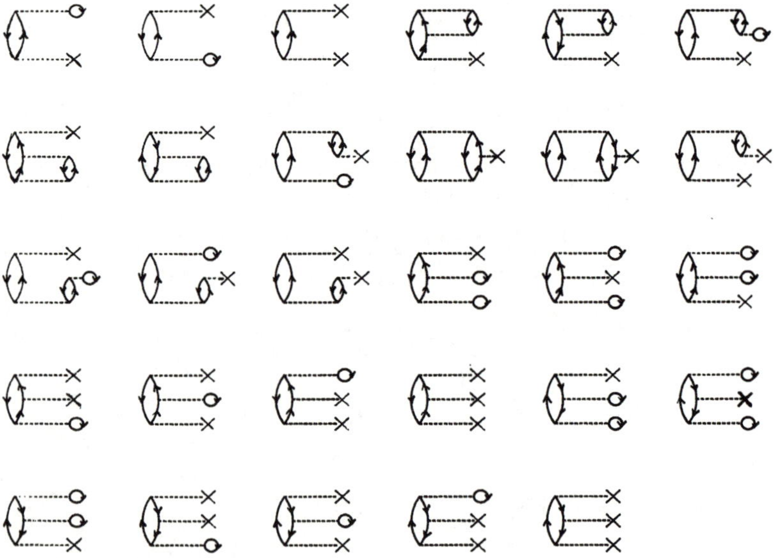

FIG. 4.16. Diagrams containing the Hartree–Fock potential which arise through third-order in the energy.

taken to be

$$\hat{\mathscr{H}}_1 = \sum_{i>j} r_{ij}^{-1} - \sum_i \hat{V}_i^{\mathrm{N}} \tag{4.8.16}$$

so that the full molecular hamiltonian is recovered on adding the two components of the hamiltonian. The presence of the Hartree–Fock potential in the perturbation (4.8.16) leads to additional diagrams in the diagrammatic perturbation series. These additional diagrams are shown through third order in the energy in Fig. 4.16 in which the diagram element

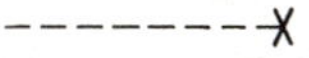

(4.8.17)

is employed to represent the Hartree–Fock potential, $-V^{\mathrm{N}}$. However, diagrammatic perturbation theory with respect to a Hartree–Fock reference function is much simplified by a cancellation of terms which may be represented diagrammatically as follows

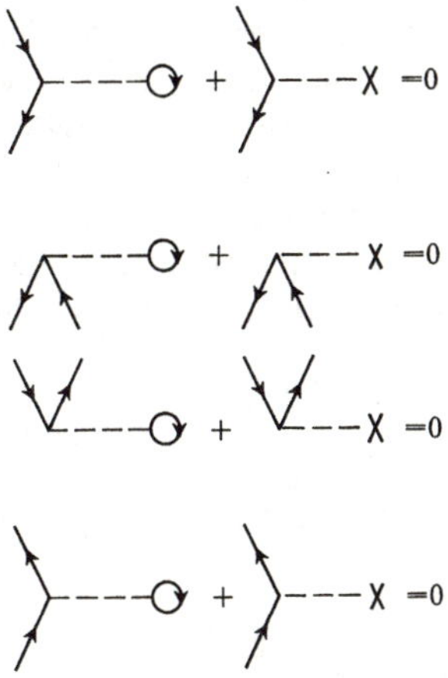

(4.8.18)

The Hartree–Fock potential terms are equal in magnitude but opposite in sign to the self-energy or 'bubble' terms. The diagrams which remain in the perturbation expansion when taken through third order after this cancellation has been effected are collected together in Fig. 4.17.

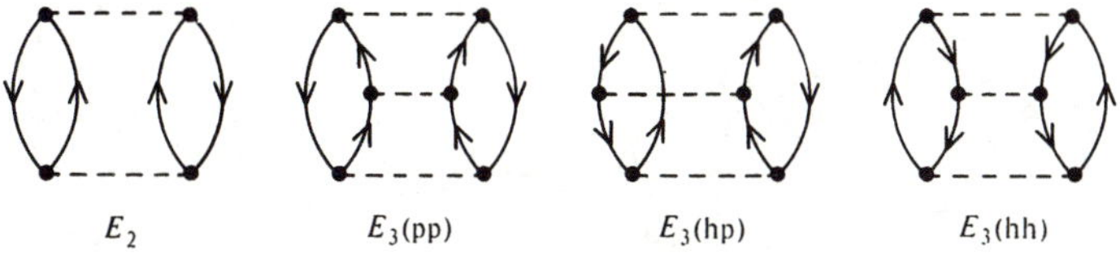

FIG. 4.17. Second-order and third-order energy diagrams which arise when the Hartree–Fock model is used to obtain a reference function.

The zero-order and first-order terms in the diagrammatic perturbation theory for closed-shell molecules, described by a single determinant in zero order, add up to the expectation value of the total hamiltonian operator for that wave function. Thus for the Hartree–Fock reference function, the sum of the zero-order and first-order energies is equal to the Hartree–Fock energy. The correlation energy is given by the sum of the second and higher order terms.

When the Hartree–Fock reference function is used, the second-order component of the correlation energy is described by one diagram if antisymmetrized vertices are employed. In third-order there are three diagrams which may be distinguished by the nature of the central interaction line: the particle–particle diagram component E_3(pp), the hole–particle diagram component E_3(hp), and the hole–hole diagram component E_3(hh). The second-order diagram and the third-order (p–p) diagram have only two hole lines. The correlation effects which they describe are, therefore, exclusively of a two-body type. The third-order (h–p) diagram contains three hole lines and can, therefore, describe three-body effects. However, if two of the hole lines are associated with the same single particle state then it describes two-body correlations. Finally, the third-order (h–h) diagram has four hole lines and can thus describe two-body, three-body, and four-body interactions. In Fig. 4.18 the possible cases which can arise through third order in the energy are illustrated.

Explicit expressions corresponding to the four diagrams given in Fig.

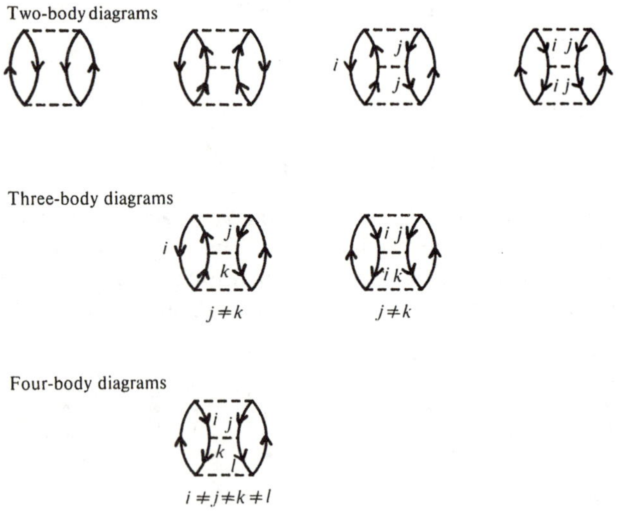

FIG. 4.18. Two-body, three-body, and four-body energy diagrams which can arise through third-order.

4.17 can be written down by following the rules presented in Section 4.7. The second-order energy expression has the form

$$E_2 = \frac{1}{4}\sum_{ij}\sum_{ab}\frac{\mathcal{I}_{ijab}\mathcal{I}_{abij}}{(\epsilon_i+\epsilon_j-\epsilon_a-\epsilon_b-d_{ijab})} \tag{4.8.19}$$

where the d_{ijab} will be discussed further below but in the immediate discussion should be taken to be zero. The third-order (p–p) energy has the form

$$E_3(\mathrm{pp}) = \frac{1}{8}\sum_{ij}\sum_{abcd}\theta_{ac;bd}\theta_{bc;ad}\frac{\mathcal{I}_{ijab}\mathcal{I}_{abcd}\mathcal{I}_{idij}}{(\epsilon_i+\epsilon_j-\epsilon_a-\epsilon_b-d_{ijab})(\epsilon_i+\epsilon_j-\epsilon_c-\epsilon_a-d_{ijcd})} \tag{4.8.20}$$

(θ should be taken to be unity in the present discussion). The third-order (h–p) energy is

$$E_3(\mathrm{hp}) = \sum_{ijk}\sum_{abc}\theta_{ik;ac}\frac{\mathcal{I}_{ijab}\mathcal{I}_{akic}\mathcal{I}_{bcjk}}{(\epsilon_i+\epsilon_j-\epsilon_a-\epsilon_b-d_{ijab})(\epsilon_j+\epsilon_k-\epsilon_b-\epsilon_c-d_{jkbc})} \tag{4.8.21}$$

and the third-order (h–h) energy may be written as follows

$$E_3(\mathrm{hh}) = \frac{1}{8}\sum_{ijkl}\sum_{ab}\theta_{ik;jl}\theta_{il;jk}\frac{\mathcal{I}_{ijab}\mathcal{I}_{klij}\mathcal{I}_{abkl}}{(\epsilon_i+\epsilon_j-\epsilon_a-\epsilon_b-d_{ijab})(\epsilon_k+\epsilon_l-\epsilon_a-\epsilon_b-d_{klab})}. \tag{4.8.22}$$

Note that all of the above expressions, eqns (4.8.19–22), are written in terms of single-electron functions and no reference is made to many-electron wave functions. This is a fundamental characteristic of the diagrammatic many-body perturbation theoretic approach to the correlation problem and there is, of course, an intimate connection between this observation and the direct proportionality of the theory to the number of electrons in the system.

In the above discussion, the N-electron Hartree–Fock hamiltonian, $\hat{\mathcal{H}}_0$, was used as a zero-order operator. This leads to the perturbation series first discussed, through second order in the energy, by Moeller and Plesset (1934; see also Binkley and Pople 1975). However, it is clear that any operator, $\hat{A}$, obeying the commutation relation

$$[\hat{\mathcal{H}}_0, \hat{A}] = 0 \Rightarrow \hat{A} = \sum_k |k\rangle\langle k|\,\hat{X}\,|k\rangle\langle k| \tag{4.8.23}$$

where $|k\rangle$ is an eigenfunction of $\hat{\mathcal{H}}_0$ may be used to develop a perturbation expansion. The operator

$$\hat{\mathcal{H}}_{\text{shifted}} = \sum_k |k\rangle\langle k|\,\hat{\mathcal{H}}\,|k\rangle\langle k| \tag{4.8.24}$$

is a special case of relation (4.8.23) in which A is set equal to the total hamiltonian. The choice (4.8.24) leads to the shifted denominator perturbation series which was first considered by Epstein (1926) and by Nesbet (1955). The resulting perturbation expansion may be given the same diagrammatic representation as the expansion based on the Hartree–Fock model hamiltonian. The corresponding algebraic expressions are given by eqns (4.8.19–22) with

(i) In third order the diagonal scattering terms are omitted, that is, the summations are modified by putting

$$\theta_{pq;rs} = 2^{-\Gamma_{pq}\Gamma_{rs}}(\Gamma_{pq} + \Gamma_{rs})$$
$$\Gamma_{pq} = 1 - \delta_{pq} \begin{array}{l} = 1 \quad \text{if} \quad p \neq q \\ = 0 \quad \text{if} \quad p = q \end{array} \tag{4.8.25}$$

(ii) The denominators are 'shifted' by

$$d_{ijab} = \mathcal{J}_{ijij} + \mathcal{J}_{abab} + \mathcal{J}_{iaai} + \mathcal{J}_{ibbi} + \mathcal{J}_{jaaj} + \mathcal{J}_{jbbj} \tag{4.8.26}$$

The use of 'shifted' denominators may also be interpreted as the summation through infinite-order of certain types of terms which occur in the perturbation series based on the Hartree–Fock model hamiltonian. This infinite-order summation of diagonal scattering terms is illustrated in Fig. 4.19.

The sum of the terms in the perturbation series through infinite order is, of course, independent of the choice of zero-order hamiltonian if all contributions are included. Assuming that the perturbation series are convergent, the Hartree–Fock model perturbation expansion, that is the Moeller–Plesset theory, and the shifted denominator, that is the Epstein–Nesbet theory, or indeed any other perturbation series which can be obtained by making different choices of the operator A in relation (4.8.23), will lead to identical results at sufficiently high order.

Most methods currently employed in the study of electron correlation effects in atoms and molecules, for example, limited configuration interaction, coupled pair many-electron theory, may be regarded as third-order theories in that they neglect, or approximate, fourth-order and higher-order terms in the perturbational analysis of the energy. Diagrammatic perturbation theory offers a very systematic approach for extending such calculations. Perturbation theory provides a clearly defined order parameter which indicates the relative importance of the various terms and thus forms the basis of a balanced treatment of the correlation problem. There is, therefore, considerable interest in the evaluation of the fourth-order terms in the diagrammatic perturbation theory since these terms represent, at least in part, the dominant corrections to most

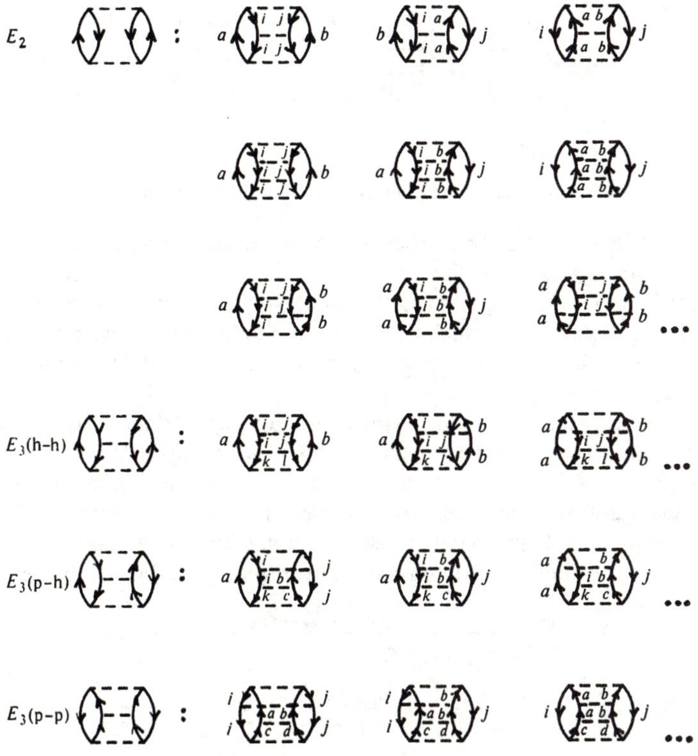

FIG. 4.19. Infinite-order summation of diagonal scattering terms by the use of denominator shifts.

of the techniques currently being used in the calculation of atomic and molecular correlation energies and correlation effects.

From the discussion of Section 4.5, it should be recalled that the fourth-order Rayleigh–Schrödinger perturbation theory contribution to the electron correlation energy of a closed-shell system when a Hartree–Fock reference function is employed may be written in the form

$$E_4 = \sum_{D_1, D_2, X} \left(\frac{\langle \Phi_0 | \hat{\mathcal{H}}_1 | \Phi_{D_1} \rangle \langle \Phi_{D_1} | \hat{\mathcal{H}}_1 | \Phi_X \rangle \langle \Phi_X | \hat{\mathcal{H}}_1 | \Phi_{D_2} \rangle \langle \Phi_{D_2} | \hat{\mathcal{H}}_1 | \Phi_0 \rangle}{(E_0 - E_{D_1})(E_0 - E_X)(E_0 - E_{D_2})} \right)_L \tag{4.8.27}$$

where Φ_0 denotes the matrix Hartree–Fock function, Φ_D denotes an N-electron state which is doubly excited with respect to the reference function, and Φ_X denotes an intermediate state obtained by further excitation. E_I denotes the zero-order energy of the Ith state. The

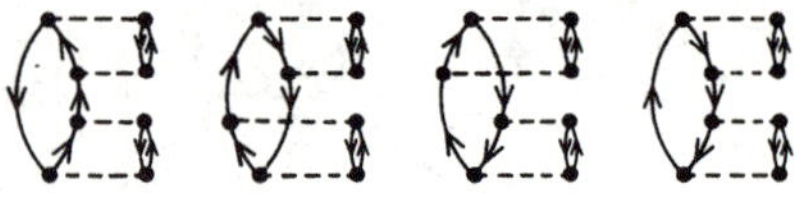

FIG. 4.20. Fourth-order energy diagrams involving singly excited intermediate states.

subscript L is used to indicate that only terms which correspond to linked diagrams are included in the summation. Unlike the second-order and third-order energy components, the fourth-order terms can involve an intermediate state Φ_X which is singly excited, doubly excited, triply excited, or quadruply excited with respect to the single determinantal reference function.

The fourth-order linked diagrams involving singly, doubly, triply, and quadruply excited intermediate states are shown in Figs. 4.20, 4.21, 4.22, and 4.23, respectively. Below, the algebraic expressions corresponding to these diagrams will be given for the case in which the Hartree–Fock model zero-order hamiltonian is used. The algebraic expressions corresponding to the single-excitation diagrams displayed in Fig. 4.20 are

$$E_4(A_S) = \tfrac{1}{4} \sum \frac{\mathscr{I}_{ijab}\mathscr{I}_{abcj}\mathscr{I}_{ckde}\mathscr{I}_{deik}}{(\epsilon_i + \epsilon_j - \epsilon_a - \epsilon_b)(\epsilon_i - \epsilon_c)(\epsilon_i + \epsilon_k - \epsilon_d - \epsilon_e)} \tag{4.8.28a}$$

$$E_4(B_S) = -\tfrac{1}{4} \sum \frac{\mathscr{I}_{ijab}\mathscr{I}_{kbij}\mathscr{I}_{alcd}\mathscr{I}_{cdkl}}{(\epsilon_i + \epsilon_j - \epsilon_a - \epsilon_b)(\epsilon_k - \epsilon_a)(\epsilon_k + \epsilon_l - \epsilon_c - \epsilon_d)}. \tag{4.8.28b}$$

Diagrams A_S and D_S are related by time reversal (Wilson and Silver 1979*a*) and thus the expression for D_S is obtained by interchanging hole and particle labels in the expression for A_S. Diagrams B_S and C_S are related by complex conjugation (Wilson and Silver 1979*a*) and the expressions corresponding to these two diagrams are equal if real orbitals are used. The twelve fourth-order diagrams which involve only doubly excited intermediate states give rise to the following algebraic expressions

$$E_4(A_D) = \tfrac{1}{16} \sum \frac{\mathscr{I}_{ijab}\mathscr{I}_{abcd}\mathscr{I}_{cdef}\mathscr{I}_{efij}}{(\epsilon_i + \epsilon_j - \epsilon_a - \epsilon_b)(\epsilon_i + \epsilon_j - \epsilon_c - \epsilon_d)(\epsilon_i + \epsilon_j - \epsilon_e - \epsilon_f)} \tag{4.8.29a}$$

$$E_4(B_D) = \tfrac{1}{16} \sum \frac{\mathscr{I}_{ijab}\mathscr{I}_{klij}\mathscr{I}_{abcd}\mathscr{I}_{cdkl}}{(\epsilon_i + \epsilon_j - \epsilon_a - \epsilon_b)(\epsilon_k + \epsilon_l - \epsilon_a - \epsilon_b)(\epsilon_k + \epsilon_l - \epsilon_c - \epsilon_d)} \tag{4.8.29b}$$

$$E_4(E_D) = -\tfrac{1}{2} \sum \frac{\mathscr{I}_{ijab}\mathscr{I}_{kbic}\mathscr{I}_{acde}\mathscr{I}_{dekj}}{(\epsilon_i + \epsilon_j - \epsilon_a - \epsilon_b)(\epsilon_j + \epsilon_k - \epsilon_a - \epsilon_c)(\epsilon_j + \epsilon_k - \epsilon_d - \epsilon_e)} \tag{4.8.29c}$$

$$E_4(I_D) = \sum \frac{\mathscr{I}_{ijab}\mathscr{I}_{akcj}\mathscr{I}_{cldk}\mathscr{I}_{dbil}}{(\epsilon_i + \epsilon_j - \epsilon_a - \epsilon_b)(\epsilon_i + \epsilon_k - \epsilon_b - \epsilon_c)(\epsilon_i + \epsilon_l - \epsilon_b - \epsilon_d)} \tag{4.8.29d}$$

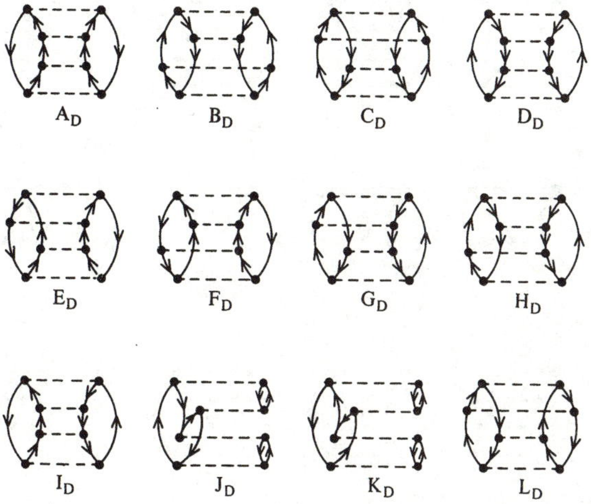

FIG. 4.21. Fourth-order energy diagrams involving only doubly excited intermediate states.

$$E_4(J_D) = -\sum \frac{\mathscr{I}_{ijab}\mathscr{I}_{kbcj}\mathscr{I}_{clid}\mathscr{I}_{adkl}}{(\epsilon_i+\epsilon_j-\epsilon_a-\epsilon_b)(\epsilon_i+\epsilon_k-\epsilon_a-\epsilon_c)(\epsilon_k+\epsilon_l-\epsilon_a-\epsilon_d)} \quad (4.8.29e)$$

$$E_4(L_D) = \sum \frac{\mathscr{I}_{ijab}\mathscr{I}_{akcj}\mathscr{I}_{lbid}\mathscr{I}_{cdlk}}{(\epsilon_i+\epsilon_j-\epsilon_a-\epsilon_b)(\epsilon_i+\epsilon_k-\epsilon_b-\epsilon_c)(\epsilon_k+\epsilon_l-\epsilon_c-\epsilon_d)}. \quad (4.8.29f)$$

Diagrams B_D and C_D, E_D and F_D, and G_D and H_D are related by complex conjugation. Diagrams A_D and D_D, E_D and H_D, F_D and G_D, and J_D and K_D are related by time reversal. The sixteen diagrams in fourth order which involve triply excited intermediate states are shown in Fig. 4.22 and lead to the following algebraic expressions

$$E_4(A_T) = -\tfrac{1}{2}\sum \frac{\mathscr{I}_{ijab}\mathscr{I}_{akcd}\mathscr{I}_{cbek}\mathscr{I}_{edij}}{(\epsilon_i+\epsilon_j-\epsilon_a-\epsilon_d)(\epsilon_i+\epsilon_j+\epsilon_k-\epsilon_b-\epsilon_c-\epsilon_d)(\epsilon_i+\epsilon_j-\epsilon_d-\epsilon_e)} \quad (4.8.30a)$$

$$E_4(B_T) = -\tfrac{1}{2}\sum \frac{\mathscr{I}_{ijab}\mathscr{I}_{akcd}\mathscr{I}_{cdej}\mathscr{I}_{ebik}}{(\epsilon_i+\epsilon_j-\epsilon_a-\epsilon_b)(\epsilon_i+\epsilon_j+\epsilon_k-\epsilon_b-\epsilon_c-\epsilon_d)(\epsilon_i+\epsilon_k-\epsilon_b-\epsilon_e)} \quad (4.8.30b)$$

$$E_4(E_T) = \sum \frac{\mathscr{I}_{ijab}\mathscr{I}_{akcd}\mathscr{I}_{cbej}\mathscr{I}_{edik}}{(\epsilon_i+\epsilon_j-\epsilon_a-\epsilon_b)(\epsilon_i+\epsilon_j+\epsilon_k-\epsilon_b-\epsilon_c-\epsilon_d)(\epsilon_i+\epsilon_k-\epsilon_d-\epsilon_e)} \quad (4.8.30c)$$

$$E_4(F_T) = \tfrac{1}{4}\sum \frac{\mathscr{I}_{ijab}\mathscr{I}_{akcd}\mathscr{I}_{cdek}\mathscr{I}_{ebij}}{(\epsilon_i+\epsilon_j-\epsilon_a-\epsilon_b)(\epsilon_i+\epsilon_j+\epsilon_k-\epsilon_b-\epsilon_c-\epsilon_d)(\epsilon_i+\epsilon_j-\epsilon_b-\epsilon_e)} \quad (4.8.30d)$$

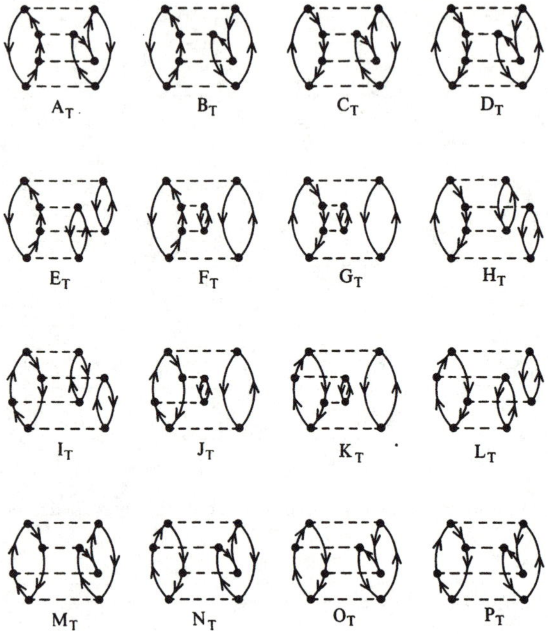

FIG. 4.22. Fourth-order energy diagrams involving triply excited intermediate states.

$$E_4(I_T) = -\tfrac{1}{4}\sum \frac{\mathcal{I}_{ijab}\mathcal{I}_{klic}\mathcal{I}_{abdj}\mathcal{I}_{dckl}}{(\epsilon_i+\epsilon_j-\epsilon_a-\epsilon_b)(\epsilon_j+\epsilon_k+\epsilon_l-\epsilon_a-\epsilon_b-\epsilon_c)(\epsilon_k+\epsilon_l-\epsilon_c-\epsilon_d)} \tag{4.8.30e}$$

$$E_4(J_T) = -\sum \frac{\mathcal{I}_{ijab}\mathcal{I}_{klic}\mathcal{I}_{acdl}\mathcal{I}_{dbkj}}{(\epsilon_i+\epsilon_j-\epsilon_a-\epsilon_b)(\epsilon_j+\epsilon_k+\epsilon_l-\epsilon_a-\epsilon_b-\epsilon_c)(\epsilon_j+\epsilon_k-\epsilon_b-\epsilon_d)} \tag{4.8.30f}$$

$$E_4(M_T) = \tfrac{1}{2}\sum \frac{\mathcal{I}_{ijab}\mathcal{I}_{klic}\mathcal{I}_{abdl}\mathcal{I}_{dckj}}{(\epsilon_i+\epsilon_j-\epsilon_a-\epsilon_b)(\epsilon_j+\epsilon_k+\epsilon_l-\epsilon_a-\epsilon_b-\epsilon_c)(\epsilon_j+\epsilon_k-\epsilon_c-\epsilon_d)}. \tag{4.8.30g}$$

Time-reversal operations can be seen to relate diagrams A_T, B_T, E_T, and F_T with diagrams D_T, C_T, H_T, and G_T, respectively. Diagrams M_T and N_T, and N_T and O_T are also related by time reversal, whilst M_T and N_T, and O_T and P_T are complex conjugates. The algebraic expressions arising from the seven diagrams given in Fig. 4.23 may be cast in the form

$$E_4(A_Q) = \tfrac{1}{2}\sum \frac{\mathcal{I}_{ijab}\mathcal{I}_{klcd}\mathcal{I}_{cbil}\mathcal{I}_{adkj}}{(\epsilon_i+\epsilon_j-\epsilon_a-\epsilon_b)(\epsilon_i+\epsilon_l-\epsilon_b-\epsilon_c)(\epsilon_j+\epsilon_l-\epsilon_a-\epsilon_d)} \tag{4.8.31a}$$

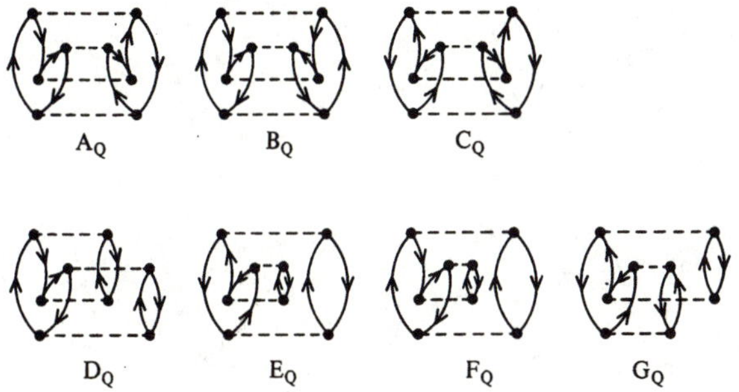

FIG. 4.23. Fourth-order energy diagrams involving quadruply excited intermediate states.

$$E_4(B_Q+C_Q)=\frac{1}{16}\sum\frac{\mathscr{I}_{ijab}\mathscr{I}_{klcd}\mathscr{I}_{cdij}\mathscr{I}_{abkl}}{(\epsilon_i+\epsilon_j-\epsilon_a-\epsilon_b)(\epsilon_i+\epsilon_j-\epsilon_c-\epsilon_d)(\epsilon_k+\epsilon_l-\epsilon_a-\epsilon_b)} \tag{4.8.31b}$$

$$E_4(D_Q+E_Q)=-\frac{1}{4}\sum\frac{\mathscr{I}_{ijab}\mathscr{I}_{klcd}\mathscr{I}_{cbij}\mathscr{I}_{adkl}}{(\epsilon_i+\epsilon_j-\epsilon_a-\epsilon_b)(\epsilon_i+\epsilon_j-\epsilon_b-\epsilon_c)(\epsilon_k+\epsilon_l-\epsilon_a-\epsilon_d)}. \tag{4.8.31c}$$

Time reversal relates diagrams B_Q and C_Q, D_Q and G_Q, and E_Q and F_Q.

The fourth-order, triple-excitation energy diagrams correspond to connected second-order wave function diagrams of the type

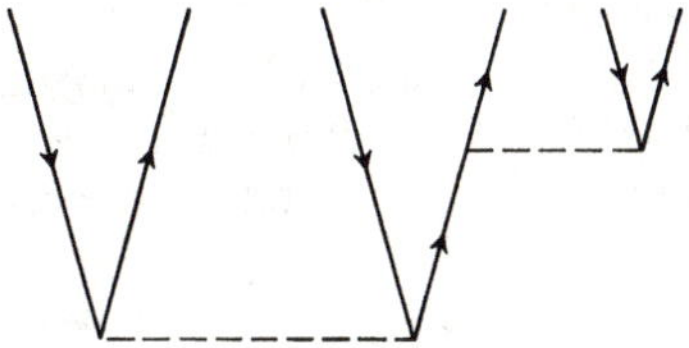

(4.8.32)

On the other hand, the fourth-order, quadruple-excitation energy diagrams, shown in Fig. 4.23, correspond to disconnected second-order wave function diagrams of the form

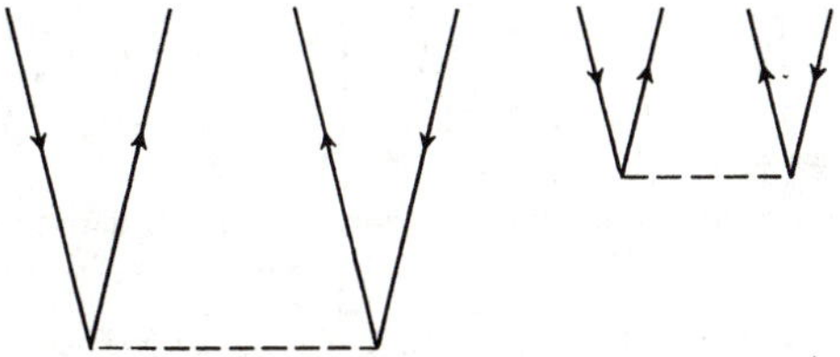

(4.8.33)

The second-order quadruple-excitation wave function diagrams can,

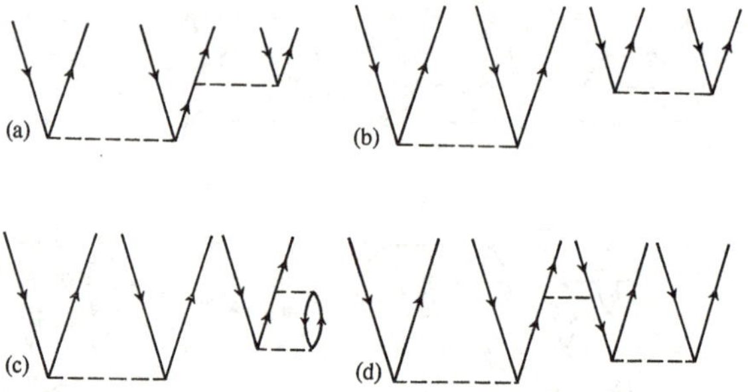

FIG. 4.24. Connected and disconnected wave function diagrams: (a) second-order connected triply excited diagram; (b) second-order disconnected quadruply excited diagram; (c) third-order disconnected triply excited diagram; (d) third-order connected quadruply excited diagram.

therefore, lead to both linked and unlinked diagrams components of the fourth-order energy whereas the second-order, triple-excitation wave function diagrams lead only to linked diagram components of the fourth-order energy.

Disconnected wave function diagrams involving triply excited intermediate states and connected wave function diagrams involving quadruply excited intermediate states first arise in third-order of the perturbation series for the wave function, as illustrated in Fig. 4.24. In Fig. 4.25, examples of the linked and unlinked energy diagrams which can be derived from the disconnected third-order wave function diagrams shown in Fig. 4.24 are displayed.

In Fig. 4.26, examples of linked energy diagrams are given which involve the interaction of intermediate states arising from different degrees of excitation. Figure 4.26(a) is a fifth-order diagram which includes

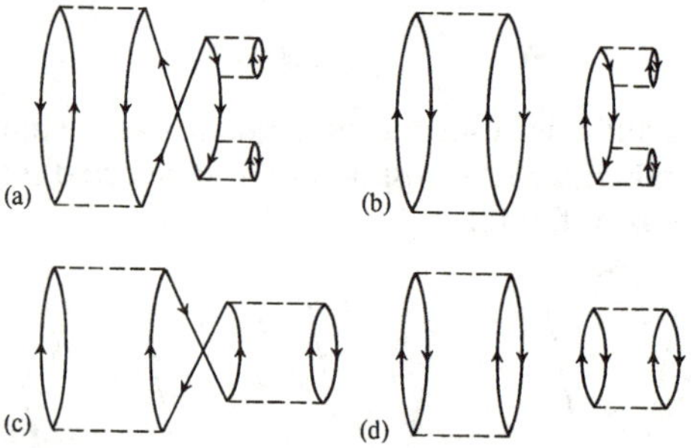

FIG. 4.25. Examples of linked and unlinked energy diagrams derived from disconnected wave-function diagrams involving triply excited and quadruply excited intermediate states: (a) linked energy diagram and (b) unlinked energy diagram derived from a disconnected wave-function diagram involving a triply excited state; (c) linked energy diagram and (d) unlinked energy diagram derived from a disconnected wave-function diagram involving a quadruply excited intermediate state.

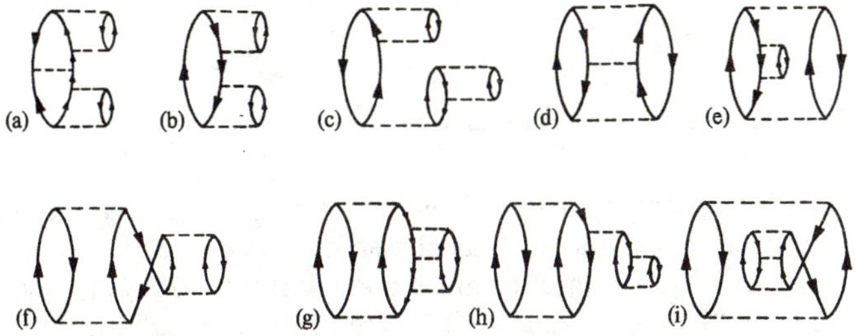

FIG. 4.26. Examples of linked energy diagrams containing the interaction between intermediate states involving different degrees of excitation; see text for details.

an interaction between two intermediate states which are singly excited with respect to the reference state. The diagram shown in Fig. 4.26(b) describes an interaction between a singly excited state and a doubly excited state, whilst Fig. 4.26(c) involves a singly excited state and a triply excited state. Figures 4.26(d), (e), and (f) show the interaction of a doubly excited intermediate state with states which are doubly excited, triply excited, and quadruply excited, respectively. The interaction of two triply excited states is illustrated in Fig. 4.26(g). Figure 4.26(h) illustrates the interaction of a triply excited state and a quadruply excited state. Finally, Fig. 4.26(i) shows an interaction between two quadruply excited states. The diagrams shown in Fig. 4.26 are typical of those which can arise in fifth-order and higher-orders of the perturbation expansion for the energy.

4.9 Diagrammatic perturbation theory for open-shell systems

Certain important classes of open-shell systems can be described within the restricted Hartree–Fock model by a single determinant. In this section, the development of a diagrammatic perturbation expansion for open-shell molecules which can be described in zero order by a restricted Hartree–Fock function will be discussed. The general description of open-shell systems requires the use of reference functions which go beyond the single determinant approximation. The use of multi-determinantal reference function will be discussed in Section 4.10 and the use of valence bond function will be described in Section 4.11.

It is, of course, possible to develop a perturbation series for correlation effects with respect to an unrestricted Hartree–Fock wave function. However, as described in Section 2.4, this theory does possess a number of serious deficiencies, such as unphysical features in calculated potential energy curves, which preclude its use in quantitative studies.

The most important open-shell systems which can be described by the restricted Hartree–Fock ansatz (Roothaan 1960) are the half-closed-shell

FIG. 4.27. Second-order energy diagrams for open-shell systems.

systems in which the open-shell consists of singly occupied, completely degenerate sets of orbitals with all the spins being parallel. Examples of half-closed-shell systems are the ground (triplet) state of the O_2 molecule, the atomic N ^4S state and the a $^4\Sigma^-$ state of the CH radical.

Hubač and Čársky (1980) demonstrated that if, following Cederbaum and Schirmer (1974), the restricted Hartree–Fock operator is cast in the form

$$f_R = f + u \tag{4.9.1}$$

where u is a one-electron operator which depends on the particular electronic configuration being considered, then a diagrammatic perturbation series analogous to that described in Section 4.8 for closed-shell systems can be obtained.

In order to give the diagrammatic representation of the perturbation series for the correlation energy in an open-shell system described by a restricted Hartree–Fock function, it is necessary to introduce a diagrammatic representation of the one-electron operator $\hat{u}$.

Following Hubač and Čársky (1980), the Hugenholtz diagrammatic convention is employed to enumerate the diagrammatic terms. The

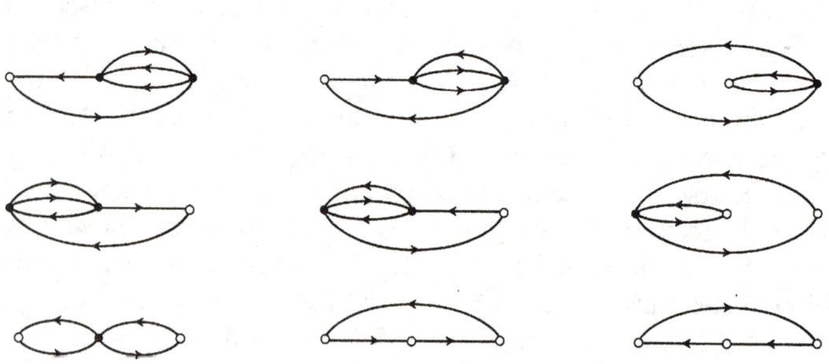

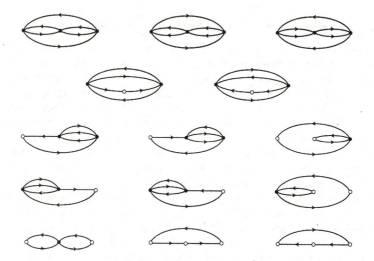

FIG. 4.28. Third-order energy diagrams for open-shell systems.

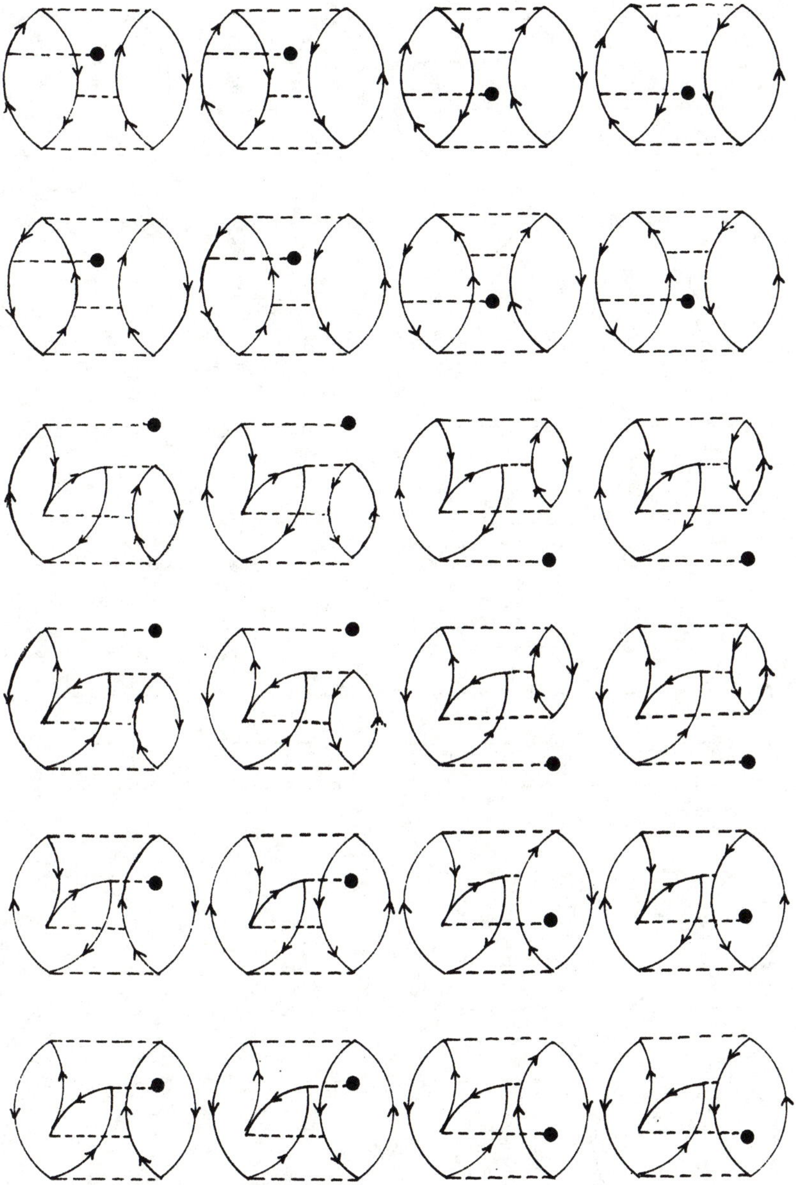

FIG. 4.29. Fourth-order energy diagrams for open-shell systems involving one single particle insertion.

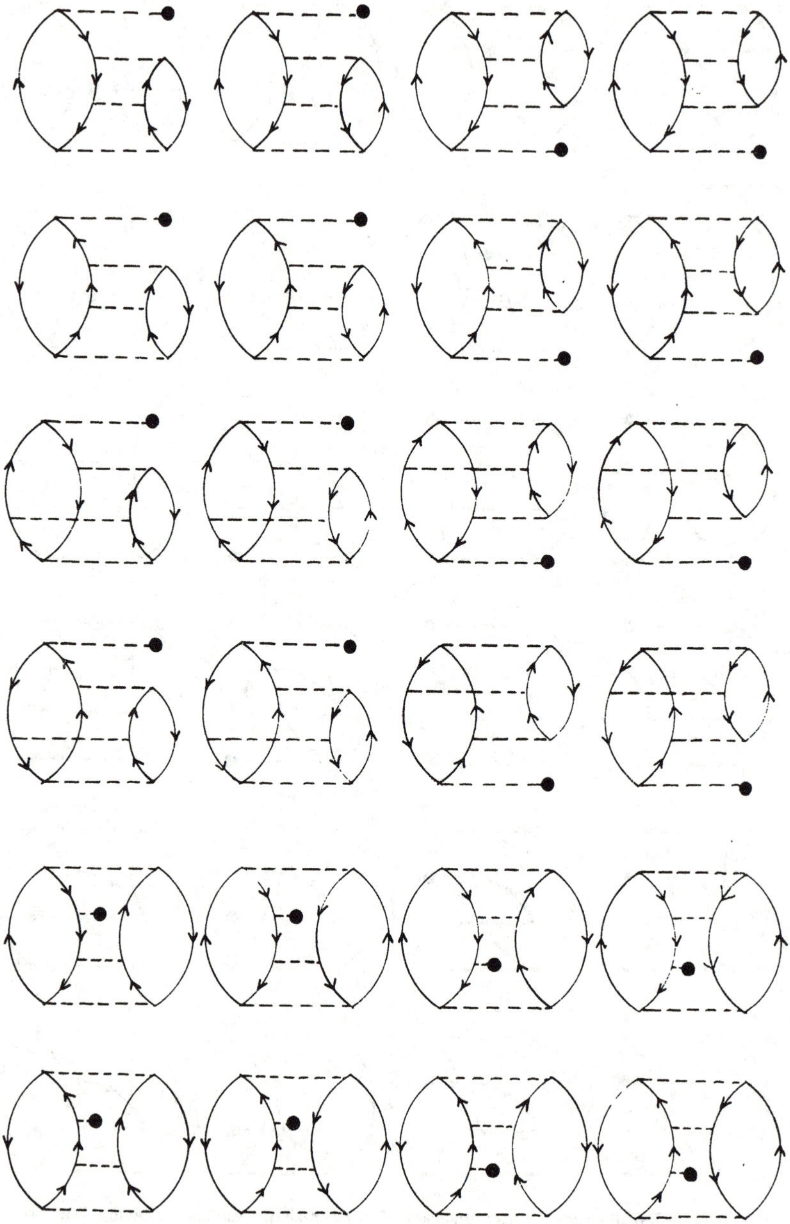

FIG. 4.29. (continued)

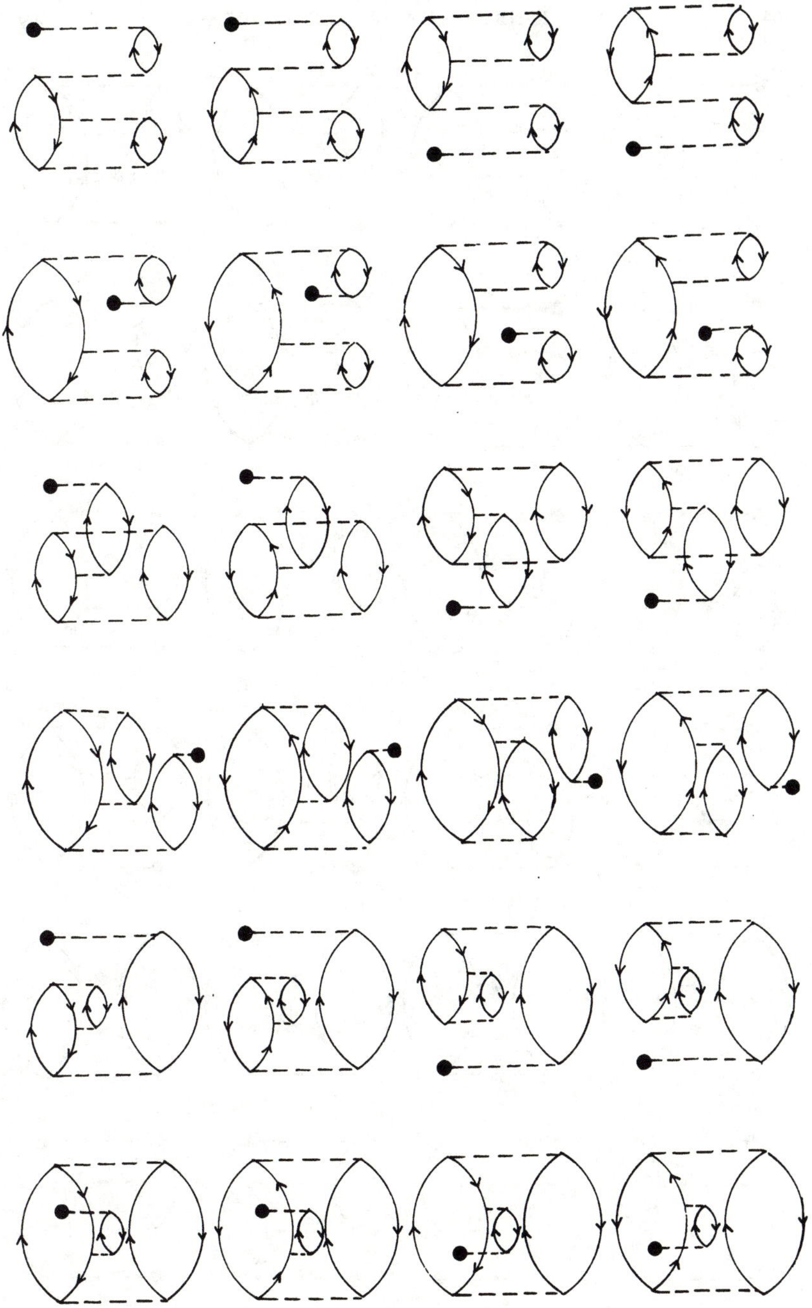

FIG. 4.29. (continued)

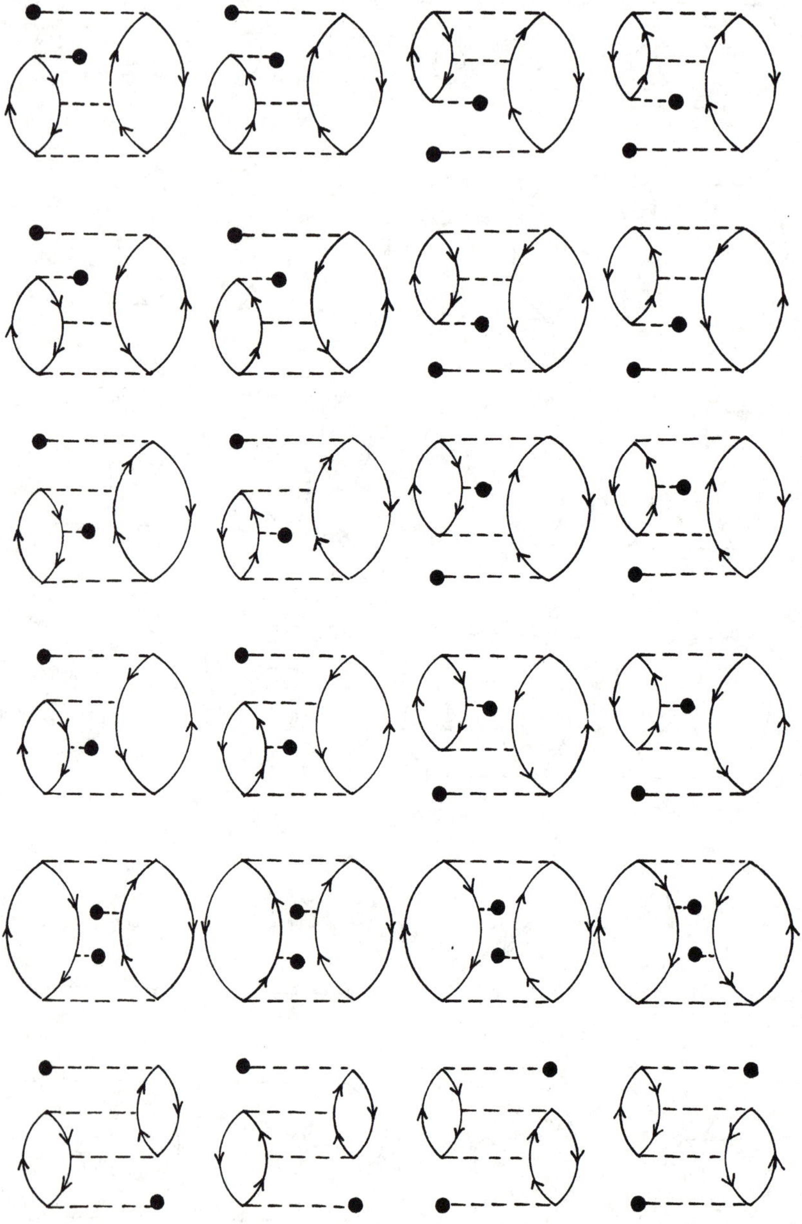

FIG. 4.30. Fourth-order energy diagrams for open-shell systems involving two single particle insertions.

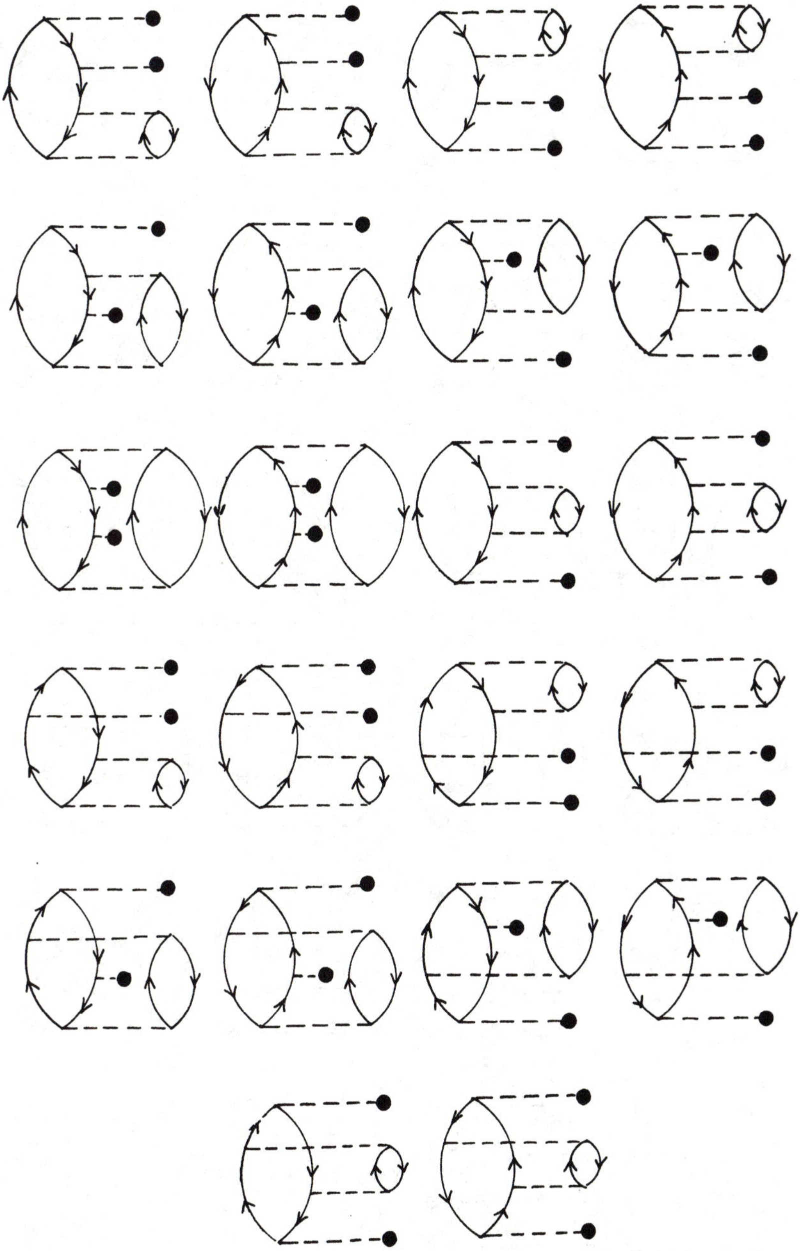

FIG. 4.30. (continued)

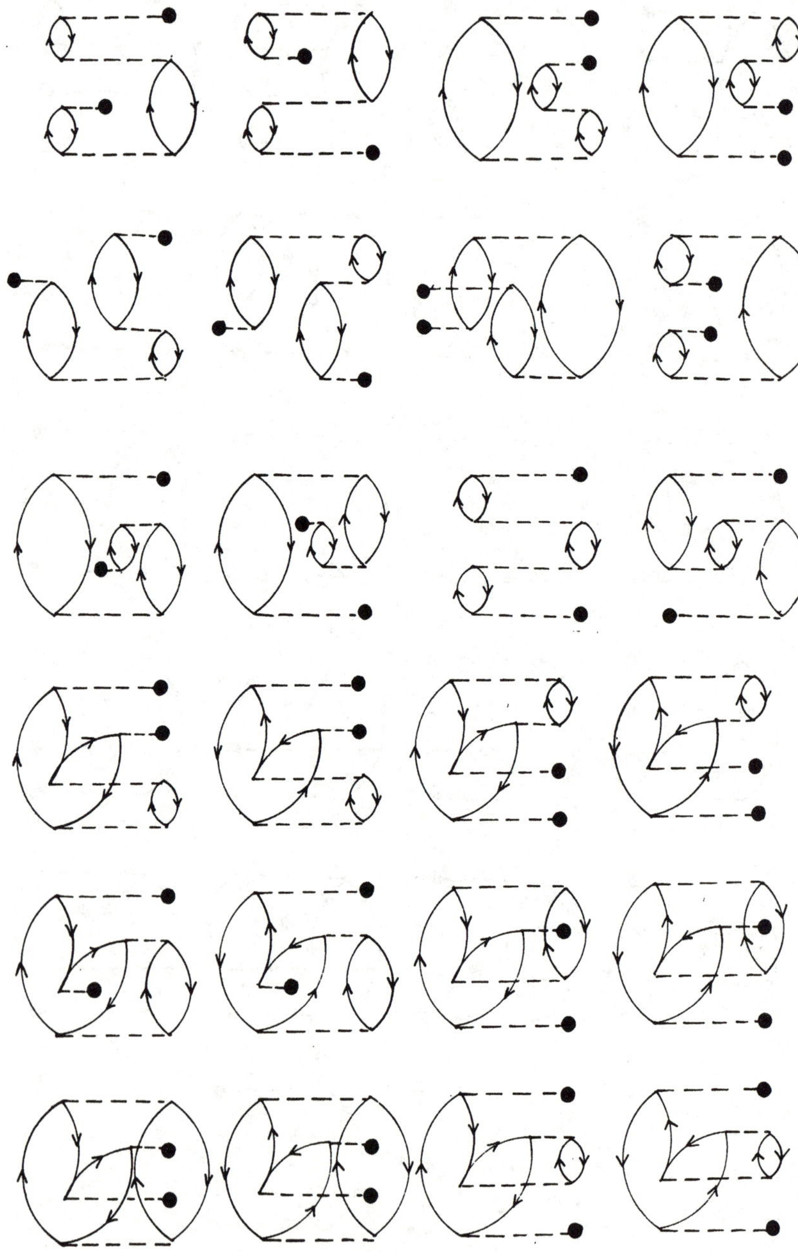

FIG. 4.30. (continued)

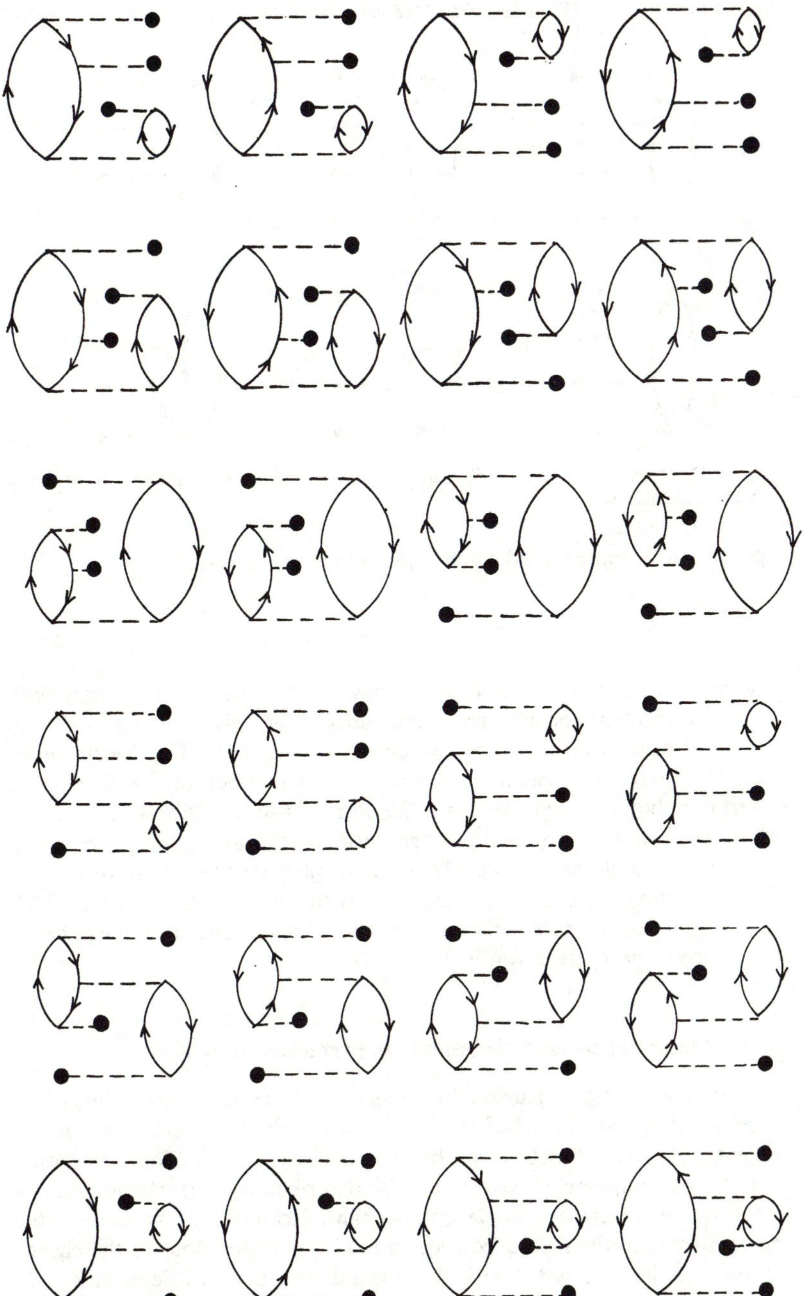

FIG. 4.31. Fourth-order energy diagrams for open-shell systems involving three single particle insertions.

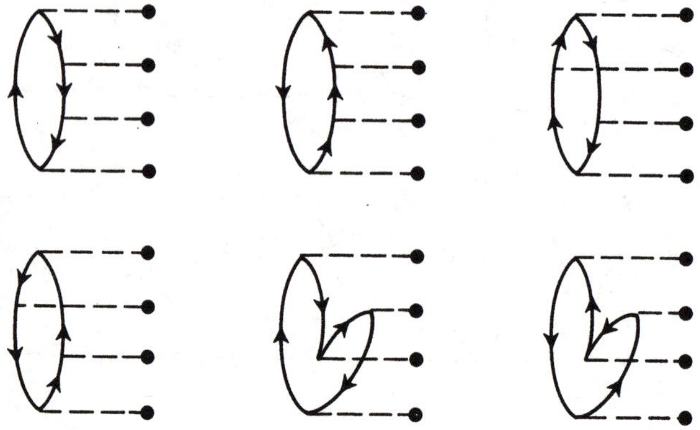

FIG. 4.32. Fourth-order energy diagrams for open-shell systems involving four single particle insertions.

operator u is represented by an open circle as follows:

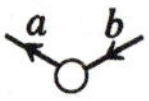

$$-\langle a|\, u\, |b\rangle N[a_a^\dagger a_b] \tag{4.9.2}$$

The second-order Hugenholtz diagrams for the correlation energy with respect to a restricted Hartree–Fock function are given in Fig. 4.27 and the third-order diagrams are displayed in Fig. 4.28. The fourth-order diagrams may be divided according to the number of single particle insertions they obtained. In Fig. 4.29 all fourth-order Brandow diagrams which contain one single particle insertion are given. The diagrams which contain two single particle insertions are displayed in Fig. 4.30, those with three insertions in Fig. 4.31, and those with four insertions in Fig. 4.32. Of course, fourth-order diagrams with no insertions also have to be considered; these are given in Figs. 4.20–23.

4.10 Quasi-degenerate diagrammatic perturbation theory

The diagrammatic perturbation theory of Brueckner and Goldstone, considered in Sections 4.8 and 4.9, can only be applied when the unperturbed wave function can be described by a single Slater determinant. A large number of systems can be described by perturbation expansions with respect to a single determinant. However, if there are other determinants in the reference spectrum with energies close to the chosen zero-order determinant, that is there is a degree of quasi-degeneracy, the convergence properties of the expansion may deteriorate. One method for avoiding this problem is to employ a reference function which consists of a linear combination of determinants

$$\Psi = \sum_{i \in S} C_i \Phi_i \tag{4.10.1}$$

where S denotes the space spanned by the set of determinants Φ_i. The zero-order space is multidimensional and quasi-degenerate diagrammatic perturbations theory must be used. The hamiltonian is diagonalized within a space of low dimensionality, referred to as 'model' space, and interactions with the complementary space are calculated perturbatively.

It should be noted that a wide range of problems which do involve some degree of quasi-degeneracy can be treated by the non-degenerate perturbation theory of Sections 4.6, 4.8, and 4.9. The formation of Padé approximants to the perturbation series has been shown to be most useful in this respect (Wilson 1979). This aspect of quasi-degeneracy effects in diagrammatic perturbation theory will be considered further in Section 7.6.

The problem of obtaining a many-body perturbation theory with respect to a multideterminantal reference function has been addressed by a number of authors (Bloch 1958; Des Cloizeaux 1960; Primas 1961, 1963; Morita 1963; Brandow 1967, 1977; Sanders 1969; Johnson and Baranger 1971; Kuo, Lee, and Ratcliff 1971; Lindgren 1974, 1978; Kvasnicka 1974; Levy 1978; Hose and Kaldor 1979, 1980; Lindgren and Morrison 1982). Brandow (1967) gave the first complete derivation of a quasi-degenerate diagrammatic perturbation theory and this has been applied to molecular systems by Kaldor and his coworkers (Kaldor 1975*a*,*b*; Stern and Kaldor 1976). In Brandow's formalism, and, indeed in most other formalisms, the orbital basis set is partitioned into core orbitals, valence or active orbitals, and virtual orbitals as illustrated in Fig. 4.33. The core orbitals are doubly occupied in all of the determinants, Φ_i,

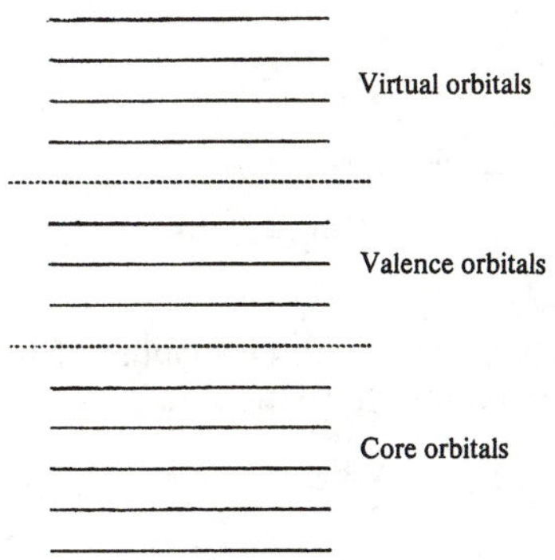

FIG. 4.33. Partitioning of orbitals for quasi-degenerate diagrammatic perturbation theory.

included in the model space, S; the valence orbitals are sometimes occupied and sometimes unoccupied and the virtual orbitals are always unoccupied.

The total molecular hamiltonian, with eigenproblem $\hat{\mathcal{H}}\psi = E\psi$, is written in the form

$$\hat{\mathcal{H}} = \hat{\mathcal{H}}_0 + \hat{\mathcal{H}}_1 \tag{4.10.2}$$

with

$$\hat{\mathcal{H}}_0 \Phi_i = E_i \Phi_i. \tag{4.10.3}$$

The projection operator onto the model space S is

$$\hat{P} = \sum_{i \in S} |\Phi_i\rangle\langle\Phi_i| \tag{4.10.4}$$

and its complement is

$$\begin{aligned} \hat{Q} &= \hat{I} - \hat{P} \\ &= \sum_{i \notin S} |\Phi_i\rangle\langle\Phi_i|. \end{aligned} \tag{4.10.5}$$

For an exactly degenerate model space with energy E_{m}, the energy correction is defined by

$$\Delta E = E - E_{\mathrm{m}} \tag{4.10.6}$$

where E is the total non-relativistic energy of the system. For a quasi-degenerate model space $\hat{\mathcal{H}}_0$ and $\hat{\mathcal{H}}_1$ are shifted by a diagonal operator

$$\hat{V} = \sum_{i \in S} |\Phi_i\rangle(E_{\mathrm{m}} - E_i)\langle\Phi_i| \tag{4.10.7}$$

giving

$$\hat{\mathcal{H}}_0' = \hat{\mathcal{H}}_0 + V \tag{4.10.8}$$

and

$$\hat{\mathcal{H}}_1' = \hat{\mathcal{H}}_1 - V. \tag{4.10.9}$$

The modified zero-order eigenproblem

$$\mathcal{H}_0' \Phi_i = E_{\mathrm{m}} \Phi_i \tag{4.10.10}$$

is exactly degenerate.

The energy correction, ΔE, is an eigenvalue of the reaction operator, $\hat{\Lambda}'$, (Löwdin 1962*a,b*; Feshbach 1958, 1962)

$$\hat{\Lambda}' P\psi = \Delta E P\psi \tag{4.10.11}$$

where

$$\hat{\Lambda}' = \hat{P}\hat{\mathcal{H}}_1\hat{P} + \hat{P}\hat{\mathcal{H}}_1\hat{\mathcal{G}}\hat{\mathcal{H}}_1\hat{P}; \tag{4.10.12}$$

$\hat{\mathscr{G}}$ is the Green function of $\hat{\mathscr{H}}$

$$\hat{\mathscr{G}} = \mathscr{G}_0 + \hat{\mathscr{G}}_0 \hat{\mathscr{H}}_1 \hat{\mathscr{G}} \tag{4.10.13}$$

where $\hat{\mathscr{G}}_0$ is the Green function of $\hat{\mathscr{H}}_0$. Equation (4.10.12) can be iterated to give the Lennard-Jones–Brillouin–Wigner perturbation series in a similar fashion to the non-degenerate case discussed in Section 4.2. In the case of quasi-degenerate perturbation theory, the operator

$$\hat{\Lambda} = \hat{\Lambda}' + V_1 \tag{4.10.14}$$

can be defined, satisfying

$$\hat{\Lambda}\hat{P}\psi = (\Delta E + V_1)\hat{P}\psi \tag{4.10.15}$$

Before discussing the diagrammatic expansion for the matrix elements of $\hat{\Lambda}$, the construction of the model space S will be considered. Almost all formulations of quasi-degenerate many-body perturbation theory require the use of a model space consisting of determinants corresponding to all possible occupations of the valence orbitals. This is termed a '*complete*' *model space*. However, this can lead to the situation in which determinants which are not in the model space have an energy significantly below the energies corresponding to some of the highest states included in S. These Φ_ps are called *intruder states* and their presence can impair or even completely destroy the convergence of the quasi-degenerate many-body perturbation theory expansion. There appear to be three possible solutions to this problem:

(i) By forming matrix Padé approximants to the expansion for the reaction operator or effective interaction operator. For example, in third-order the expansion

$$\hat{\mathbf{\Lambda}} = \mathbf{\Lambda}_0 + \mathbf{\Lambda}_1 + \mathbf{\Lambda}_2 + \mathbf{\Lambda}_3 \tag{4.10.16}$$

is replaced by

$$\mathbf{\Lambda}[2/1] = \mathbf{\Lambda}_0 + \mathbf{\Lambda}_1 + \mathbf{\Lambda}_2(1 - \mathbf{\Lambda}_2^{-1}\mathbf{\Lambda}_3)^{-1} \tag{4.10.17}$$

Such Padé approximants have been found to be useful in studies of effective interactions in nuclei in which the problem of intruder states also arises (Schucan and Weidenmuller 1972, 1973; Krenciglowa and Kuo 1974; Hoffmann, Lee, Richert, and Weidenmuller 1974; Leinaas and Kuo 1978; Richert, Schucan, Scrubel, and Weidenmuller 1976).

(ii) By including the intruder states in the model space. This, of course, leads to an 'incomplete' model space which can also be much larger than the initial complete model space and which may lead to a formalism which is not size-consistent.

(iii) By excluding the higher energy configurations from the model space which again may lead to a formalism which is not size-consistent.

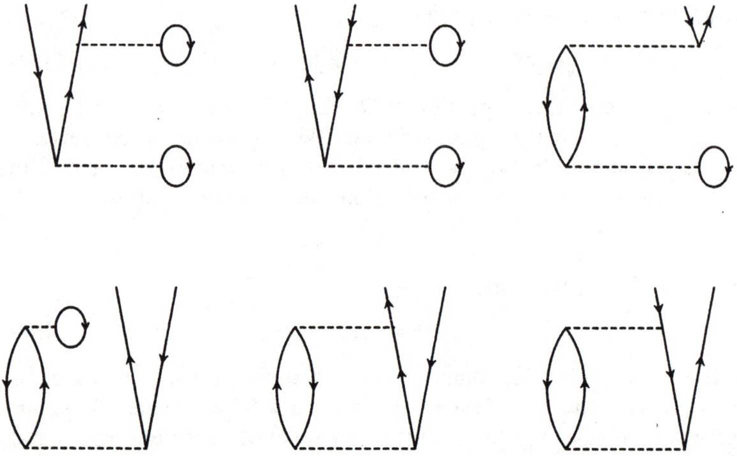

FIG. 4.34. Second-order diagrams in quasi-degenerate diagrammatic perturbation theory in which Φ_i and Φ_j differ by one orbital.

The first approach is more balanced in the sense that the model space is well defined once the orbitals have been partitioned. There is, however, no guarantee that the formation of matrix Padé approximants will overcome the convergence problems created by intruder states in all cases.

Since the use of a 'complete' model space is a special case of the formalism which is necessary to handle an incomplete model space, let us continue by considering the diagrams which arise in the calculation of the

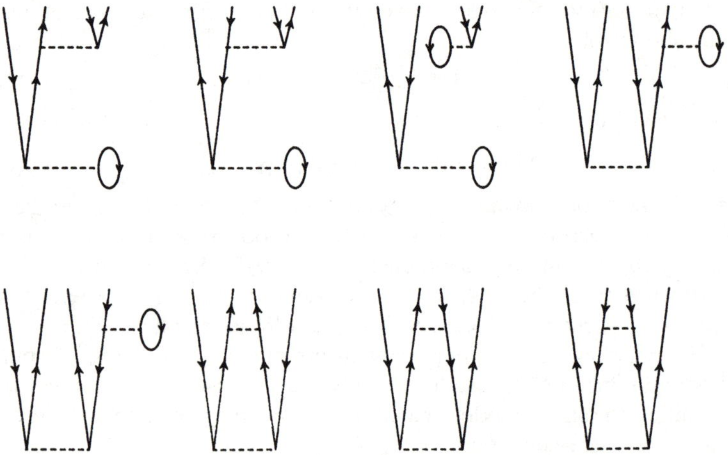

FIG. 4.35. Second-order diagrams in quasi-degenerate diagrammatic perturbation theory in which Φ_i and Φ_j differ by two orbitals.

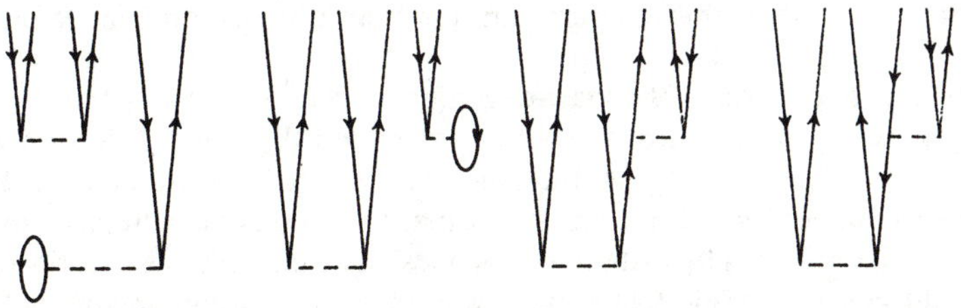

FIG. 4.36. Second-order diagrams in quasi-degenerate diagrammatic perturbation theory in which Φ_i and Φ_j differ by three orbitals.

matrix elements of $\hat{\Lambda}$, $\langle\Phi_i|\hat{\Lambda}|\Phi_j\rangle$. The diagrammatic rules and conventions of Hose and Kaldor (1979) will be followed. The orbitals are divided into holes, which are occupied in Φ_j, and particles which are unoccupied. The diagonal elements $\langle\Phi_i|\hat{\Lambda}|\Phi_j\rangle$ lead to the usual Goldstone diagrams and to folded diagrams which may be derived from them. Folded diagrams will be discussed further below. Off-diagonal matrix elements $\langle\Phi_i|\hat{\Lambda}|\Phi_j\rangle$, $i \neq j$ lead to open diagrams, which can also be folded. Open diagrams contain incoming and outgoing lines describing the transition from Φ_j to Φ_i. Open diagrams may be derived by cutting Goldstone diagrams in all possible ways. Figure 4.34 gives all of the second-order diagrams to be evaluated if Φ_i and Φ_j differ by one orbital. In Figs. 4.35, 4.36, and 4.37, the second-order diagrams in which Φ_i and Φ_j differ by two, three, and four orbitals, respectively, are displayed. Internal lines are summed over hole and particle orbitals, defined with respect to Φ_j. No model state, or linear combination of model states, may appear as an intermediate state.

If a 'complete' model space is used or if the model space is exactly degenerate, there are no unlinked diagrams in the expansion. When an incomplete quasi-degenerate model space is employed then unlinked diagrams do occur, but these diagrams may not cancel with any other diagrams and may not lead to a size-consistency problem (Hose and Kaldor 1979). The third diagram in Fig. 4.35, for example, is unlinked, but none

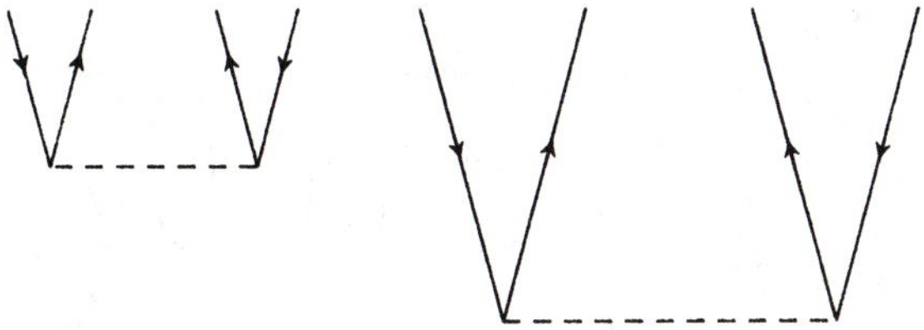

FIG. 4.37. Second-order diagrams in quasi-degenerate diagrammatic perturbation theory in which Φ_i and Φ_j differ by four orbitals.

of the two parts of this diagram can represent a transition connecting Φ_i and Φ_j.

In all derivations of quasi-degenerate diagrammatic perturbation theory, terms arise which correspond to folded diagrams. Johnston and Baranger (1971) have given a detailed analysis of folded diagrams. In a time-dependent formalism, they show that folded diagrams 'replace time-delayed interactions by instantaneous ones'. The interactions described by the full non-relativistic hamiltonian are, of course, instantaneous; in the effective hamiltonian in model space, however, interactions are often 'time-delayed'. In the diagrammatic scheme of Hose and Kaldor, the folded diagrams are generated from the unfolded ones as follows:

> "A group of lines (particles, holes or both) is designated 'folded' if a simultaneous cut through them produces two vertically overlapping parts, each of which is a legitimate open diagram in the model space. This process may be repeated".

The type of the folded lines, holes, or particles, is determined by the requirement that particle lines enter the portion of the diagram containing the highest interaction line and hole lines enter the other part. Thus there are equal numbers of hole and particle lines in a fold and the line directions are as in the parent diagram. A double arrow notation is employed with the upper arrow giving the line type and the lower arrow the line direction. In Fig. 4.38, an example of the folding of a closed diagram is presented. In Fig. 4.39, the folding of an open diagram is illustrated. When evaluating folded diagrams, the usual summation rules apply for the unfolded lines. For the folded lines, the hole and particle lines are summed over transition hole and particle lines corresponding to

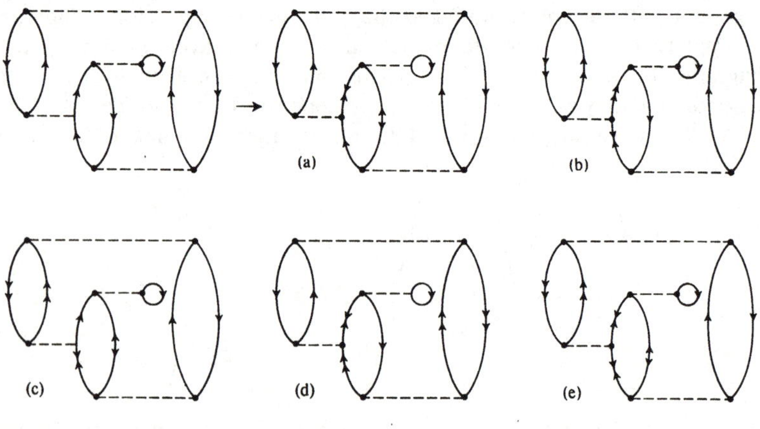

FIG. 4.38. Example of the folding of a closed diagram.

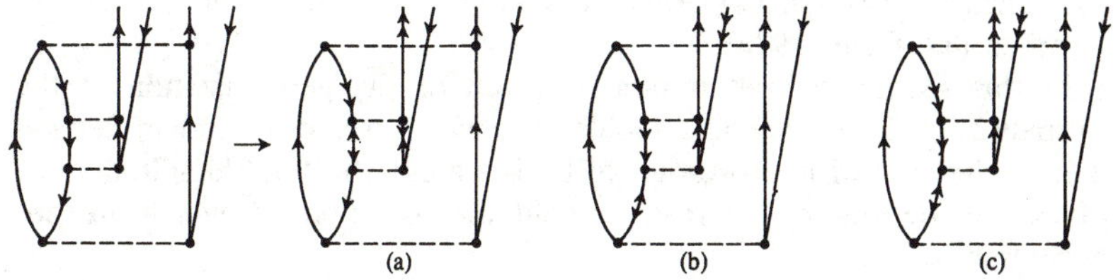

FIG. 4.39. Example of the folding of an open diagram.

all possible transitions from the vacuum state of the relevant matrix element to all other model states. In the evaluation of the denominator factors the rules given in Section 4.7 again apply; folded lines are classified according to their upper arrow and external lines coming into the diagram are always downgoing. The overall sign factor is

$$(-1)^{h+l+f} \tag{4.10.18}$$

where h is the number of hole lines, including folded lines, l is the number of loops after any open incoming lines are connected to outgoing lines in ascending order of orbital labels, and f is the number of folds.

The rules, outlined above, enable the matrix of $\hat{\Lambda}$ to be evaluated in the model space. The calculation is then complete by diagonalizing this matrix, which may, of course, destroy the size-consistency if the model space, S, is not complete.

4.11 Diagrammatic perturbation theory for the valence bond model

It was shown in Section 2.6 that a qualitatively correct description of many dissociative processes can be obtained by using an orbital product wave function in which the orbitals are not required to be orthogonal. This suggests that, by employing a valence bond reference function in diagrammatic perturbation theory, a description of molecular electronic structure can be obtained which would be useful in the calculation of potential energy curves and surfaces. Furthermore, this could be achieved without having to resort to multideterminantal reference wave functions and thereby suffering the attendant loss of physical interpretability of the correlated wave function and energy expansion. The work of Gerratt (Gerratt and Lipscomb 1968; Gerratt 1971, 1974; Wilson and Gerratt 1975; Pyper and Gerrett 1977), of Gallup (1968, 1969*a,b*, 1973), and of Goddard (1967*a,b*) suggests a possible reference function for use in such a scheme. Newman (1969, 1970) has examined a many-body perturbation theory formalism in which biorthogonal sets of orbitals are employed. The use of reference functions constructed from non-orthogonal orbitals in perturbation theory has also been discussed by Kvasnicka (1977*c*),

Moshinsky and Seligman (1971), Gouyet (1973), Cantu *et al.* (1975), and Kirtman and Cole (1978).

In this section a brief outline is given of the generalization of the formalism of Brueckner and Goldstone, which was discussed in Section 4.5, to the case of non-orthogonal basis states. The total hamiltonian is, of course, hermitian, but may be split into components which are not hermitian

$$\hat{\mathcal{H}} = \hat{\mathcal{H}}_0 + \hat{\mathcal{H}}_1; \qquad \hat{\mathcal{H}}_0 \neq \hat{\mathcal{H}}_0^\dagger. \tag{4.11.1}$$

The left and right eigenfunctions of the zero-order hamiltonian must be distinguished

$$\hat{\mathcal{H}}_0 \,|\phi_n\rangle = E_n \,|\phi_n\rangle; \qquad \langle\phi^n|\, \hat{\mathcal{H}}_0 = \langle\phi^n|\, E_n \tag{4.11.2}$$

with

$$\langle\phi^m \mid \phi_n\rangle = \delta_n^m. \tag{4.11.3}$$

If the true ground-state wave function, Ψ_0, satisfies

$$\hat{\mathcal{H}}\,|\Psi_0\rangle = \mathbb{E}_0\,|\Psi_0\rangle \tag{4.11.4}$$

and the intermediate normalization condition

$$\langle\Psi_0 \mid \phi_0\rangle = 1 \tag{4.11.5}$$

is assumed, then

$$E_0 = \langle\Psi_0|\, \hat{\mathcal{H}}_0\,|\phi_0\rangle \tag{4.11.6}$$

and

$$\begin{aligned}\mathbb{E}_0 &= \langle\Psi_0|\, \hat{\mathcal{H}}\,|\phi_0\rangle \\ &= \langle\Psi_0|\, \hat{\mathcal{H}}_0\,|\phi_0\rangle + \langle\Psi_0|\, \hat{\mathcal{H}}_1\,|\phi_0\rangle\end{aligned} \tag{4.11.7}$$

and the level shift is given by

$$\begin{aligned}\Delta E_0 &= \mathbb{E}_0 - E_0 \\ &= \langle\Psi_0|\, \hat{\mathcal{H}}_1\,|\phi_0\rangle.\end{aligned} \tag{4.11.8}$$

Defining the projection operators

$$\hat{P}_0 = |\phi_0\rangle\langle\phi^0| \tag{4.11.9}$$

and

$$\hat{Q}_0 = \hat{I} - \hat{P}_0 = \sum_{i=1}^{\infty} |\phi_i\rangle\langle\phi^i| \tag{4.11.10}$$

and following a treatment analogous to that given for the case where the zero-order hamiltonian is hermitian as given in Section 4.3, leads to an expression for the wave function

$$\langle\Psi_0| = \langle\phi^0| \sum_{n=0}^{\infty} \left\{(\hat{\mathcal{H}}_1 - \Delta E_0)\frac{\hat{Q}}{E_0 - \mathcal{H}_0}\right\}^n \tag{4.11.11}$$

and for the level shift

$$\Delta E_0 = \langle\phi^0| \sum_{n=0}^{\infty} \left\{(\hat{\mathcal{H}}_1 - \Delta E_0)\frac{\hat{Q}}{E_0 - \hat{\mathcal{H}}_0}\right\}^n \hat{\mathcal{H}}_1 |\phi_0\rangle. \tag{4.11.12}$$

Further iteration can then be performed for the level shift

$$\Delta E_0 = E_0^{(1)} + E_0^{(2)} + E_0^{(3)} + \ldots \tag{4.11.13}$$

which then leads to linked-diagram expressions which are denoted by the subscript L

$$E_0^{(0)} = \langle\phi^0| \hat{\mathcal{H}}_0 |\phi_0\rangle \tag{4.11.14a}$$

$$E_0^{(1)} = \langle\phi^0| \hat{\mathcal{H}}_1 |\phi_0\rangle_L \tag{4.11.14b}$$

$$E_0^{(2)} = \left(\sum_{k\neq 0} \frac{\langle\phi^0| \hat{\mathcal{H}}_1 |\phi_k\rangle\langle\phi^k| \hat{\mathcal{H}}_1 |\phi_0\rangle}{E_0 - E_k}\right)_L \tag{4.11.14c}$$

$$E_0^{(3)} = \left(\sum_{k,l\neq 0} \frac{\langle\phi^0| \hat{\mathcal{H}}_1 |\phi_k\rangle\langle\phi^k| \hat{\mathcal{H}}_1 |\phi_l\rangle\langle\phi^l| \hat{\mathcal{H}}_1 |\phi_0\rangle}{(E_0 - E_k)(E_0 - E_l)}\right)_L \tag{4.11.14d}$$

In order to evaluate the matrix elements which occur in expressions (4.11.14) it is necessary to describe the use of non-orthogonal orbitals in a second-quantized formalism. In the second-quantized formalism with non-orthogonal orbitals, the creation and annihilation operators, which are hermitian conjugates, no longer satisfy the usual anticommutation relations. If, however, a biorthogonal basis set is introduced, the creation and annihilation operators satisfy the usual anticommutation relations but are no longer hermitian conjugates. Given a basis set $\boldsymbol{\chi}$ its biorthogonal complement is defined by

$$\boldsymbol{\omega} = \mathbf{S}^{-1}\boldsymbol{\chi}; \qquad S_{ij} = \langle\chi_i \mid \chi_j\rangle. \tag{4.11.15}$$

The covariant creation and annihilation operators obey the following anticommutation relations

$$\begin{aligned} [a_i, a_j]_+ &= 0 \\ [a_i^\dagger, a_j^\dagger]_+ &= 0 \\ [a_i^\dagger, a_j]_+ &= S_{ij}. \end{aligned} \tag{4.11.16}$$

The corresponding contravariant operators are

$$\begin{aligned} a^{i\dagger} &= \sum_j S^{ij} a_j \\ a^i &= \sum_j S^{ij} a_j; \qquad (S^{ij}) = (S_{ij})^{-1} \end{aligned} \tag{4.11.17}$$

which obey the following anticommutation relations

$$[a^{i\dagger}, a_j]_+ = [a^i, a_j^\dagger]_+ = \delta_i^j$$
$$[a^{i\dagger}, a^j]_+ = S^{ij}. \tag{4.11.18}$$

In terms of these operators the total molecular hamiltonian may be written

$$\hat{\mathcal{H}} = \sum_{ij} \langle \phi^i | \, h \, | \phi_j \rangle a_i^\dagger a^j$$
$$+ \tfrac{1}{2} \sum_{ijkl} \langle \phi^i \phi^j | \, g \, | \phi_k \phi_l \rangle a_i^\dagger a_j^\dagger a^l a^k. \tag{4.11.19}$$

The diagrammatic expansion of the Rayleigh–Schrödinger theory with a non-hermitian partition of the total hamiltonian operators is topologically equivalent to the usual one. In the algebraic expressions corresponding to these diagrams each summation is over one covariant and one contravariant index For example, in second-order the diagram

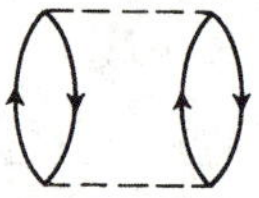

(4.11.20)

leads to the algebraic expression

$$\tfrac{1}{4} \sum_{ijab} \frac{\langle \phi^i \phi^j | \, g \, | \phi_a \phi_b \rangle \langle \phi^a \phi^b | \, g \, | \phi_i \phi_j \rangle}{\epsilon_i + \epsilon_j - \epsilon_a - \epsilon_b} \tag{4.11.21}$$

4.12 Appendix to Chapter 4

In this appendix, two aspects of the derivation of the linked diagram theorem which were briefly mentioned in Section 4.6 are described in more detail.

The factorization theorem

The denominators of disconnected parts of a diagram may be evaluated separately if the sum of diagrams with all possible orderings of the interactions are considered. For example, the diagrams

and

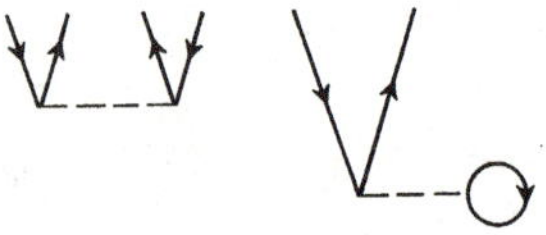

have identical numerator terms and have denominators

$$\mathscr{D}_1=(\epsilon_i+\epsilon_j-\epsilon_a-\epsilon_b)^{-1}(\epsilon_i+\epsilon_j+\epsilon_k-\epsilon_a-\epsilon_b-\epsilon_c)^{-1}$$

and

$$\mathscr{D}_2=(\epsilon_k-\epsilon_c)^{-1}(\epsilon_i+\epsilon_j+\epsilon_k-\epsilon_a-\epsilon_b-\epsilon_c)^{-1}$$

respectively. It can readily be shown that

$$\mathscr{D}_1+\mathscr{D}_2=(\epsilon_i+\epsilon_j-\epsilon_a-\epsilon_b)^{-1}(\epsilon_k-\epsilon_c)^{-1}.$$

Further details may be found in the original work of Frantz and Mills (1960) and in the recent monograph by Lindgren and Morrison (1982).

Exclusion principle violating diagrams

Formally, linked diagrams which are said to be 'exclusion principle violating' should be included in the diagrammatic many-body perturbation theory expansion for the correlation energy. Such 'exclusion principle violating diagrams' are related to unlinked diagrams with restrictions on the orbital indices. For example, it can be demonstrated that

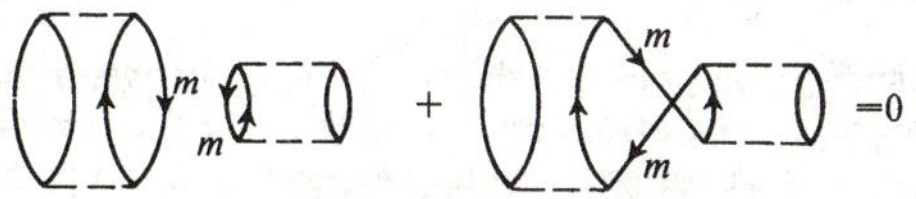

A detailed discussion of 'exclusion principle violating diagrams' can be found in the work of Hubač (1980) and in the recent monograph by Lindgren and Morrison (1982).

5

GROUP THEORETICAL ASPECTS

5.1 Symmetry properties

Group theory enables a number of properties of a molecular system to be examined qualitatively without having to resort to actual calculation. When calculations are made, however, the use of group theoretical techniques can result in substantial simplifications. The symmetry properties of a molecular system with respect to spatial transformations has long been recognized to be an important aspect of the quantum mechanics of molecules. On the other hand, the symmetry properties of systems with respect to permutation of the electrons or with respect to linear transformations amongst the orbitals employed in independent electron models has only been fully exploited more recently, although Weyl recognized their importance as early as 1928 when he wrote, in *Gruppentheorie und Quantenmechanik*:

> "I have particularly emphasized the 'reciprocity' between the representations of the symmetric permutation group and those of the complete linear group; this reciprocity has as yet been unduly neglected in the physical literature, in spite of the fact that it follows most naturally from the conceptual structure of quantum mechanics".

In the second section of this chapter, the use of spatial symmetry in the treatment of electron correlation effects in molecules is briefly discussed while the remaining sections are devoted to the use of the symmetric group and the unitary group in molecular electronic structure theory. Exploitation of these closely related groups leads to powerful methods for the analysis of the correlation problem in molecules even when the particular molecule under consideration has no spatial symmetry whatsoever. In Section 3, spin functions are considered in some detail since there is an intimate connection between the symmetry properties of molecular spin functions and those of the spatial wave function for physically realizable total wave functions. An introduction to the graphical methods of spin algebras is provided in Section 4. Graphical methods form the basis of an elegant and highly effective approach to the construction of many-electron spin functions and have proved valuable in applications of the symmetric group and of the unitary group to both the spin and space part of molecular wave functions. An outline of the theory of the symmetric group S_N is given in Section 5. In Section 6, the importance of S_N to the study of molecular spin functions is described, whilst in Section 7 the use of S_N in constructing spatial wave functions with appropriate symmetry properties is discussed. A basic introduction to the theory of the unitary group $U(n)$ is presented in Section 8. $SU(2)$, a special unitary group, is used in Section 9 to elaborate some further properties of molecular spin functions. The importance of the unitary group $U(n)$ in the construction of spatial wave functions used in the study of electron correlation effects is outlined in Section 10. In the final section of this chapter, the group theoretical aspects of the particle–hole formalism are described. The particle–hole formalism is employed in most approaches to the correlation problem based on the linked diagram expansion, which was discussed in Chapter 4.

5.2 Point symmetry groups

The importance of the point symmetry groups in the study of the electronic structure of molecules has long been recognized. The electronic wave function reflects the spatial symmetry properties of the molecule.

Every subgroup of the orthogonal group O_n is called a point group. The orthogonal group is the set of all real unitary transformations. The transformations of the point symmetry groups leave at least one point of the molecule under consideration fixed when they are applied. The rotation group R_3 is a continuous point group. The elements of the finite point groups consist of combinations of rotations through definite angles and reflections in planes. The point group symmetry operations are summarized in Table 5.1.

There are several notations for the point groups. Here, following Landau and Lifshitz (1965), the notation of Schonflies will be followed. The point groups are classified in Table 5.2. They are subdivided into the discrete axial point groups for which there is a single axis of symmetry whose order is greater than two-fold, the cubic point groups which possess several symmetry axes of order greater than two-fold, and the continuous point groups which contain an axis of infinite order.

Table 5.1

Point group symmetry operations

E	Identity operator
C_n	Rotation by an angle of $2\pi/n$. The axis with the highest n is called the principal axis. Rotations about other axes are denoted by C'_m, C''_m, etc.
σ	Reflection
σ_h	Reflection in a plane perpendicular to the principal axis
σ_v	Reflection in a plane containing the principal axis
σ_d	Reflection in a plane containing the principal axis and bisecting two C'_2 axes
S_n	Rotation–reflection, $S_n = C_n\sigma_h = \sigma_h C_n$
i	Inversion, $i = S_2 = C_2\sigma_h$

Table 5.2

Point group	Elements	Example
The discrete axial point groups		
C_n	$C_n, C_n^2, \ldots, C_n^{n-1}, C_n^n(=E)$	$H_2C{=}CCl_2$ (C_2)
S_{2n}	$S_{2n}, S_{2n}^2, \ldots, S_{2n}^{2n-1}, S_{2n}^{2n}(=E)$	*trans*-ClBrHC—CHBrCl (S_2)
C_{nh}	$C_n \times C_h$; $C_h = E, \sigma_h$	*trans*-ClHC=CHCl (C_{2h})
C_{nv}	$C_n, n\sigma_v$	H_2O (C_{2v})
D_n	C_n, nC_2'	—
D_{nh}	$D_n \times C_h$	BF_3 (D_{3h})
D_{nd}	$D_n, n\sigma_d, S_{2n}^{2k+1}$	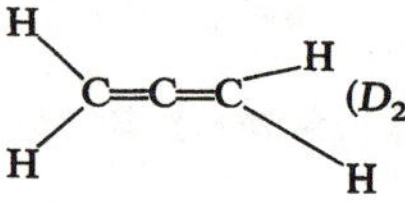 (D_{2d})
The cubic point groups		
T	$E, 3C_2, 4C_3, 4C_3^2$	—
T_d	$E, 8C_3, 3C_2, 6S_4, 6\sigma_d$	CH_4
T_h	$T \times C_i$; $C_i = E, i$	—
O	$E, 8C_3, 3C_2, 6C_4, 6C_2'$	—
O_h	$O \times C_i$; $C_i = E, i$	SF_6
The continuous point groups		
O_3	$R_3 \times C_i$; $C_i = E, i$ (R_3: rotations about any axis)	H
$C_{\infty v}$	$C_\infty, \infty\sigma_v$	LiH
$D_{\infty h}$	$C_{\infty v} \times C_i$; $C_i = E, i$	H_2

Molecular point symmetry can be exploited in almost all stages of a molecular electronic structure calculation. If a basis set of symmetry adapted functions is employed, then the number of integrals which have to be evaluated is significantly reduced. The two-electron integral

$$\langle \phi_i(\Gamma^{(i)})\phi_j(\Gamma^{(j)})| \frac{1}{r_{12}} |\phi_k(\Gamma^{(k)}\phi_l(\Gamma^{(l)})\rangle \tag{5.2.1}$$

will only be non-zero if the direct product

$$\Gamma^{(i)} \times \Gamma^{(j)} \times \Gamma^{(k)} \times \Gamma^{(l)} \tag{5.2.2}$$

contains the totally symmetric A_1 irreducible representation where $\Gamma^{(i)}$ is used to denote the irreducible representation with respect to which the orbital ϕ_i transforms. By using integrals over symmetry adapted basis functions, the number of integrals which have to be processed in the construction of the Fock matrix is reduced (see, for example, Winter, Ermler, and Pitzer 1973; Bagus and Wahlgren 1976; Dupuis and King 1977; Dacre 1970; Elder 1973). Furthermore, molecular symmetry allows the Fock matrix to be cast into block diagonal form for which the eigenvalue problem may be solved block by block. The integral evaluation phase of a molecular calculation depends on the fourth power of the number of basis function, the construction of the Fock matrix also depends on the fourth power of the number of basis function, and the diagonalization depends on the third power. The transformation of the two-electron integrals over the basis set to two electron integrals over molecular orbitals, depends on the fifth power of the number of basis function, but here again molecular symmetry can significantly reduce the amount of computation required to effect this transformation. Clearly, the greatest efficiency will be achieved when the basis functions and the molecular orbitals are symmetry-adapted with respect to the molecular point symmetry group. In general, we have

$$\langle \psi_i(\Gamma^{(i)})\psi_j(\Gamma^{(j)})| \frac{1}{r_{12}} |\psi_k(\Gamma^{(k)})\psi_l(\Gamma^{(l)})\rangle$$

$$= \sum_{pqrs} C_{ip}(\Gamma^{(i)})C_{jq}(\Gamma^{(j)})C_{kr}(\Gamma^{(k)})C_{ls}(\Gamma^{(l)})$$

$$\times \langle \phi_p(\Gamma^{(i)})\phi_q(\Gamma^{(j)})| \frac{1}{r_{12}} |\phi_r(\Gamma^{(k)})\phi_s(\Gamma^{(l)}).\rangle \quad (5.2.3)$$

where ϕ denotes a basis function and ψ denotes a molecular orbital. The irreducible representations, with respect to which these functions and orbitals transform, must satisfy the condition (5.2.2). Finally, it should be noted that since symmetry adaption also leads to a shorter list of integrals over molecular orbitals, this will lead to greater efficiency in the calculation of correlation effects.

5.3 Spin functions

Although we are concerned in this book with the study of molecular systems within the framework of non-relativistic quantum mechanics, relativistic effects cannot entirely be ignored. Dirac (1929) first provided a theoretical explanation of the relativistic property of electrons first observed by Uhlenbeck and Goudsmit (1925, 1929), namely, their intrinsic angular momentum or spin. There is a spin associated with every electron which has no analogy in classical mechanics. The wave function for any molecule within non-relativistic quantum mechanics consists of a product of a spin function and a spatial function. Since the symmetry properties of these two functions are intimately connected, the spin functions cannot be neglected even in problems which do not involve spin-dependent operators (see, for example, Kotani *et al.* 1963; Matsen 1964, 1976; Kaplan 1975; Pauncz 1979).

The spin wave function $\theta(\sigma)$ for a single electron is a function of the

z-component σ of the spin angular momentum. σ may take one of two values: $+\frac{1}{2}$ or $-\frac{1}{2}$ in units of $h/2\pi$. α and β are usually taken to denote the linearly independent spin functions such that

$$\begin{aligned} \alpha(+\tfrac{1}{2}) &= 1 \qquad \beta(+\tfrac{1}{2}) = 0 \\ \alpha(-\tfrac{1}{2}) &= 0 \qquad \beta(-\tfrac{1}{2}) = 1; \end{aligned} \tag{5.3.1}$$

α and β are eigenfunctions of $\hat{s}_z$

$$\hat{s}_z\alpha = +\tfrac{1}{2}\alpha \qquad \hat{s}_z\beta = -\tfrac{1}{2}\beta \tag{5.3.2}$$

and of $\hat{s}^2$

$$\hat{s}^2\theta = s(s+1)\theta, \qquad \theta = \alpha \quad \text{or} \quad \beta, \qquad s = \tfrac{1}{2}; \tag{5.3.3}$$

α and β are usually taken to be normalized to unity.

For an N-electron system, each electron may have a spin function α or β. For the whole system there will be 2^N linearly independent spin functions

$$\Theta_k^N = \theta(\sigma_1)\theta(\sigma_2)\ldots\theta(\sigma_N); \quad k = 1, 2, \ldots, 2^N \tag{5.3.4}$$

where

$$\theta(\sigma_i) = \alpha_i(\sigma_i) \quad \text{or} \quad \beta_i(\sigma_i). \tag{5.3.5}$$

The functions Θ_k^N are not, in general, eigenfunctions of the total spin operator $\hat{S}^2$ for the N-electron system. However, linear combinations of the Θ_k^N which are eigenfunctions of $\hat{S}^2$ can always be found. Let $\Theta_{SM;k}^N$ denote one such combination, then

$$\hat{S}^2\Theta_{SM;k}^N = S(S+1)\Theta_{SM;k}^N \tag{5.3.6}$$

and

$$\hat{S}_z\Theta_{SM;k}^N = M\Theta_{SM;k}^N. \tag{5.3.7}$$

If the N-electron spin functions $\Theta_{SM;k}^N$ are to be linearly independent then

$$k = 1, 2, \ldots, f_S^N \tag{5.3.8}$$

where (Wigner 1931, 1959)

$$f_S^N = \{(2S+1)N!\}/\{(\tfrac{1}{2}N+S+1)!\,(\tfrac{1}{2}N-S)!\}. \tag{5.3.9}$$

Note that f_S^N is independent of M. Since there are $2S+1$ substates for a given state S

$$\sum_{S=0\text{ or }\frac{1}{2}}^{\frac{1}{2}N} (2S+1)f_S^N = 2^N. \tag{5.3.10}$$

The functions $\Theta_{SM;k}^N$ span a subspace of dimension f_S^N of the total spin space.

5.4 Graphical methods of spin algebras

One of the aims of this chapter is to demonstrate that there is a close connection between the applications of the theory of the symmetric group to the problem of molecular electronic structure and applications of the unitary group. This connection becomes particularly transparent when a formalism based on the graphical methods of spin algebras is employed. In this section, a brief overview of the graphical methods of spin algebras is given.

The graphical methods of spin algebras have been used for almost twenty years to simplify angular momentum calculations in atomic, nuclear, and particle physics (Yutsis, Levinson, and Vanagas 1962; Yutsis and Bandzaitus 1977; Sanders 1969, 1971; El Baz and Castel 1972). Although it is possible to use angular momentum theory in purely algebraic terms, the graphical methods are more intuitive and have advantages arising from their universality and transparency in handling various coupling schemes. Even in very complicated problems, the graphical methods of spin algebra can provide a useful qualitative insight. When used in conjunction with diagrammatic techniques based on second-quantization and Wick's theorem (see Chapter 4), they are of value whenever a spin-adapted formalism is required.

A brief introduction to the graphical methods of spin algebra will be given in this section. Detailed descriptions may be found in the original work of Yutsis and his collaborators (1962, 1977) the articles of Sanders (1969, 1971) and in the monograph by El Baz and Castel (1972). Some familiarity with the standard Racah algebra will be assumed (Edmonds 1957; Brink and Satchler 1968).

Graphical elements

The graphical representation of the 'ket' (contravariant) vector is

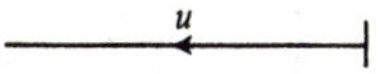

$|u\rangle$ (5.4.1)

while the 'bra' (covariant) vector is represented by

$\langle u|$ (5.4.2)

The direction of the arrows represents the sign of the magnetic quantum numbers.

The 3-*jm* symbols are represented by

$\begin{Bmatrix} j_1 & j_2 & j_3 \\ m_1 & m_2 & m_3 \end{Bmatrix} =$

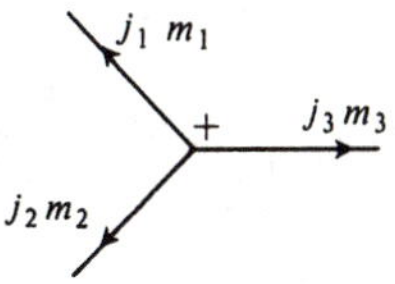

$= (-)^{j_1+j_2+j_3}$

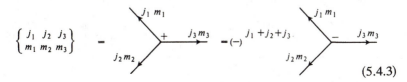

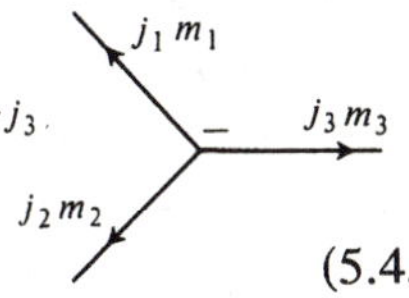

(5.4.3)

where the plus (minus) sign indicates that the moments should be read in anticlockwise (clockwise) direction.

Elimination of zero-spin lines

Lines corresponding to zero spin, which are often represented by a dashed line, may be eliminated as follows

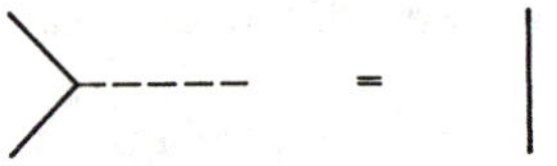

(5.4.4)

Orthogonality

The well known orthogonality condition for 3-*jm* coefficients may be expressed graphically as follows

(5.4.5)

Summation over the m quantum number

If an arbitrary part of an angular momentum graph is represented by a rectangle and only 'external' lines are shown explicitly, then the following rule for summation over the *m* quantum number may be given

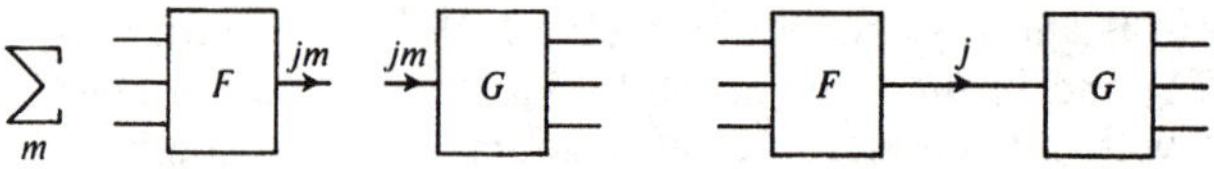

(5.4.6)

Separation rules

A graph is said to be separable over *n* lines in that it has the following form

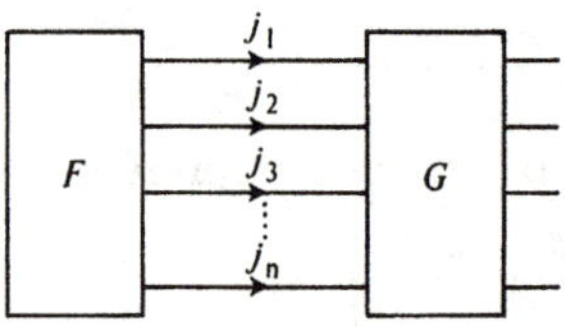

(5.4.7)

The following are examples of the powerful separation rules developed initially by Yutsis and his collaborators (1962) for simplifying such graphs.

(a) Separation over one line

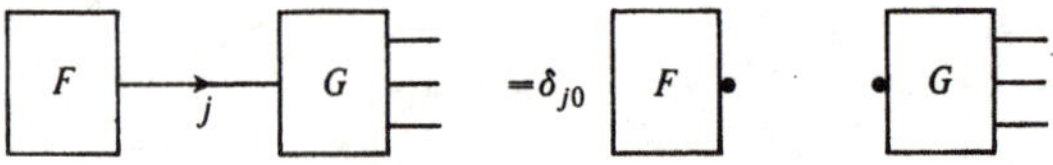

(5.4.8)

(b) Separation over two lines

$$= (2j_1 + 1)^{-1}\, \delta_{j_1 j_2} \times$$

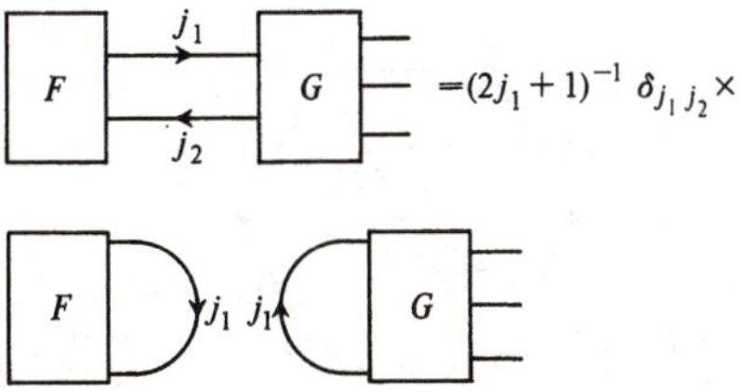

(5.4.9)

(c) Separation over three lines

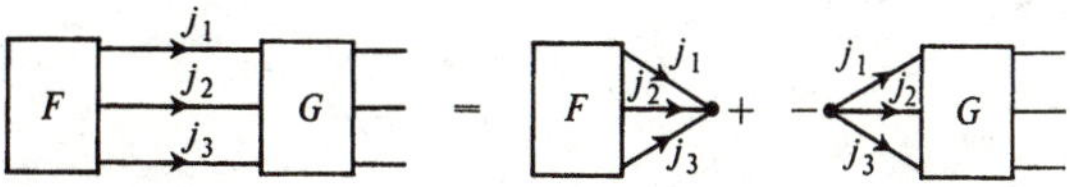

— (5.4.10)

(d) Separation over four lines

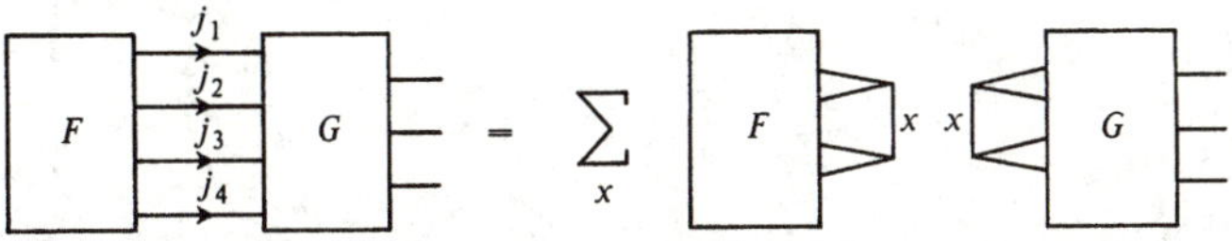

(5.4.11)

These rules, and similar separation rules, often enable a very complicated spin graph to be separated into components which can be readily evaluated. The use of graphical techniques enables simplifications which can be made by using these separation rules to be easily recognized even in complicated problems.

Some components of spin graphs

Typically, a given spin graph can be decomposed into components of the following form

(a) A 3-j coefficient or 'triangle delta'

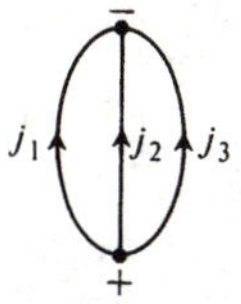

(5.4.12)

(b) A 6-j coefficient

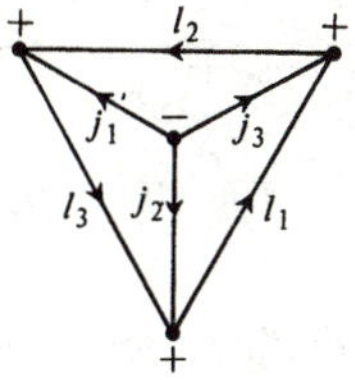

(5.4.13)

(c) A 9-j coefficient

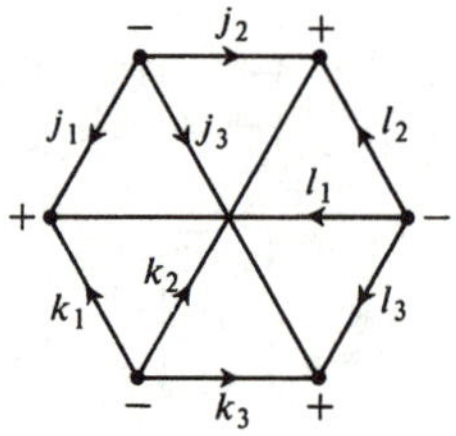

(5.4.14)

(d) A 3n-j coefficient

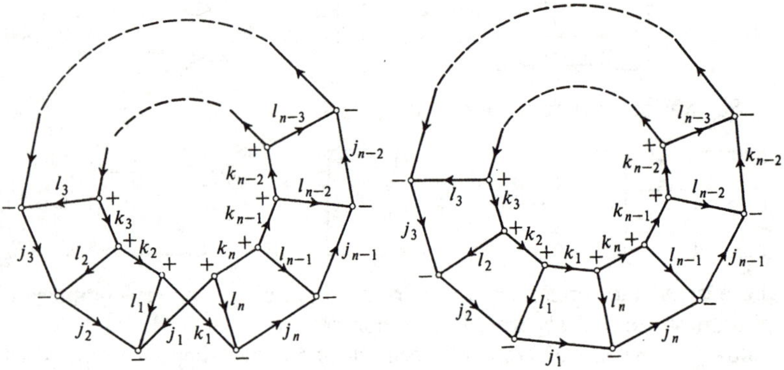

(5.4.15)

Let us now consider an application of the graphical methods of spin algebra. In Section 5.6, different methods for coupling the one-electron spin functions to give the total spin function for a molecule will be considered. Two such schemes are the Young–Yamanouchi method which may be represented graphically as

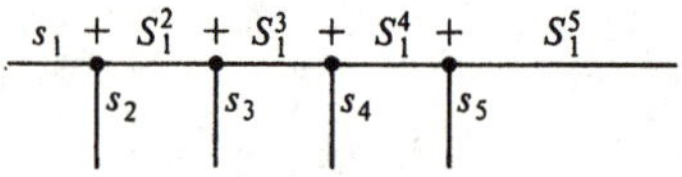

(5.4.16)

and the Serber method which may be represented graphically as

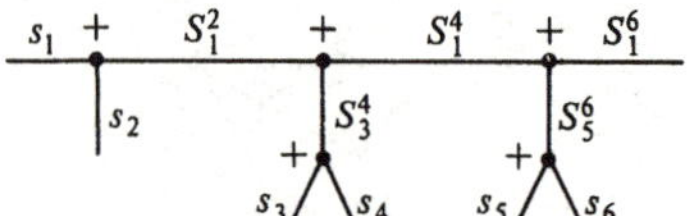

(5.4.17)

The transformation matrix relating these two coupling schemes may be

written as (Wilson 1977*c,d*)

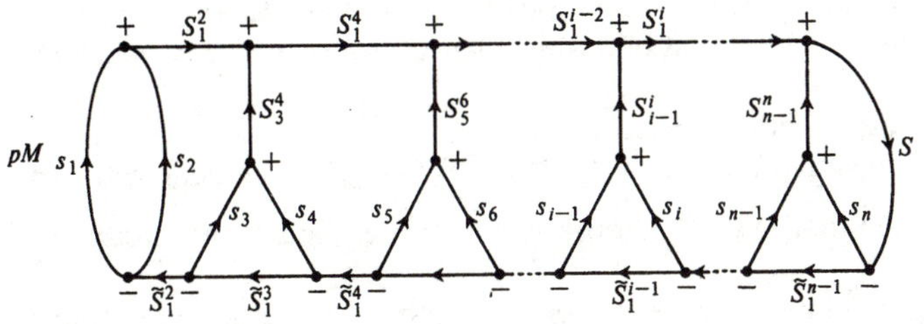

(5.4.18)

where

$$M^2 = \prod_{i=2}^{i=n-2} \{(2S_1^i+1)(2S_{i+1}^{i+2}+1)\} \prod_{i=2}^{i=n-1} (2\bar{S}_1^{i}+1) \qquad (5.4.19)$$

and p is a phase factor.

This graph may be simplified by using the rule for separation over two lines and the following result is obtained

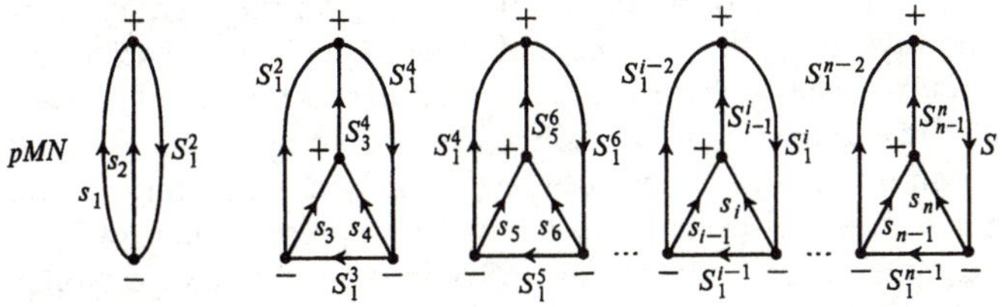

(5.4.20)

where

$$N = \prod_{i=2,4,\ldots}^{n-2} \delta(S_1^i, \bar{S}_1^i)(2S_1^i+1)^{-1}. \qquad (5.4.21)$$

The required transformation matrix is essentially a single product of 6-j coefficients.

5.5 The symmetric group S_N

The symmetric group S_N is the group of the $N!$ permutations P of N objects (Rutherford 1948; Robinson 1961; Hamermesh 1962; Coleman 1966; Kaplan 1975). A typical permutation may be written

$$P = \begin{pmatrix} 1 & 2 & \ldots & N \\ i_1 & i_2 & \ldots & i_N \end{pmatrix} \qquad (5.5.1)$$

where the indices $i_1, i_2, \ldots, i_N$ denote the numbers $1\,2 \ldots N$ in some order. It can readily be shown that (see, for example, Kaplan 1975)

(a) The product of two permutations is also a permutation.
(b) The identity permutation, which leaves every number unchanged, exists.
(c) Every permutation has an inverse.
(d) The associative law, i.e. $P_1(P_2P_3)=(P_1P_2)P_3$, can be demonstrated.

The $N!$ permutations of N objects, therefore, satisfy the group postulates.

Every permutation can be written as a product of cycles. For example,

$$\begin{pmatrix}1 & 2 & 3 & 4 & 5 & 6\\ 2 & 4 & 5 & 1 & 3 & 6\end{pmatrix}=\begin{pmatrix}1 & 2 & 4\\ 2 & 4 & 1\end{pmatrix}\begin{pmatrix}3 & 5\\ 5 & 3\end{pmatrix}\begin{pmatrix}6\\ 6\end{pmatrix}$$
$$=(1\ \ 2\ \ 4)(3\ \ 5)(6) \qquad (5.5.2)$$

All permutations which are related by the equation

$$P_i=QP_jQ^{-1}, \qquad Q\in S_N \qquad (5.5.3)$$

are, by definition, members of the same class. P_i is obtained by the action of Q on P_j (in the sense of a permutation acting on the arguments of a function) and thus the cyclic structures of P_i and P_j coincide. Each class of S_N is, therefore, characterized by the number of different ways of partitioning the number N into positive integral components. Since the number of non-equivalent irreducible representations of a group is equal to the number of its classes, the non-equivalent representations of S_N are also defined by the different partitions of N into positive integer components. These partitions are usually written as

$$[\lambda]=[\lambda^{(1)}\lambda^{(2)}\ldots\lambda^{(m)}] \qquad (5.5.4)$$

with

$$\lambda^{(1)}+\lambda^{(2)}+\ldots+\lambda^{(m)}=N \qquad (5.5.5a)$$

and

$$\lambda^{(1)}\geqslant\lambda^{(2)}\geqslant\ldots\geqslant\lambda^{(m)} \qquad (m\leqslant N) \qquad (5.5.5b)$$

These partitions can be depicted by Young diagrams. In these diagrams, each $\lambda^{(i)}$ is represented by a row of i cells as illustrated in Fig. 5.1 for the partition $[\lambda]=[6,4^2,2,1]$. The irreducible representations of S_4 are, for example, labelled by the following partitions and Young diagrams:

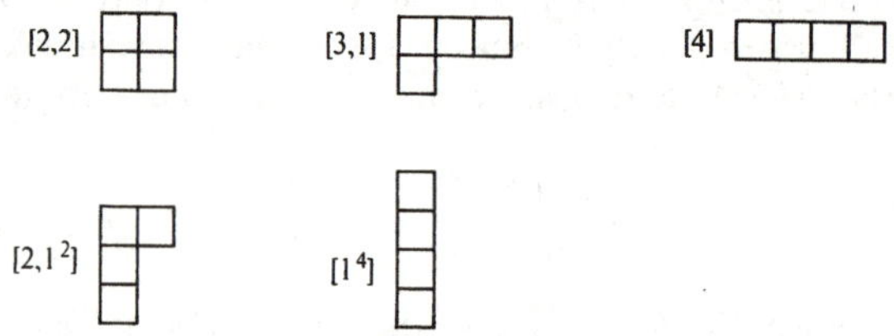

(5.5.6)

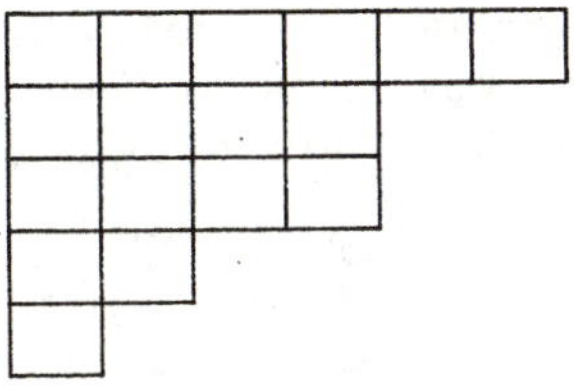

FIG. 5.1. Young diagram corresponding to the partition $[6, 4^2, 2, 1]$.

In general, an irreducible representation of S_N becomes reducible on passing to its subgroup S_{N-1}. However, the basis functions for S_N may be chosen in such a way that they become bases for irreducible representations on passing to the subgroups $S_{N-1}, S_{N-2}, \ldots, S_1$. This choice of basis functions may be described graphically as follows:

Each basis functions is associated with a Young diagram $[\lambda]$ the cells of which each contain an integer in the range 1 to N. This is termed a Young tableau. The numbers are arranged amongst the cells in such a way that if the cell containing the number N is removed one obtains the Young diagram of the irreducible representation of S_{N-1} according to which the basis function transforms on passing to this subgroup. This process can be repeated until a diagram with a single cell is reached. Only those Young tableaux in which the numbers increase from left to right and down the columns are allowed. These are referred to as standard Young tableaux. Some of the standard Young tableaux for the partition $[\lambda]=[6, 4^2, 2, 1]$, which was considered above in Fig. 5.1, are shown in Fig. 5.2. Each basis function of an irreducible representation of S_N can be associated with a

1	2	3	6	9	11
4	5	8	12		
7	10	13	16		
14	17				
15					

1	2	3	5	13	15
4	7	11	14		
6	10	16	17		
8	12				
9					

1	3	5	7	10	11
2	4	6	9		
8	12	14	15		
13	16				
17					

1	4	7	10	12	15
2	5	13	14		
3	6	16	17		
8	9				
11					

FIG. 5.2. Young tableaux corresponding to the partition $[6, 4^2, 2, 1]$.

Young tableau. The dimension of the irreducible representation is given by the number of standard Young tableaux. For example, for S_5 $[\lambda]=[3,2]$ there are five linearly independent basis functions which may be labelled as follows:

$$\begin{array}{|c|c|c|}\hline 1&3&5\\\hline 2&4\\\cline{1-2}\end{array}\qquad \begin{array}{|c|c|c|}\hline 1&2&5\\\hline 3&4\\\cline{1-2}\end{array}\qquad \begin{array}{|c|c|c|}\hline 1&3&4\\\hline 2&5\\\cline{1-2}\end{array}\qquad \begin{array}{|c|c|c|}\hline 1&2&4\\\hline 3&5\\\cline{1-2}\end{array}\qquad \begin{array}{|c|c|c|}\hline 1&2&3\\\hline 4&5\\\cline{1-2}\end{array}\tag{5.5.7}$$

The irreducible matrix representation for S_5 corresponding to the partition $[\lambda]=[3,2]$ is thus a set of 5×5 matrices. Consider the behaviour of the basis functions under the permutations of the subgroup S_4. The representation of the subgroup is described by the Young tableau:

$$\begin{array}{|c|c|}\hline 1&3\\\hline 2&4\\\hline\end{array}\qquad \begin{array}{|c|c|}\hline 1&2\\\hline 3&4\\\hline\end{array}\qquad \begin{array}{|c|c|c|}\hline 1&3&4\\\hline 2\\\cline{1-1}\end{array}\qquad \begin{array}{|c|c|c|}\hline 1&2&4\\\hline 3\\\cline{1-1}\end{array}\qquad \begin{array}{|c|c|c|}\hline 1&2&3\\\hline 4\\\cline{1-1}\end{array}\tag{5.5.8}$$

Different shapes are associated with different irreducible representations of S_4 and, therefore, on confining our attention to this subgroup of S_5, the matrices assume a block diagonal form—one 2×2 irreducible representation and one 3×3 representation. On passing through the chain, $S_5\rightarrow S_4\rightarrow S_3\rightarrow S_2\rightarrow S_1$, the basis functions given above for S_5 are uniquely identified by the chains:

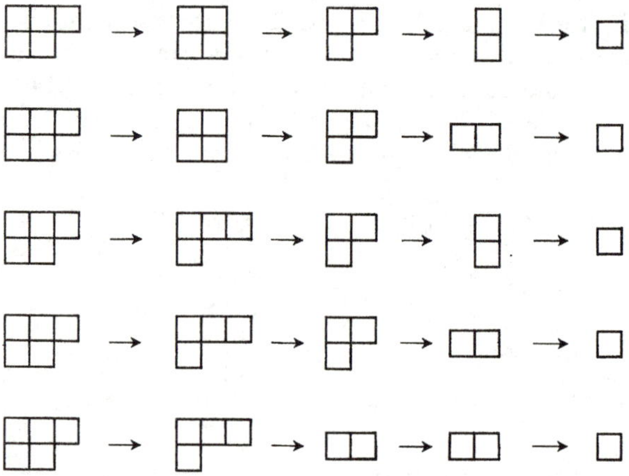

(5.5.9)

The Young tableaux uniquely identify the basis functions.

The basis functions, introduced above, form a basis for the Young–Yamanouchi representation of S_N (Rutherford 1948; Yamanouchi 1937, 1938, 1948). Young and Yamanouchi showed that this representation may be chosen to be real and gave the following simple rules for constructing the representation matrices corresponding to the permutation $P_{i-1,i}$.

The representation matrices $\mathbf{V}(P_{i-1,i})$ have non-zero elements:

(a) $V_{mm}(P_{i-1,i})=1$, if in tableau m, the numbers $i-1$ and i occur in the same row.
(b) $V_{mm}(P_{i-1,i})=-1$, if in tableau m, the numbers $i-1$ and i occur in the same column.
(c)

$$\mathbf{V}(P_{i-1,i})=\begin{array}{c} \\ m \\ \\ n \\ \\ \end{array}\begin{pmatrix} & m & & n & \\ & \vdots & & \vdots & \\ \cdots & -d^{-1} & \cdots & -(1-d^{-2})^{\frac{1}{2}} & \cdots \\ & \vdots & & \vdots & \\ \cdots & -(1-d^{-2})^{\frac{1}{2}} & \cdots & d^{-1} & \cdots \\ & \vdots & & \vdots & \end{pmatrix} \tag{5.5.10}$$

if the tableaux m and n differ only by a permutation of the numbers $i-1$ and i and if the row containing $i-1$ in tableau m is above that containing i. d denotes the axial distance between $i-1$ and i, that is, the number of horizontal and vertical steps one must take in the Young tableaux to move from $i-1$ to i.

5.6 S_N and spin functions

Let P^σ denote an operator which permutes the spin coordinates in an N-electron system

$$P^\sigma\Theta^N_{SM;k}=\begin{pmatrix}1 & 2 & \cdots & N\\ i_1 & i_2 & & i_N\end{pmatrix}\Theta^N_{SM;k}(\sigma_1,\sigma_2,\ldots,\sigma_N)$$
$$=\Theta^N_{SM;k}(\sigma_{i_1},\sigma_{i_2},\ldots,\sigma_{i_N}). \tag{5.6.1}$$

Since electrons are indistinguishable and $\hat{S}^2$ and $\hat{S}_z$ are symmetric with respect to the labelling of the electrons, the following commutation relations are obeyed

$$[P^\sigma,\hat{S}^2]=0,\qquad [P^\sigma,\hat{S}_z]=0. \tag{5.6.2}$$

The spin functions $\Theta^N_{SM;k}$ form a basis for a representation of the symmetric group. For P^σ we have

$$P^\sigma\Theta^N_{SM;k}=\sum_{j=1}^{j=f^N_S} V^{NS}_{j,k}(P^\sigma)\Theta^N_{SM;j} \tag{5.6.3}$$

where $V^{NS}_{j,k}(P^\sigma)$ is an element of the $\mathbf{V}^{NS}(P^\sigma)$. If the spin functions are orthogonal

$$\langle\Theta^N_{SM;k}\,|\,\Theta^N_{SM;l}\rangle=\delta_{k,l} \tag{5.6.4}$$

then the $\mathbf{V}^{NS}(P^\sigma)$ are orthogonal matrices. Since the one-to-one correspondence

$$P^\sigma:\mathbf{V}^{NS}(P^\sigma) \tag{5.6.5}$$

between the elements $P^\sigma \in S_N$ and the set of square matrices $\mathbf{V}^{NS}$ such that

$$\mathbf{V}^{NS}(P_1^\sigma)\mathbf{V}^{NS}(P_2^\sigma) = \mathbf{V}^{NS}(P_1^\sigma P_2^\sigma) \tag{5.6.6}$$

the matrices $\mathbf{V}^{NS}(P^\sigma)$ are said to form a matrix representation of the group S_N (see, for example, Pauncz 1979).

In order to determine the matrices $\underline{V}^{NS}(P^\sigma)$ eqn (5.6.3) is multiplied from the left by $\Theta^{N*}_{SM;j}$ and integrated over all spin coordinates. Remembering eqn (5.6.4), we obtain

$$V^{NS}_{j,k} = \int\int \ldots \int \Theta^{N*}_{SM;j}(P^\sigma \Theta^N_{SM;k})\, d\sigma_1\, d\sigma_2 \ldots d\sigma_N \tag{5.6.7}$$

Hence, the matrix elements of $\underline{V}^{NS}(P^\sigma)$ are obtained by constructing the spin functions and determination of the effect of the spin coordinate permutation operators P^σ on them.

There is a considerable degree of freedom in the mode of construction of the spin functions $\Theta^N_{SM;k}$. In general, let us divide a given N-electron system into m subsystems. Let S_{N_i} denote the symmetric group of order $N_i!$ of the permutations P_i of the ith subsystem. Assume that the total spin function for each of these subsystems is known and that they are now coupled together to give the spin function for the whole N-electron system. This may be represented as follows

$$S_N = S_{N_1} \otimes S_{N_2} \otimes \ldots \otimes S_{N_m} \tag{5.6.8}$$

Typically, the spin functions for two subsystems may be coupled according to

$$\Theta^{N_i+N_j}_{SM;S_iS_jk_ik_j} = \sum_{M_i+M_j=M} \langle S_iS_jM_iM_j \mid SM\rangle \Theta^{N_i}_{S_iM_i;k_i} \Theta^{N_j}_{S_jM_j;k_j} \tag{5.6.9}$$

where the coefficients are Clebsch–Gordan coefficients (Brink and Satchler 1968). Of course, the spin functions in each of the subsystems may be coupled according to some arbitrary scheme. This point will be discussed further below after an important property of the representation matrices constructed according to the scheme (5.6.8) has been described.

For simplicity, consider an N-electron system divided into two subsystems. Put

$$S_N = S_{N_1} \otimes S_{N_2}; \qquad N = N_1 + N_2. \tag{5.6.10}$$

Let

$$\begin{aligned} P_1 &\in S_{N_1} & P_1 &: \underline{V}^{N_1S_1}(P_1) \\ P_2 &\in S_{N_2} & P_2 &: \underline{V}^{N_2S_2}(P_2) \end{aligned} \tag{5.6.11}$$

omitting the superscript σ from now on.

Since S_{N_1} and S_{N_2} possess no common element, apart from the identity, the commutation relation

$$[P_1, P_2]=0 \tag{5.6.12}$$

is obeyed. For permutation operators of the form $P_1 \cdot P_2$ it is easily shown that

$$(P_1 \cdot P_2)\cdot(P_1' \cdot P_2')=(P_1 \cdot P_1')\cdot(P_2 \cdot P_2') \tag{5.6.13}$$

The operators of the direct product (5.6.13) satisfy the group postulates. Furthermore, if $\mathbf{V}^{S_1N_1}(P_1)$ and $\mathbf{V}^{S_2N_2}(P_2)$ form irreducible representations of S_{N_1} and S_{N_2}, respectively, then

$$V^{NS}_{k,l}(P)=V^{N_1S_1}_{k_1,l_1}(P_1)\cdot V^{N_2S_2}_{k_2,l_2}(P_2) \tag{5.6.14}$$

where

$$P=P_1\cdot P_2;\qquad k=(S_1S_2k_1k_2);\qquad l=(S_1S_2l_1l_2) \tag{5.6.15}$$

forms an irreducible representation of S_N.

All of the different spin bases which can be formed by coupling the component spin functions in different orders, e.g.

$$(S_{N_1}\otimes(S_{N_2}\otimes S_{N_3})) \quad \text{and} \quad ((S_{N_1}\otimes S_{N_2})\otimes S_{N_3})$$

are related by unitary transformations.

The most commonly employed spin basis functions are the Young–Yamanouchi orthogonal functions. This basis is obtained by putting $m=N$ in eqn (5.6.8). The Young–Yamanouchi representation is conveniently characterized by the branching diagram shown in Fig. 5.3. The N-electron spin function is built up by coupling the one-electron spin

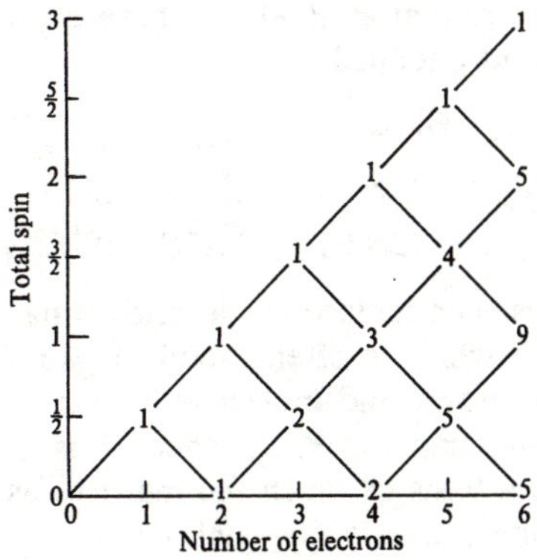

FIG. 5.3. Branching diagram for Young–Yamanouchi spin functions.

functions, one at a time, according to

$$\Theta^N_{SM;k} = \langle S+\tfrac{1}{2}\ \tfrac{1}{2}\ M-\tfrac{1}{2}\ \tfrac{1}{2} \mid SM\rangle \Theta^{N-1}_{S+\frac{1}{2}M-\frac{1}{2};k}\alpha(\sigma_N) + \langle S+\tfrac{1}{2}\ \tfrac{1}{2}\ M+\tfrac{1}{2}\ -\tfrac{1}{2} \mid SM\rangle \Theta^{N-1}_{S+\frac{1}{2}M+\frac{1}{2};k}\beta(\sigma_N) \quad (k=1,2,\ldots,f^{N-1}_{S+\frac{1}{2}}) \tag{5.6.16}$$

and

$$\Theta^N_{SM;k} = \langle S-\tfrac{1}{2}\ \tfrac{1}{2}\ M-\tfrac{1}{2}\ \tfrac{1}{2} \mid SM\rangle \Theta^{N-1}_{S-\frac{1}{2}M-\frac{1}{2};l}\alpha(\sigma_N) + \langle S-\tfrac{1}{2}\ \tfrac{1}{2}\ M+\tfrac{1}{2}\ -\tfrac{1}{2} \mid SM\rangle \Theta^{N-1}_{S-\frac{1}{2}M+\frac{1}{2};l}\beta(\sigma_N) \quad (k=f^{N-1}_{S+\frac{1}{2}}+l;\ l=1,2,\ldots,f^{N-1}_{S-\frac{1}{2}}). \tag{5.6.17}$$

The Clebsh–Gordan coefficients arising in (5.6.16) and (5.6.17) are

$$\begin{aligned}
\langle S+\tfrac{1}{2}\ \tfrac{1}{2}\ M-\tfrac{1}{2}\ \tfrac{1}{2} \mid SM\rangle &= -\left(\frac{S-M+1}{2S+2}\right)^{\frac{1}{2}} \\
\langle S+\tfrac{1}{2}\ \tfrac{1}{2}\ M+\tfrac{1}{2}\ -\tfrac{1}{2} \mid SM\rangle &= \left(\frac{S+M+1}{2S+2}\right)^{\frac{1}{2}} \\
\langle S-\tfrac{1}{2}\ \tfrac{1}{2}\ M-\tfrac{1}{2}\ \tfrac{1}{2} \mid SM\rangle &= \left(\frac{S+M}{2S}\right)^{\frac{1}{2}} \\
\langle S-\tfrac{1}{2}\ \tfrac{1}{2}\ M+\tfrac{1}{2}\ -\tfrac{1}{2} \mid SM\rangle &= \left(\frac{S-M}{2S}\right)^{\frac{1}{2}}.
\end{aligned} \tag{5.6.18}$$

The representation matrices for the Young–Yamanouchi basis can be conveniently generated by means of an inductive scheme (Gabriel 1961; Kotani *et al.* 1963; Pauncz 1979; Rettrup 1979). If all of the representation matrices are known for the first $N-1$ electrons, then only $P_{N-1,N}$ need be considered since all other matrices can be obtained by means of matrix multiplication. Kotani *et al.* (1963) have explicitly performed the $P_{N-1,N}$ interchange and obtained

$$\begin{aligned}
P_{N-1,N}\Theta^N_{SM;S_1S_2\ldots S\pm 1 S\pm\frac{1}{2}} &= \Theta^N_{SM;S_1S_2\ldots S\pm 1 S\pm\frac{1}{2}} \\
P_{N-1,N}\Theta^N_{SM;S_1S_2\ldots SS\pm\frac{1}{2}} &= \pm(2S+1)^{-1}\Theta^N_{SM;S_1S_2\ldots SS\pm\frac{1}{2}} \\
&\quad + (2S+1)^{-1}\{2S(2S+2)\}^{\frac{1}{2}}\Theta^N_{SM;S_1S_2\ldots SS\mp\frac{1}{2}}
\end{aligned} \tag{5.6.19}$$

where the index k has been replaced by a series of partial resultant spins.

A coupling scheme which is often useful in pair theories of electron correlation effects in atoms and molecules is due to Serber (1934*a*,*b*). Since the simple interchanges, $P_{i-1,i}$ where i is even, commute and $(P_{i-1,i})^2 = E$, there is a basis in which the matrices associated with these permutations are simultaneously diagonal, i.e.

$$V^{NS}_{m,n}(P_{i-1,i}) = +\delta_{m,n}. \tag{5.6.20}$$

This basis may be characterized by the chain

$$S_N = S_2 \otimes S_2 \otimes \ldots \otimes S_2.$$

The spin functions of the individual electrons are first coupled together in pairs to give either singlets

$$\Theta^2_{0,0} = (\tfrac{1}{2})^{\frac{1}{2}}(\alpha(\sigma_{i-1})\beta(\sigma_i) - \beta(\sigma_{i-1})\alpha(\sigma_i)) \tag{5.6.21}$$

or triplets

$$\begin{aligned} \Theta^2_{1,1} &= \alpha(\sigma_{i-1})\alpha(\sigma_i) \\ \Theta^2_{1,0} &= (\tfrac{1}{2})^{\frac{1}{2}}(\alpha(\sigma_{i-1})\beta(\sigma_i) + \beta(\sigma_{i-1})\alpha(\sigma_i)) \\ \Theta^2_{1,-1} &= \beta(\sigma_{i-1})\beta(\sigma_i) \end{aligned} \tag{5.6.22}$$

These pair spin functions are then coupled to give the total spin function. This coupling scheme can be conveniently represented by means of the branching diagram shown in Fig. 5.4. The index k is given by the series of partial resultant spins

$$k = (\ldots(((s_{12}, s_{34})S_4, s_{56})S_6, s_{78})S_8 \ldots S_{N-2}, s_{N-1,N}) \tag{5.6.23}$$

in which $s_{i-1,i}$ is used to denote a pair spin function. It can be shown that

$$P^{\sigma}_{i-1,i}\Theta^N_{SM;k} = (-1)^{s_{i-1,i}+1}\Theta^N_{SM;k}. \tag{5.6.24}$$

Inductive schemes can be devised in order to generate the representation matrices in the Serber basis.

Consider the portion of the branching diagram for the Serber spin basis shown in Fig. 5.5. The numbers f^N_S, as before, are the number of linearly independent spin functions which can be constructed for N electrons with a total spin of S. It will be assumed that the irreducible representation

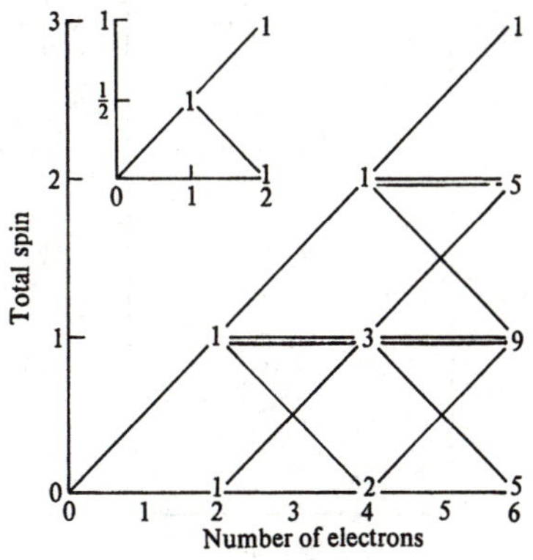

FIG. 5.4. Branching diagram for Serber spin functions.

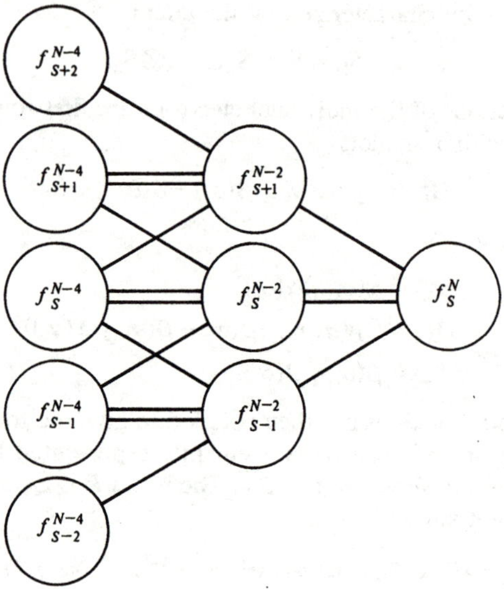

FIG. 5.5. Portion of the branching diagram for Serber spin functions.

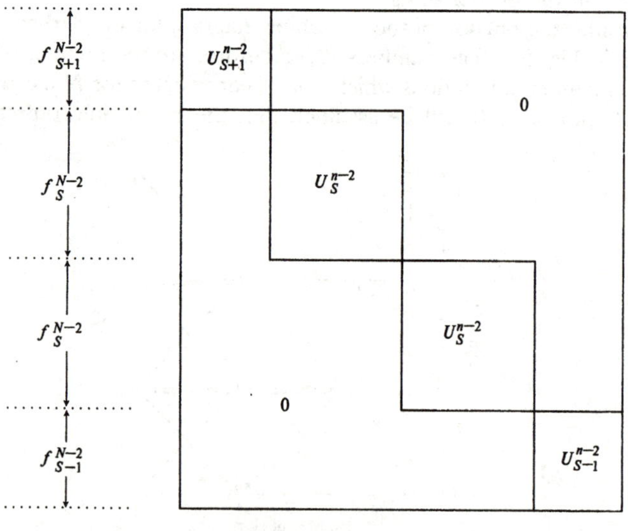

FIG. 5.6. Form of the irreducible representation matrices for S_N in the Serber basis for permutations which do not involve electrons $N-1$ and N.

matrices for the spin permutation group for $N-2$ electron systems are known. The matrix $\mathbf{V}^{NS}(P_{N-1,N})$ is a diagonal matrix of the form

$$\begin{pmatrix} \underline{I}(f^{N-2}_{S+1}) & \underline{0} & \underline{0} & \underline{0} \\ \underline{0} & \underline{I}(f^{N-2}_{S}) & \underline{0} & \underline{0} \\ \underline{0} & \underline{0} & -\underline{I}(f^{N-2}_{S}) & \underline{0} \\ \underline{0} & \underline{0} & \underline{0} & \underline{I}(f^{N-2}_{S-1}) \end{pmatrix} \tag{5.6.25}$$

where $\underline{I}(f)$ denotes the unit matrix of dimension f. The matrices for permutations which do not involve electrons $N-1$ and N have the reduced form shown in Fig. 5.6. To determine the remaining representation matrices, it is only necessary to find the matrix $\mathbf{V}^{NS}(P_{N-2,N-1})$, since all other matrices can be found by matrix multiplication. The matrix $\mathbf{V}^{NS}(P_{N-2,N-1})$ may be determined as follows:

We begin by assuming that we have all of the matrices for $N-4$ electrons. These will correspond to total spins of $S+2$, $S+1$, S, $S-1$, and $S-2$. Using the following formulae for coupling electron pair spin functions

$$\Theta^n_{SM;k} = \Theta^2_{0,0}\Theta^{n-2}_{SM;k}$$

$$\Theta^n_{S,M;k} = -\left\{\frac{(S+M)(S-M+1)}{2S(S+1)}\right\}^{\frac{1}{2}}\Theta^2_{1,1}\Theta^{n-2}_{SM-1;k} + \frac{M}{\{2S(S+1)\}^{\frac{1}{2}}}\Theta^2_{1,0}\Theta^{n-2}_{SM;k} + \left\{\frac{(S-M)(S+M+1)}{2S(S+1)}\right\}^{\frac{1}{2}}\Theta^2_{1,-1}\Theta^{n-2}_{SM+1;k}$$

$$\Theta^n_{S,M;k} = \left\{\frac{(S-M+1)(S-M+2)}{(2S+2)(2S+3)}\right\}^{\frac{1}{2}}\Theta^2_{1,1}\Theta^{n-2}_{S+1,M-1;k} - \left\{\frac{(S-M+1)(S+M+1)}{(2S+2)(2S+3)}\right\}^{\frac{1}{2}}\Theta^2_{1,0}\Theta^{n-2}_{S+1,M;k} + \left\{\frac{(S+M+1)(S+M+2)}{(2S+2)(2S+3)}\right\}^{\frac{1}{2}}\Theta^2_{1,-1}\Theta^{n-2}_{S+1,M+1;k}$$

$$\Theta^n_{SM;k} = \left\{\frac{(S+M)(S+M-1)}{2S(2S-1)}\right\}^{\frac{1}{2}}\Theta^2_{1,1}\Theta^{n-2}_{S-1,M-1;k} + \left\{\frac{(S+M)(S-M)}{2S(2S-1)}\right\}^{\frac{1}{2}}\Theta^2_{1,0}\Theta^{n-2}_{S-1,M;k} + \left\{\frac{(S-M)(S-M-1)}{2S(2S-1)}\right\}^{\frac{1}{2}}\Theta^2_{1,-1}\Theta^{n-2}_{S-1,M+1,k} \tag{5.6.26}$$

we can couple the spin function for electrons $N-3$ and $N-2$ to the $N-4$ electron spin function to obtain an $N-2$ electron spin function. We then use the above formulae again in order to couple the spin function for the $(N-1)$th and Nth electrons to obtain an N electron spin function. Thus an N electron spin function is obtained in which the coordinates of electrons $N-3$, $N-2$, $N-1$, and N occur explicitly. Be performing the permutation $P_{N-2,N-1}$ any using (5.6.7) the required matrix may be shown to have the form shown in Fig. 5.7, where only the non-zero elements are shown. The quantities A and X in Fig. 5.7 are defined in terms of the spin quantum number S in Table 5.3.

	f^{N-2}_{S+1}			f^{N-2}_{S}				f^{N-2}_{S}				f^{N-2}_{S-1}			
f^{N-4}_{S+2}	f^{N-4}_{S+1}	f^{N-4}_{S+1}	f^{N-4}_{S}	f^{N-4}_{S+1}	f^{N-4}_{S}	f^{N-4}_{S}	f^{N-4}_{S-1}	f^{N-4}_{S+1}	f^{N-4}_{S}	f^{N-4}_{S}	f^{N-4}_{S-1}	f^{N-4}_{S}	f^{N-4}_{S-1}	f^{N-4}_{S-1}	f^{N-2}_{S-2}
A_1															
	A_2	X_1		X_2				X_3							
		A_3		X_4				X_5							
			A_4		X_6	X_7			X_8	X_9					
				A_5				X_{10}							
					A_6	X_{11}			X_{12}	X_{13}		X_{14}			
						A_7			X_{15}			X_{16}			
							A_8				X_{17}		X_{18}	X_{19}	
								A_9							
									A_{10}			X_{20}			
										A_{11}		X_{21}			
											A_{12}		X_{22}	X_{23}	
												A_{13}			
													A_{14}	X_{24}	
														A_{15}	
															A_{16}

FIG. 5.7. Form of the irreducible representation matrices for S_N in the Serber basis for permutations which involve electrons $N-1$ and N.

Table 5.3
Non-zero elements of the matrix $\mathbf{V}^{NS}(P_{N-1,N-2})$ *in the Serber basis*

(a) *Diagonal elements*

$A_1=1$	$A_9=\frac{1}{2}$
$A_2=\frac{S}{2(S+1)}$	$A_{10}=\frac{1}{2}$
$A_3=\frac{1}{2}$	$A_{11}=\frac{1}{2}$
$A_4=\frac{-1}{2(S+1)}$	$A_{12}=\frac{1}{2}$
$A_5=\frac{S+2}{2(S+1)}$	$A_{13}=\frac{1}{2S}$
$A_6=\frac{S^2+S-1}{2S(S+1)}$	$A_{14}=\frac{(S+1)}{2S}$
$A_7=\frac{1}{2}$	$A_{15}=\frac{1}{2}$
$A_8=\frac{(S-1)}{2S}$	$A_{16}=1$

(b) *Off-diagonal elements*

$X_1=\sqrt{\left\{\frac{(S+2)}{4(S+1)}\right\}}$	$X_{13}=-\frac{1}{2}$
$X_2=\sqrt{\left\{\frac{S(S+2)}{4(S+1)}\right\}}$	$X_{14}=\sqrt{\left\{\frac{(S+1)^2(2S-1)}{4S(2S+1)}\right\}}$
$X_3=\sqrt{\left\{\frac{(S+2)}{4(S+1)}\right\}}$	$X_{15}=-\frac{1}{2}$
$X_4=-\sqrt{\left\{\frac{S}{4(S+1)}\right\}}$	$X_{16}=-\sqrt{\left\{\frac{(2S-1)(S+1)}{4S(2S+1)}\right\}}$
$X_5=-\frac{1}{2}$	$X_{17}=\sqrt{\left\{\frac{(S+1)}{4S}\right\}}$
$X_6=\sqrt{\left\{\frac{S^2(2S+3)}{4(S+1)^2(2S+1)}\right\}}$	$X_{18}=\sqrt{\left\{\frac{(S+1)(S-1)}{4S^2}\right\}}$
$X_7=\sqrt{\left\{\frac{S(2S+3)}{4(S+1)(2S+1)}\right\}}$	$X_{19}=\sqrt{\left\{\frac{(S+1)}{4S}\right\}}$
$X_8=\sqrt{\left\{\frac{S(2S+3)}{4(S+1)(2S+1)}\right\}}$	$X_{20}=-\sqrt{\left\{\frac{(2S-1)(S+1)}{4S(2S+1)}\right\}}$
$X_9=\sqrt{\left\{\frac{(2S+3)}{4(2S+1)}\right\}}$	$X_{21}=\sqrt{\left\{\frac{(2S-1)}{4(2S+1)}\right\}}$
$X_{10}=-\sqrt{\left\{\frac{S}{4(S+1)}\right\}}$	$X_{22}=-\sqrt{\left\{\frac{(S-1)}{4S}\right\}}$
$X_{11}=\frac{1}{2}(S(S+1))^{-\frac{1}{2}}$	$X_{23}=-\frac{1}{2}$
$X_{12}=\frac{1}{2}(S(S+1))^{-\frac{1}{2}}$	$X_{24}=-\sqrt{\left\{\frac{(S-1)}{4S}\right\}}$

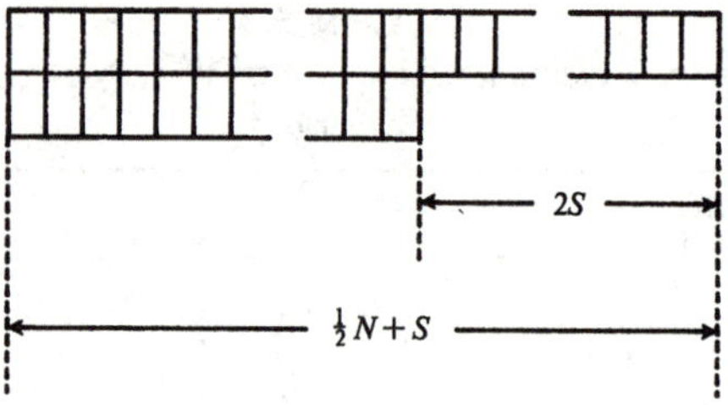

FIG. 5.8. Young diagram for the spin coordinate permutation group of an N electron system with total spin S.

The theory of the symmetric group, described in Section 5.5, can be used to determine the representation matrices $\mathbf{V}^{NS}(P^0)$. The general theory given in the preceding section yields all of the irreducible representations of the symmetric group. For spin coordinate permutations only those representations which correspond to partitions of N into two summands, i.e.

$$(N-k)+k \qquad k \leqslant \tfrac{1}{2}N \tag{5.6.27}$$

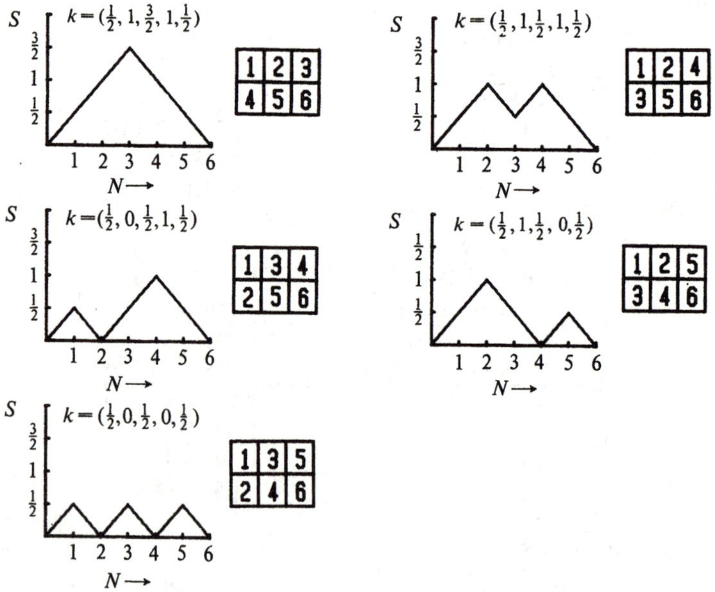

FIG. 5.9. Illustration of the correspondence between branching diagram functions and the standard Young tableaux.

are of interest. This is a consequence of the fact that the electron spin can assume one of two values. The irreducible representations of the spin coordinate permutation group are those corresponding to Young diagrams containing at most two rows. Physically, the partition of N is determined by the total spin of the molecule under consideration as illustrated in Fig. 5.8.

There is a one-to-one correspondence between the standard Young tableaux and the branching diagram functions for the Young–Yamanouchi coupling scheme introduced above. This correspondence is illustrated in Fig. 5.9 for a system of six electrons with a total spin of zero.

The graphical methods of spin algebra, described in Section 5.4, provide a powerful method for the direct calculation of the representation matrices for the spin permutation group. Muftakhova (1976) has demonstrated that the representation matrix for any permutation may be written as a transformation matrix relating different coupling schemes. We have seen in Section 5.4 that these transformation matrices may be related to $3n$-j symbols which can be reduced to sums of products of 6-j symbols.

As an example, consider the matrix elements for transpositions in the Young–Yamanouchi basis. The Young–Yamanouchi spin functions may be represented by

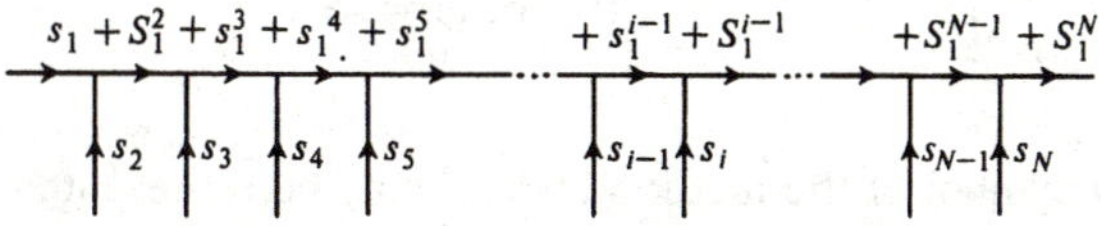

(5.6.28)

where s_i denotes the spin of the ith electron and S_i^j the resultant spin obtained by coupling $s_i \ldots s_j$. The matrix element $\mathbf{V}_{k,k}^{NS}(P_{i,j})$ may be obtained by changing the signs of the nodes and the sense of the arrows in (5.6.28), interchanging s_i and s_j and 'contracting' the resulting graph with the original, giving

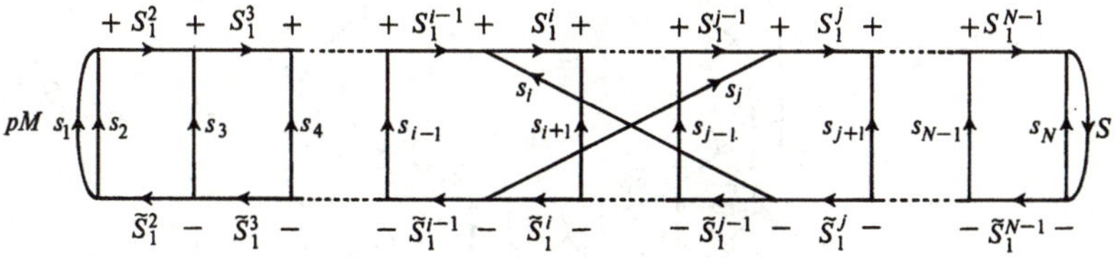

(5.6.29)

where p is a phase factor and

$$M^2 = \prod_{n=2}^{n=N-1} (2S_1^n + 1) \prod_{n=2}^{n=N+1} (2\tilde{S}_1^n + 1). \tag{5.6.30}$$

Using the rule (5.4.9), the required matrix element may be reduced to the form

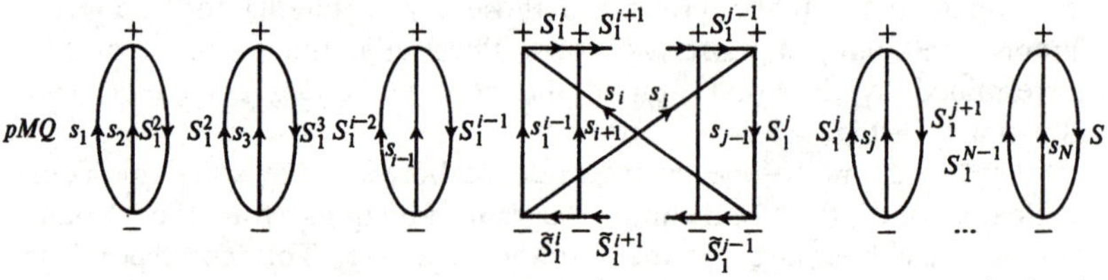

(5.6.31)

where the factor Q is given by

$$Q=\prod_{n=2}^{n=i-1}\delta(S_1^n,\tilde{S}_1^n)(2S_1^n+1)^{-1}\prod_{n=j}^{n=N-1}\delta(S_1^n,\tilde{S}_1^n)(2S_1^n+1)^{-1}. \quad (5.6.32)$$

The matrix element is thus a product of triangle deltas

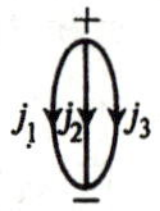

$$=\{J_1J_2J_3\}=\begin{cases}1 & \text{if } |J_1-J_2|\leqslant J_3\leqslant J_1+J_2\\ 0 & \text{otherwise}\end{cases} \quad (5.6.33)$$

and a $3n$-j symbol of the first kind which may be represented graphically by a Möbius band

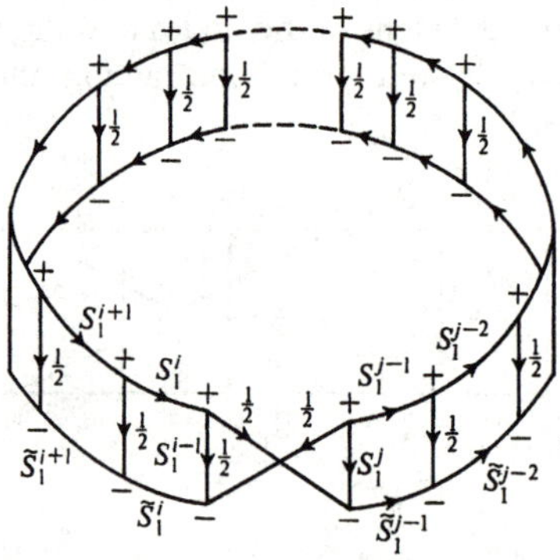

(5.6.34)

The $3n$-j symbol can then be written as a sum of two products of 6-j

symbols

$$\left\{\begin{matrix} \frac{1}{2} & S_1^i & S_1^{i+1} & \dots & S_1^{j-2} & S_1^{j-1} \\ S_1^{i-1} & \frac{1}{2} & \frac{1}{2} & \dots & \frac{1}{2} & S_1^j \\ \frac{1}{2} & \tilde{S}_1^i & \tilde{S}_1^{i+1} & \dots & \tilde{S}_1^{j-2} & \tilde{S}_1^{j-1} \end{matrix}\right| 1 \Bigg\}$$

$$= (-1)^A \begin{Bmatrix} \frac{1}{2} & 0 & \frac{1}{2} \\ S_1^i & S_1^{i-1} & \tilde{S}_1^i \end{Bmatrix} \begin{Bmatrix} S_1^i & S_1^{i+1} & \frac{1}{2} \\ \tilde{S}_1^{i+1} & \tilde{S}_1^i & 0 \end{Bmatrix} \dots \begin{Bmatrix} S_1^{j-2} & S_1^{j-1} & \frac{1}{2} \\ \tilde{S}_1^{j-1} & \tilde{S}_1^{j-2} & 0 \end{Bmatrix}$$

$$\times \begin{Bmatrix} \frac{1}{2} & \frac{1}{2} & 0 \\ \tilde{S}_1^{j-1} & S_1^{j-1} & S_1^j \end{Bmatrix} + (-1)^B 3 \begin{Bmatrix} \frac{1}{2} & 1 & \frac{1}{2} \\ S_1^i & S_1^{i-1} & \tilde{S}_1^i \end{Bmatrix} \begin{Bmatrix} S_1^i & S^{i+1} & \frac{1}{2} \\ \tilde{S}_1^{i+1} & \tilde{S}_1^i & 1 \end{Bmatrix} \dots$$

$$\times \begin{Bmatrix} S_1^{j-2} & S_1^{i-1} & \frac{1}{2} \\ \tilde{S}_1^{j-1} & \tilde{S}_1^{j-2} & 1 \end{Bmatrix} \begin{Bmatrix} \frac{1}{2} & \frac{1}{2} & 1 \\ \tilde{S}_1^{j-1} & S_1^{j-1} & S_1^j \end{Bmatrix} \qquad (5.6.35)$$

where

$$A = \tfrac{1}{2}(j-1+3) + S_1^{i-1} + S_i^j + \sum_{j=1}^{n=j-1} (S_1^n + \tilde{S}_1^n)$$

$$B = A + j - 1. \qquad (5.6.36)$$

The matrix elements for the spin permutation group in other coupling schemes, such as the Serber scheme discussed above, can be obtained by means of similar methods (Wilson and Gerratt 1979).

5.7 S_N and spatial functions

The non-relativistic molecular electronic hamiltonian, $\hat{\mathcal{H}}$, is symmetric with respect to any permutation of the spatial coordinates of the electrons P^r and, therefore

$$[\hat{\mathcal{H}}, P^r] = 0 \qquad (5.7.1)$$

For an N-electron system the $N!$ possible permutations of the electronic coordinates are constants of the motion. The eigenfunctions of $\hat{\mathcal{H}}$ may, therefore, be chosen to form a basis for an irreducible representation of the symmetric group S_N. The indistinguishability of electrons leads to the so-called exchange degeneracy. This degeneracy may not, however, be removed, since there no physical perturbations which distinguish between electrons.

It is experimentally observed that electrons are fermions and must, therefore, conform to the Fermi–Dirac statistics. The Pauli exclusion principle summarizes this requirement for electrons. If the total wave function, a function of both space and spin coordinates

$$\Psi = \Psi(\mathbf{r}_1\sigma_1\, \mathbf{r}_2\sigma_2 \dots \mathbf{r}_N\sigma_N) \qquad (5.7.2)$$

corresponds to a physically realizable state, then it is antisymmetric with respect to simultaneous permutation of space and spin coordinates, i.e.

$$P\Psi = \epsilon_P \Psi \tag{5.7.3}$$

where $\epsilon = +1$ for an even permutation and -1 for an odd permutation.

Since the spin functions $\Theta^N_{SM;k}$ form a complete set in the spin space for an N-electron system with a total spin of S, the exact wave function may be expanded as follows (Wigner 1931, 1959)

$$\Psi_{SM} = (f^N_S)^{-\frac{1}{2}} \sum_k^{f_S^N} \Phi^N_{Sk}(\mathbf{r}_1\, \mathbf{r}_2 \ldots \mathbf{r}_N)\Theta^N_{SM;k}(\sigma_1\, \sigma_2 \ldots \sigma_N) \tag{5.7.4}$$

where the coefficients Φ^N_{Sk} depend on the spatial coordinates of the N electrons. The factor $(f^N_S)^{-\frac{1}{2}}$ maintains normalization. It immediately follows from (5.7.1) that the Φ^N_{Sk} also form a basis for a representation of S_N. Thus

$$P^r \Phi^N_{Sk} = \sum_{j=1}^{j=f_S^N} U^{NS}_{j,k}(P^r)\Phi^N_{Sj} \tag{5.7.5}$$

where the $\mathbf{U}^{NS}(P)$ form a matrix representation of the symmetric group. From (5.7.3)

$$P = P^r \cdot P^\sigma = \epsilon_P = \pm 1 \tag{5.7.6}$$

and thus

$$\mathbf{U}^{NS}(P^r)\mathbf{V}^{NS}(P^\sigma) = \pm \mathbf{I}. \tag{5.7.7}$$

If the spin functions form a basis for an irreducible representation $\mathbf{V}^{NS}(P)$ of S_N then the spatial functions form a basis for the irreducible representation $\mathbf{U}^{NS}(P)$—the two representations being related by eqn (5.7.7). These representations are said to be mutually dual or conjugate

$$\mathbf{U}^{NS}(P) \leftrightarrow \mathbf{V}^{NS}(P) \tag{5.7.8}$$

The $\mathbf{U}^{NS}(P)$ are sometimes said to form the associated representation of $\mathbf{V}^{NS}(P)$. In terms of Young diagrams, associated representations are obtained by interchanging the rows and columns in a given diagram. This is illustrated in Fig. 5.10 for a Young diagram, which arises in electronic structure theory. In general, the associated representation of a given representation is the direct product of that representation and the one-dimensional antisymmetric representation (see, for example, Wigner 1959, p. 127).

Generally, the spatial wave function Φ^N_{Sk} is not known. In practice, we have an approximate spatial wave function, Φ^N say, from which a wave function with the correct symmetry properties must be constructed. This

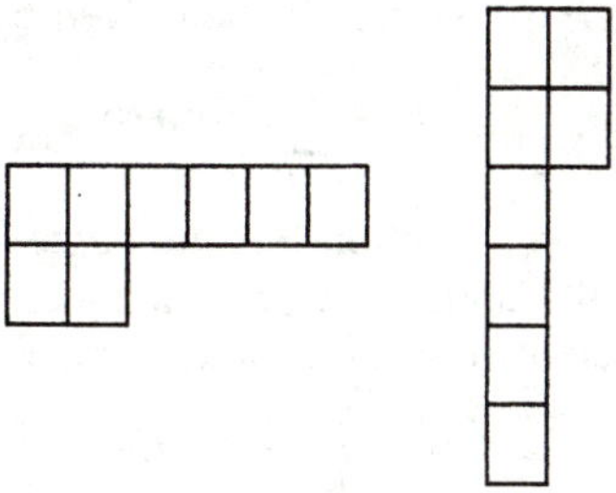

FIG. 5.10. Conjugate Young diagrams.

may be achieved by applying the Young–Wigner operator

$$\hat{\omega}^{NS}_{l,k} = (f^N_S/N!)^{\frac{1}{2}} \sum_{P \in S_N} U^{NS}_{l,k}(P)P^r. \tag{5.7.9}$$

The $(f^N_S)^2$ operators $\hat{\omega}^{NS}_{l,k}$ have the property

$$\hat{\omega}^{NS}_{i,j}\hat{\omega}^{N'S'}_{k,l} = (f^N_S/N!)^{\frac{1}{2}}\delta_{NS,N'S'}\delta_{i,k}\omega^{NS}_{j,l}. \tag{5.7.10}$$

It can be shown that the Young–Wigner operators themselves form a basis for a representation of S_N.

$$P\hat{\omega}^{NS}_{i,j} = \sum_k U^{NS}_{k,i}(P)\hat{\omega}^{NS}_{k,j} \tag{5.7.11}$$

$\hat{\omega}^{NS}$ are essentially the characteristic units of the Frobenius algebra of the symmetric group (see, for example, Littlewood 1950, Chapter 2).

The functions $(\hat{\omega}^{NS}_{i,j}\Phi^N)$ have the desired permutational symmetry properties

$$P^r(\hat{\omega}^{NS}_{i,j}\Phi^N) = \sum_{k=1}^{f_S N} U^{NS}_{k,i}(P)(\hat{\omega}^{NS}_{k,j}\Phi^N) \tag{5.7.12}$$

This is easily proved as follows

$$\begin{aligned} Q^r(\hat{\omega}^{NS}_{i,j}\Phi^N) &= (f^N_S/N!)^{\frac{1}{2}} \sum_{P \in S_N} U^{NS}_{i,j}(P)(Q^r P^r \Phi^N) \\ &= (f^N_S/N!)^{\frac{1}{2}} \sum_{P^r \in S_N} U^{NS}_{i,j}\{(Q)^{-1}\tilde{P}\}(\tilde{P}^r\Phi^N) \\ &= \sum_k U^{NS}_{ik}\{(Q)^{-1}\}\left(\frac{f^N_S}{N!}\right)^{\frac{1}{2}} \sum_{\tilde{P} \in S_N} U^{NS}_{k,j}(\tilde{P})(\tilde{P}^r\Phi^N) \\ &= \sum_k U^{NS}_{k,i}(Q)(\hat{\omega}^{NS}_{k,j}\Phi^N). \end{aligned} \tag{5.7.13}$$

The functions $(\hat{\omega}^{NS}_{i,j}\Phi^N) i = 1, 2, \ldots, f^N_S$ form, for each k, a basis for a representation of S_N. An approximate wave function with the required permutational symmetry properties is then given by

$$\Psi^N_{SM;k}=(f^N_S)^{-\frac{1}{2}}\sum_{l=1}^{f_S N}(\hat{\omega}^{NS}_{l,k}\Phi^N)\Theta^N_{SM;l} \tag{5.7.14}$$

In general, there are f^N_S functions of the type (5.7.14) for a given approximate function Φ^N and, therefore, the most general wave function which can be constructed from the primitive spatial function Φ^N is

$$\Psi^N_{SM}=\sum_k b^N_{Sk}\Psi^N_{SM;k} \tag{5.7.15}$$

where the b^N_{Sk} are variational parameters.

5.8 The unitary group $U(n)$

Linear transformations on an n-dimensional vector space form a group; the general linear group, $GL(n)$. These groups form a particular case of the Lie groups (see, for example, Hamermesh 1962).

Consider a vector space of dimension n subjected to a linear transformation, the matrix of which contains elements a_{ij}. Every vector $\boldsymbol{\psi}$ in the space is transformed into a new vector $\boldsymbol{\psi}'$ according to

$$\psi'_i=\sum_j a_{ij}\psi_j \tag{5.8.1}$$

If the matrix $\mathbf{a}$ is required to be a unitary matrix then the group of unitary transformations in an n-dimensional space is obtained. This subgroup of $GL(n)$ is usually denoted by $U(n)$. Unitary transformations with determinant equal to unity form the group of unitary unimodular transformations or the special unitary group, which is denoted by $SU(n)$.

Jordan (1935) first noticed that the hamiltonian in quantum mechanical problems can be expressed in terms of the generators of the unitary group. For the exact hamiltonian the appropriate group is $U(\infty)$; however, in molecular calculations, a finite basis set is almost always employed and the $U(2n)$ spin-orbital group or its $U(n)$ orbital subgroup is appropriate (see, for example, Paldus 1976*b*). On the other hand, the $SU(2)$ special unitary group is relevant to the analysis of spin functions. The relation of the $SU(2)$ group to spin functions will be discussed in Section 5.9 and the relation of the group $U(n)$ to spatial functions will be described in Section 5.10.

As emphasized in the early work of Weyl (1928, 1964), there is a close connection between the representation theory of the symmetric group and that of the general linear groups or unitary groups. The operations, of carrying out a linear transformation and of permuting the indices, commute. An Nth rank tensor may be defined as the set of n^N components

$$T_{i_1i_2\ldots i_N}=\psi_{i_1}\psi_{i_2}\ldots\psi_{i_N} \tag{5.8.2}$$

which under the transformations of the vector space (5.8.1) transform according to

$$T'_{i_1 i_2 \ldots i_N} = \sum_{j_1 j_2 \ldots j_N} a_{i_1 j_1} a_{i_2 j_2} \ldots a_{i_N j_N} T_{j_1 j_2 \ldots j_N}. \tag{5.8.3}$$

A transformation of the n-dimensional space, therefore, induces a transformation

$$\mathbf{a} \times \mathbf{a} \times \cdots \times \mathbf{a} \tag{5.8.4}$$

in the n^N-dimensional space of Nth rank tensors (5.8.2). Since the operations, of performing a unitary transformation and of permuting the indices, commute, linear combinations of the tensor components can be formed which possess definite symmetry properties with respect to permutation of the indices. These linear combinations will, therefore, transform only among themselves. It can be shown that the irreducible representations of the unitary group $U(n)$ which occur in the decomposition of an Nth rank tensor representation are characterized by Young diagrams consisting of N cells with no more than n cells in a column (Hamermesh 1962). The latter restriction results from the fact that no more than n of the N indices in a given tensor can be distinct. In the reduction of a tensor representation, the n^N basis functions may be arranged in a plane diagram in which functions which lie on the same row transform into each other under unitary transformations; whereas functions lying in the same column transform into each other under permutation of the arguments. This is illustrated in Fig. 5.11.

If an irreducible representation of $U(n)$ is labelled by a Young diagram $[\lambda]$, then each of the basis functions is characterized by a tableau in which the indices $k_1, k_2, \ldots, k_n$ are assigned to the cells, subject to the conditions:

(i) The indices do not decrease along the rows.
(ii) The indices increase down the columns.

The resulting labels for the basis functions are known as Weyl tableaux. Typical Weyl tableaux are shown in Fig. 5.12. The Weyl tableaux

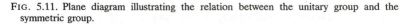

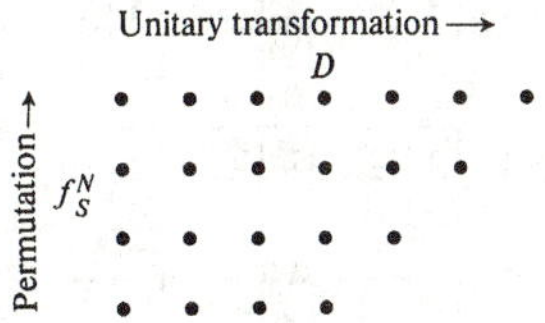

FIG. 5.11. Plane diagram illustrating the relation between the unitary group and the symmetric group.

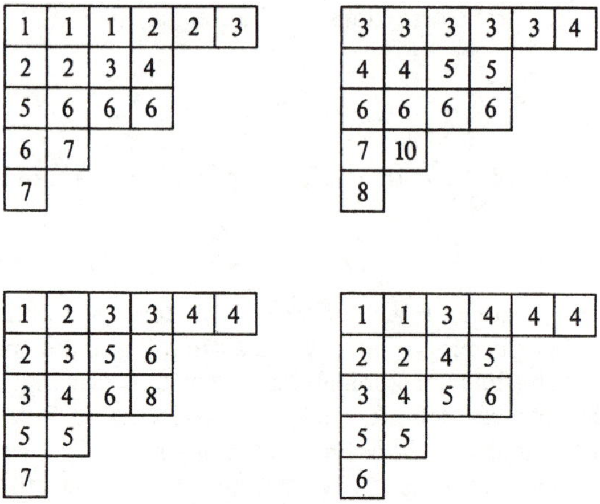

FIG. 5.12. Some of the Weyl tableaux for the partition $[\lambda]=[6, 4^2, 2, 1]$.

corresponding to a given n, N, and $[\lambda]$ can be put into a standard order, the lexically ordered Weyl tableaux. The lexically ordered Weyl tableaux for $n=4$, $N=3$, and $[\lambda]=[2, 1]$ are shown in Fig. 5.13.

As in the case of the symmetric group, the individual basis functions for the unitary group $U(n)$ are characterized by the chain of irreducible representations to which they belong on passing to the subgroups $U(n-1)$, $U(n-2), \ldots, U(1)$. Thus, for example, the irreducible representation of $U(4)$ shown in Fig. 5.13, is reduced into a sum of irreducible representations of $U(3)$ as illustrated in Fig. 5.14. The 20-by-20 matrices of the $U(4)$ irreducible representation is reduced to an 8-by-8, a 6-by-6, and two 3-by-3 irreducible representations of $U(3)$.

The Weyl tableaux exhibit the properties of the basis functions of $U(n)$ transparently. They serve to emphasize the close connection between the theory of the symmetric group and that of the unitary group. However,

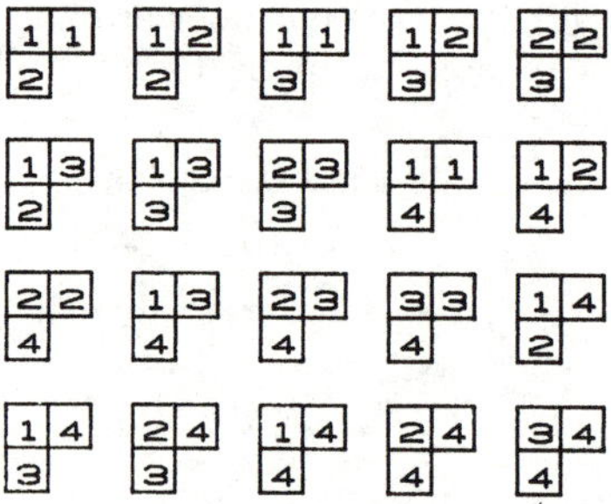

FIG. 5.13. Lexically ordered Weyl tableaux for $[\lambda]=[2, 1]$, $n=4$ and $N=3$.

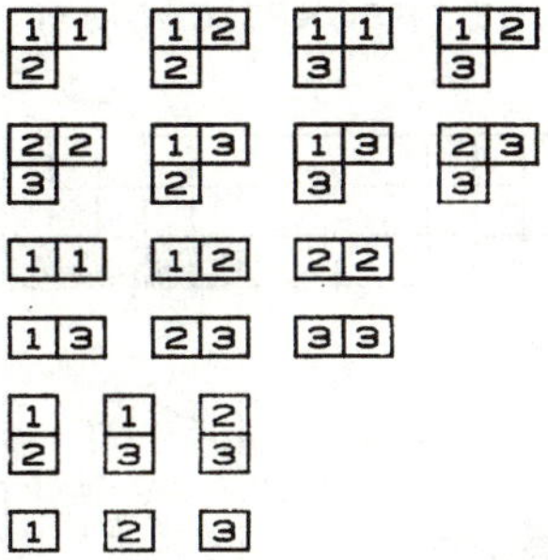

FIG. 5.14. Reduction of the irreducible representations of $U(4)$ to a sum of irreducible representations of $U(3)$.

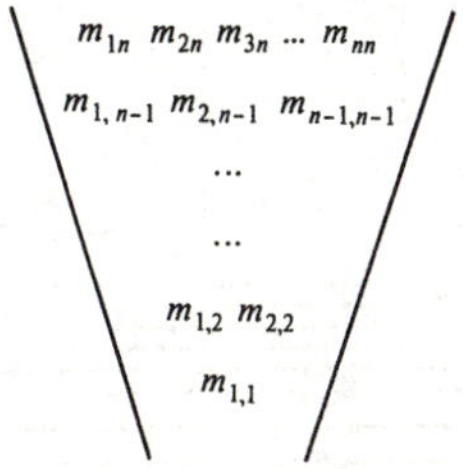

FIG. 5.15. A Gelfand tableau.

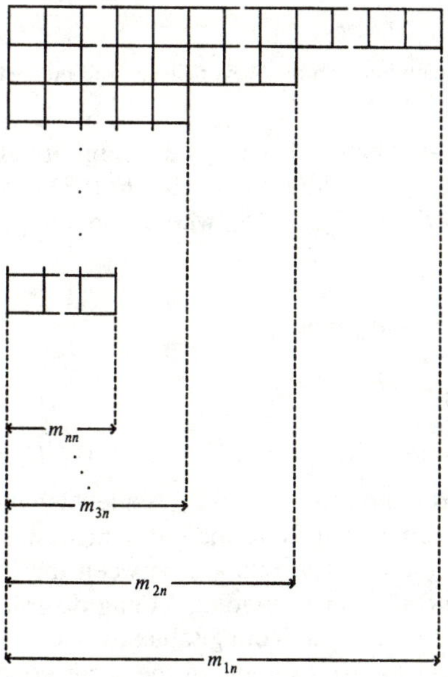

FIG. 5.16. Relation between the highest weight of a Gelfand tableau and the corresponding Young diagram.

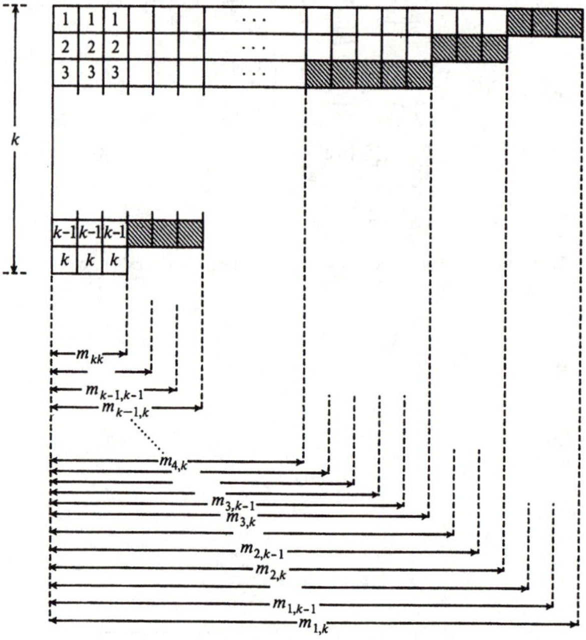

FIG. 5.17. Relation between the Weyl tableaux and the Gelfand tableaux.

the Gelfand tableaux have also proved useful in electronic structure theory (see, for example, Paldus 1974, 1976*b*, 1981). A Gelfand tableaux has the form shown in Fig. 5.15, where the integers m_{ij} satisfy the inequalities

$$m_{i,j} \geqslant m_{i+1,j} \qquad \begin{array}{l} i=1,\ldots,j \\ j=1,\ldots,n \end{array} \tag{5.8.5}$$

and the conditions

$$m_{i,j} \geqslant m_{i,j-1} \geqslant m_{i+1,j} \qquad i=1,\ldots,n-1; \quad j=2,\ldots,n. \tag{5.8.6}$$

An irreducible representation of $U(n)$ is uniquely labelled by the first row of a Gelfand tableau which is termed the highest weight of a given irreducible representation. The relation between the highest weight of a Gelfand tableau and the corresponding Young diagram is shown in Fig. 5.16. The relation between the Young tableaux and the Gelfand tableaux is illustrated in Fig. 5.17. An example of the correspondence between the scheme for labelling basis functions using Gelfand tableaux and using Weyl tableaux is given in Fig. 5.18.

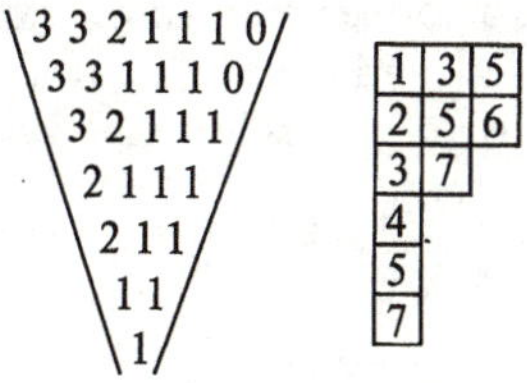

FIG. 5.18. An example of the correspondence between Weyl tableaux and Gelfand tableaux.

5.9 *SU*(2) and the spin functions

The spin functions for an N-electron system

$$\Theta_N(\theta_i) = \theta_1\,\theta_2 \ldots \theta_N; \qquad \theta = \alpha \text{ or } \beta \tag{5.9.1}$$

form a 2^N-dimensional space of Nth rank tensors. They form a basis for irreducible representations of the special unitary group, $SU(2)$. If the one-electron spin functions are written in the vector form

$$\boldsymbol{\theta} = \begin{pmatrix} \alpha \\ \beta \end{pmatrix} \tag{5.9.2}$$

then the spin operators can be expressed in terms of the Pauli spin matrices

$$\boldsymbol{\sigma}_x = \begin{pmatrix} 0 & 1 \\ 1 & 0 \end{pmatrix} \quad \boldsymbol{\sigma}_y = \begin{pmatrix} 0 & -i \\ i & 0 \end{pmatrix} \quad \boldsymbol{\sigma}_z = \begin{pmatrix} 1 & 0 \\ 0 & -1 \end{pmatrix} \tag{5.9.3}$$

which obey the following commutation relations

$$[\sigma_x, \sigma_y] = -2i\sigma_z; \qquad [\sigma_y, \sigma_z] = -2i\sigma_x; \qquad [\sigma_z, \sigma_x] = -2i\sigma_y. \tag{5.9.4}$$

The spin operators can then be written in the following form

$$\hat{S}_x = \tfrac{1}{2}\boldsymbol{\sigma}_x \tag{5.9.5a}$$

$$\hat{S}_y = \tfrac{1}{2}\boldsymbol{\sigma}_y \tag{5.9.5b}$$

$$\hat{S}_z = \tfrac{1}{2}\boldsymbol{\sigma}_z \tag{5.9.5c}$$

$$\hat{S}^2 = \tfrac{3}{4}(\boldsymbol{\sigma}_x^2 + \boldsymbol{\sigma}_y^2 + \boldsymbol{\sigma}_z^2) = \frac{3}{2}\begin{pmatrix} 1 & 0 \\ 0 & 1 \end{pmatrix} \tag{5.9.5d}$$

$$\hat{S}_+ = \hat{S}_x + i\hat{S}_y = \begin{pmatrix} 0 & 1 \\ 0 & 0 \end{pmatrix} \tag{5.9.5e}$$

$$\hat{S}_- = \hat{S}_x - i\hat{S}_y = \begin{pmatrix} 0 & 0 \\ 1 & 0 \end{pmatrix}. \tag{5.9.5f}$$

It is clear that any operator of rank 2 can be expressed in terms of the Pauli matrices $\boldsymbol{\sigma}_x$, $\boldsymbol{\sigma}_y$, and $\boldsymbol{\sigma}_z$ and the unit matrix of dimension 2. Indeed, the Pauli spin matrices are very closely related to the generators of the $SU(2)$ group

$$\begin{aligned} E_{12} &= \hat{S}_+ \\ E_{21} &= \hat{S}_- \end{aligned} \tag{5.9.6}$$

From the discussion of Section 5.8, it can be seen that the irreducible representations of the $SU(2)$ group for which the N-electron spin functions form a basis may be characterized by Young diagrams consisting of N cells but with no more than two cells to a column. This is precisely the type of Young diagram which was considered in Section 5.6 when applying the theory of the symmetric group to molecular spin functions. The spin functions are labelled by Weyl tableaux containing the indices 1 and 2 corresponding to α and β spins.

Consider, as a simple example, a two-electron system. The 2^2-dimensional space generated by the total spin functions for this system may be represented by

$$|1\rangle|1\rangle, \qquad |1\rangle|2\rangle, \qquad |2\rangle|1\rangle, \qquad |2\rangle|2\rangle \tag{5.9.7}$$

In this simple case, it is known that the simultaneous eigenstates of the spin operators S^2 and S_z are (neglecting normalization)

$$S = M = 0: \qquad |1\rangle|2\rangle - |2\rangle|1\rangle \tag{5.9.8a}$$

$$S = 1 \begin{cases} M = 1: & |1\rangle|1\rangle & (5.9.8b) \\ M = 0: & |1\rangle|2\rangle + |2\rangle|1\rangle & (5.9.8c) \\ M = -1: & |2\rangle|2\rangle. & (5.9.8d) \end{cases}$$

This 4-dimensional space can also be regarded as a representation space for the symmetric group S_2. It can thus be decomposed into irreducible representations characterized by Young diagrams

$$\begin{array}{|c|}\hline \\ \hline \\ \hline \end{array} \qquad \begin{array}{|c|c|}\hline & \\ \hline \end{array} \tag{5.9.9}$$

These same diagrams also label the irreducible representations of the unitary group $U(2)$ and the special unitary group $SU(2)$. The individual basis vectors can be uniquely labelled by the Weyl tableaux corresponding to these Young diagrams. Thus, in the two-electron example, for $S = 0$ the basis function can be labelled by the Weyl tableau

$$\begin{array}{|c|}\hline 1 \\ \hline 2 \\ \hline \end{array} \tag{5.9.10}$$

and, for $S = 1$, the labels are

$$\boxed{1|1} \qquad \boxed{1|2} \qquad \boxed{2|2} \tag{5.9.11}$$

Let us now consider a three-electron system. The possible Young diagrams, containing three cells and no more than two rows, are

$$\tag{5.9.12}$$

These diagrams correspond to $S = \frac{1}{2}$ and $S = \frac{3}{2}$, respectively. For the first case $(S = \frac{1}{2})$, the basis functions are labelled by the following $2S + 1 = 2$ Weyl tableaux

$$\tag{5.9.13}$$

while in the second case $(S = \frac{3}{2})$ there are $2S + 1 = 4$ Weyl tableaux

$$\boxed{1|1|1} \qquad \boxed{1|1|2} \qquad \boxed{1|2|2} \qquad \boxed{2|2|2} \tag{5.9.14}$$

As a third example, a six-electron system is considered in Fig. 5.19. There are four possible Young diagrams corresponding to $S = 0, 1, 2,$ and 3. The total spin associated with each diagram is given by one half of the

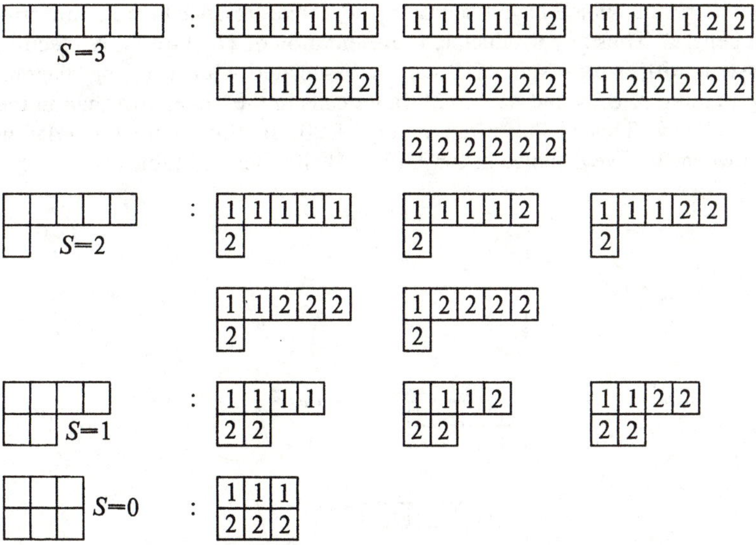

FIG. 5.19. Labelling of electron spin functions using Young diagrams and Weyl tableaux for a six electron system.

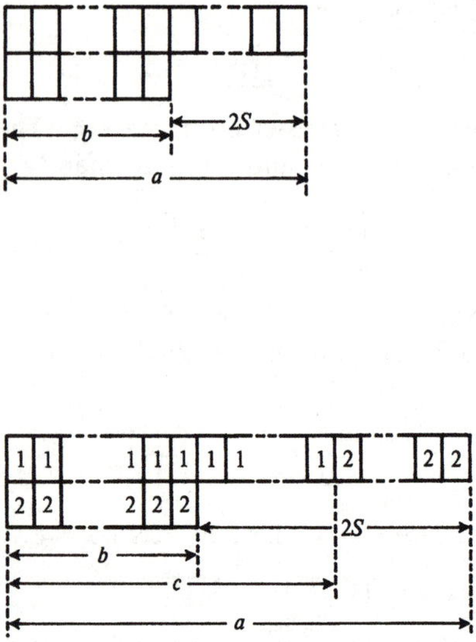

FIG. 5.20. Young diagram for an N electron system with total spin S.

difference between the number of cells in the first and second rows. There are $2S+1$ Weyl tableaux associated with a given Young diagram.

The results which have been obtained above for specific cases are true in general. Thus any irreducible representation of $U(2)$ for an N-electron system with a total spin of S can be associated with a Young diagram containing N cells and having $2S$ more cells in the upper row than in the lower row. This is illustrated in Fig. 5.20. In this figure the relation between the Weyl tableaux for $U(2)$ and the Gelfand tableaux.

$$\left\backslash \begin{matrix} b & c \\ & a \end{matrix} \right/ \tag{5.9.15}$$

$$\boxed{\begin{matrix} 1 & 1 \\ 2 \end{matrix}} \qquad \left\backslash \begin{matrix} 2 & 1 \\ & 2 \end{matrix} \right/$$

$$\boxed{\begin{matrix} 1 & 1 & 1 & 2 & 2 \\ 2 & 2 \end{matrix}} \qquad \left\backslash \begin{matrix} 5 & 2 \\ & 3 \end{matrix} \right/$$

$$\boxed{\begin{matrix} 1 & 1 & 1 & 1 & 2 & 2 & 2 \\ 2 & 2 & 2 \end{matrix}} \qquad \left\backslash \begin{matrix} 7 & 3 \\ & 4 \end{matrix} \right/$$

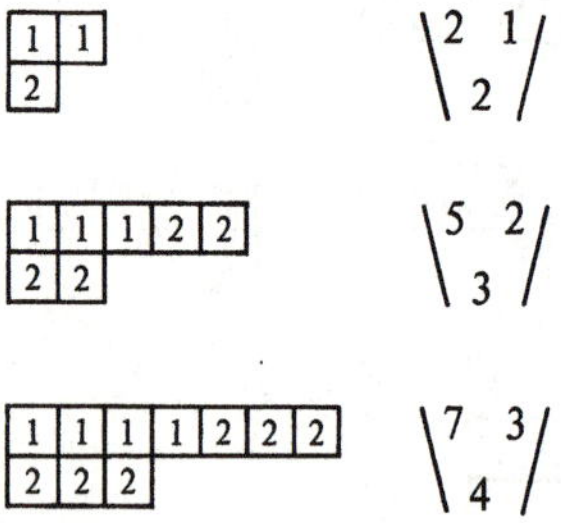

FIG. 5.21. Correspondence between Weyl tableaux and Gelfand tableaux used to label spin functions.

is also shown. There is a one-to-one correspondence between the Weyl tableaux and the Gelfand tableaux. An example of this correspondence for electron spin function functions is given in Fig. 5.21.

5.10 $U(n)$ and the spatial wave functions

If the second-quantized formalism is used (see Section 4.4), the hamiltonian operator for a system of N electrons may be written as

$$\hat{\mathcal{H}} = \sum_{i_1 i_2} \langle i_1 | \hat{f} | i_2 \rangle \hat{a}^\dagger_{i_1} \hat{a}_{i_2} + \tfrac{1}{2} \sum_{i_1 i_2 i_3 i_4} \langle i_1 i_2 | \hat{g} | i_3 i_4 \rangle \hat{a}^\dagger_{i_1} \hat{a}^\dagger_{i_2} \hat{a}_{i_4} \hat{a}_{i_3}. \quad (5.10.1)$$

The products of creation and annihilation operators which occur in this hamiltonian may be regarded as generators of the unitary group $U(2n)$

$$e_{i_1 i_2} = \hat{a}^\dagger_{i_1} \hat{a}_{i_2}. \quad (5.10.2)$$

The generators can be shown to satisfy the commutation relations

$$[e_{i_1 i_2}, e_{i_3 i_4}] = e_{i_1 i_4} \delta_{i_2 i_3} - e_{i_3 i_2} \delta_{i_1 i_4}. \quad (5.10.3)$$

The orbital and spin partial traces

$$\mathbb{E}_{I_1 I_2} = \sum_{\sigma} e_{I_1 \sigma, I_2 \sigma}; \qquad i = I\sigma \qquad E_{\sigma_1 \sigma_2} = \sum_{I} e_{I\sigma_1, I\sigma_2}; \quad (5.10.4)$$

satisfy the commutation relations (Moshinsky 1966, 1968)

$$\begin{aligned} [\mathbb{E}_{I_1 I_2}, \mathbb{E}_{I_3 I_4}] &= \mathbb{E}_{I_1 I_4} \delta_{I_2 I_3} - \mathbb{E}_{I_3 I_2} \delta_{I_1 I_4} \\ [E_{\sigma_1 \sigma_2}, E_{\sigma_3 \sigma_4}] &= E_{\sigma_1 \sigma_4} \delta_{\sigma_2 \sigma_3} - E_{\sigma_3 \sigma_2} \delta_{\sigma_1 \sigma_4} \end{aligned} \quad (5.10.5)$$

and may thus be regarded as the generators of the orbital $U(n)$ group and the spin $SU(2)$ group, respectively. Furthermore, these two sets of generators mutually commute, since the $\mathbb{E}_{I_1 I_2}$ are spin-independent and the $E_{\sigma_1 \sigma_2}$ are independent of the orbitals. Consequently, the spin-independent hamiltonian (5.10.1) can be written in terms of the generators of the unitary group $U(n)$

$$\hat{\mathcal{H}} = \sum_{I_1 I_2} \langle I_1 | f | I_2 \rangle \mathbb{E}_{I_1 I_2} + \tfrac{1}{2} \sum_{I_1 I_2 I_3 I_4} \langle I_1 I_2 | g | I_3 I_4 \rangle (\mathbb{E}_{I_1 I_2} \mathbb{E}_{I_3 I_4} - \delta_{I_2 I_3} \mathbb{E}_{I_1 I_4}) \quad (5.10.6)$$

or

$$\hat{\mathcal{H}} = \sum_{I_1 I_2} \langle I_1 | \hat{f} | I_2 \rangle \mathbb{E}_{I_1 I_2} + \tfrac{1}{2} \sum_{I_1 I_2 I_3 I_4} \langle I_1 I_2 | \hat{g} | I_3 I_4 \rangle N[\mathbb{E}_{I_1 I_3} \mathbb{E}_{I_2 I_4}] \quad (5.10.7)$$

where $N(\ldots)$ denotes the normal product which was discussed in Section 4.4.

The orbital product functions

$$|\Phi_{i_1 i_2 \dots i_N}\rangle = |\phi_{i_1}\,\phi_{i_2}\dots\phi_{i_N}\rangle$$
$$= |p_1\,p_2\dots p_n\rangle \tag{5.10.8}$$

where

$$p_i = \begin{cases} 1 & \text{if} \quad \phi_i \in \Phi \\ 0 & \text{if} \quad \phi_i \notin \Phi \end{cases} \tag{5.10.9}$$

are usually employed as a basis in which to expand N-electron spatial wave functions. Using a basis set of n one-electron functions to describe an N-electron system generates a n^N-dimensional space of Nth rank tensors. These tensors will form bases for irreducible representations of the unitary group $U(n)$, which may be labelled by Young diagrams containing N cells and no more than n cells in a column. However, if the total wave function is to correspond to a physically realizable state of the system, there is a further restriction, namely, that the direct product of the irreducible representation of $U(n)$ for the spatial wave function and the irreducible representation of $SU(2)$ for the spin wave function must contain the antisymmetric representation of $U(2n)$. This will only be the case if the irreducible representations of $U(n)$ and $SU(2)$ are associated with mutually dual Young diagrams. This requirement severely restricts the number of irreducible representations of $U(n)$ which are of interest in the study of the electronic structure of atoms and molecules. Indeed, the Young diagrams used to label the irreducible representations of $U(n)$ of interest in the electronic structure problem, can have at most two columns. This restriction corresponds to the requirement encountered in Section 5.9 for spin functions for which the Young diagrams can have at most two rows. In the Weyl tableaux used to label the basis functions of $U(n)$ the difference between the number of cells in the first column and the number in the second column is equal to twice the total spin of the system being considered.

Since the hamiltonian operator can be expressed in terms of the generators of $U(n)$, it will leave the spin part of the wave function invariant. The matrix representation of the hamiltonian in terms of the basis defined in eqn (5.10.6) can therefore be factorized into $(2S+1)$ identical matrices, each corresponding to a particular choice of M_S. For a given multiplicity, the space which has to be considered is very much smaller than the full n^N-dimensional space which can be generated from a set of n one-electron functions for an N-electron system. In fact, it can be shown (Paldus 1981) that the dimension of the space which has to be considered is given by

$$D(a, b, c) = \frac{b+1}{n+1}\binom{n+1}{a}\binom{n+1}{c}, \quad a = \tfrac{1}{2}N - S,\ b = 2S,\ c = n - \tfrac{1}{2}N - S \tag{5.10.10}$$

Table 5.4
Comparison of values of D(a, b, c) and n^N

Number of basis functions n	Number of electrons N	Total spin S	a	b	c	$D(a, b, c)$	n^N
8	4	0	2	0	6	336	4096
10	10	0	5	0	5	19 404	10^{10}
50	4	0	2	0	48	520 625	6.25×10^7

Typical values of D together with values of the dimension of the full space are given in Table 5.4 for some typical cases.

As an example of the use of Weyl tableaux to label orbital product functions, the Weyl tableaux which label the basis functions for a 5-electron system with a total spin of $\frac{1}{2}$ are given in Fig. 5.22 for the case where 5 basis functions are being used. The tableaux are displayed in

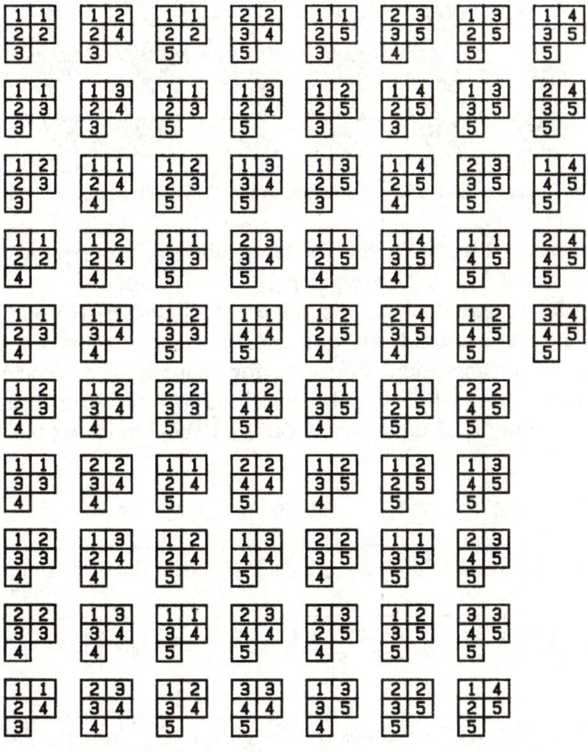

FIG. 5.22. Weyl tableaux which label the orbital product function for a five electron system with a total spin of $\frac{1}{2}$.

FIG. 5.23. Gelfand tableaux corresponding to the Weyl tableaux given in Fig. 5.22.

lexical order. In Fig. 5.23, the Gelfand tableaux corresponding to the Weyl tableaux of Fig. 5.22 are given.

It is useful to examine at this stage the characteristics of the Weyl tableaux and the Gelfand tableaux as electrons are removed from the system under consideration. Consider, for example, a 5-electron system with a total spin of $\frac{1}{2}$ which is to be described by 4 basis functions. In Fig. 5.24, the generation of the lexically ordered Weyl tableaux is illustrated.

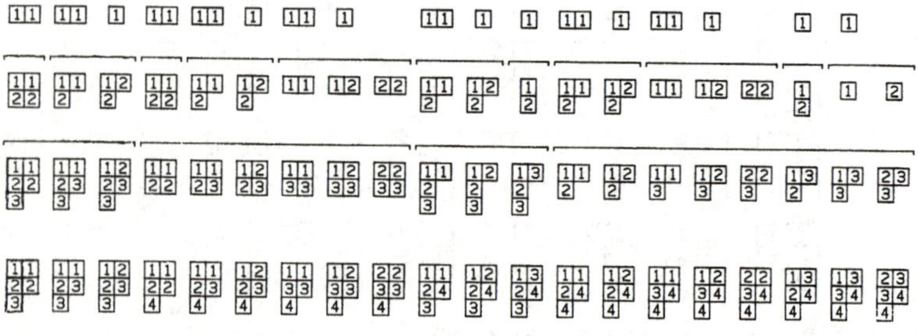

FIG. 5.24. Generation of canonical basis functions using Weyl tableaux.

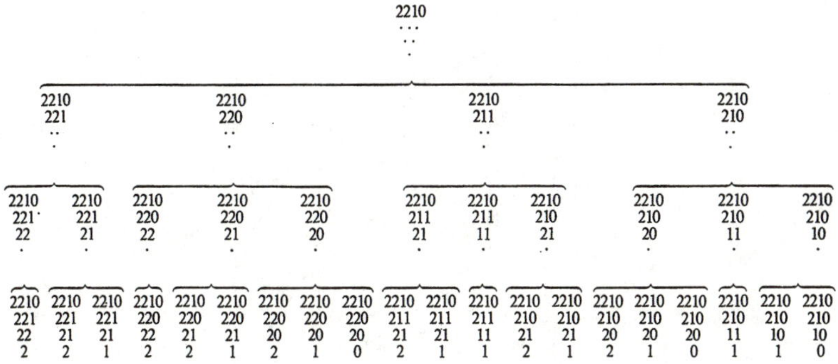

FIG. 5.25. Gelfand tableaux scheme for the example considered in Fig. 5.24.

The Gelfand labelling scheme for the same example is displayed in Fig. 5.25.

A number of different schemes have been developed (see, for example, Paldus 1981) for economically labelling the states defined by Weyl tableaux and Gelfand tableaux. In the *ABC* tableaux the number of 2's, 1's, and 0's in each row of the Gelfand tableau is recorded. This scheme and other schemes are illustrated in Fig. 5.26. In the ΔAC tableaux, the change in the number of 2's and the number of 0's on passing from one row to the next of the *ABC* tableaux is recorded. The $\Delta A\bar{C}$ tableaux are obtained from the ΔAC tableaux by replacing the ΔC values by their binary complement, i.e. $\Delta \bar{C}_k = 1 - \Delta C_k$. Finally, if the entries in the $\Delta A\bar{C}$ tableaux are read as binary numbers, the so-called case or step (Shavitt

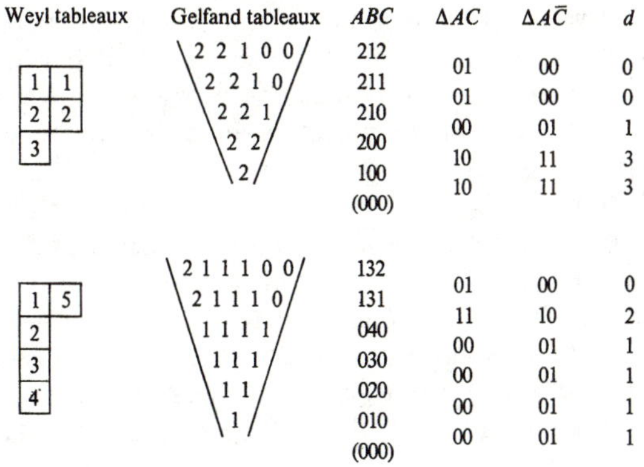
Weyl tableaux
Gelfand tableaux
ABC
ΔAC
ΔAC̄
d
1 1 / 2 2 / 3
2 2 1 0 0 / 2 2 1 0 / 2 2 1 / 2 2 / 2
212 211 210 200 100 (000)
01 01 00 10 10
00 00 01 11 11
0 0 1 3 3
1 5 / 2 / 3 / 4
2 1 1 1 0 0 / 2 1 1 1 0 / 1 1 1 1 / 1 1 1 / 1 1 / 1
132 131 040 030 020 010 (000)
01 11 00 00 00 00
00 10 01 01 01 01
0 2 1 1 1 1

FIG. 5.26. Illustration of the different labelling schemes for states defined by Weyl tableaux.

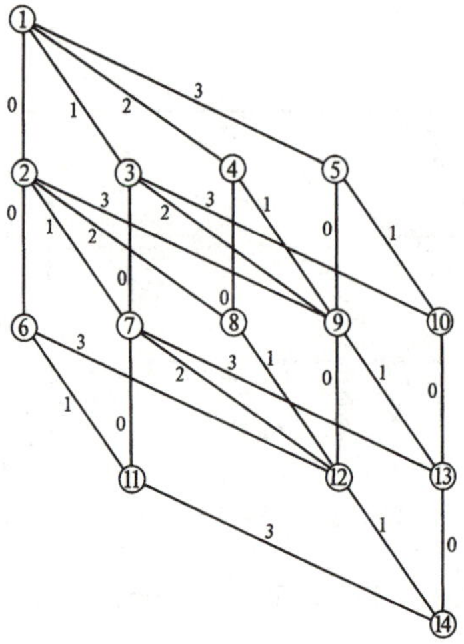

FIG. 5.27. Illustration of Shavitt's Graphical Unitary Group Approach for $n=4$, $N=3$, $S=\frac{1}{2}$. The vertices represent distinct rows and the edges represent case numbers.

1977*a*,*b*) numbers are obtained. These case numbers provide a unique label for the basis of the irreducible representation of $U(n)$. The case numbers have been used in the graphical unitary group approach to the configuration interaction method developed by Shavitt (1977*a*,*b*, 1978, 1979, 1980, 1981) and others (Brooks and Schaefer 1979; Siegbahn 1979, 1980). In this method it is only necessary to store the distinct rows of the *ABC* tableaux and the case numbers interrelating them. Shavitt developed a graphical representation of this procedure which is illustrated in Fig. 5.27. The use of the graphical unitary group approach to the configuration interaction problem will be discussed further in Section 7.2.

The graphical methods of spin algebras provide a powerful technique for the direct calculation of the matrix elements of the generators of the unitary group and, more importantly, matrix elements of products of generators. Paldus and Boyle (1980*a*,*b*) observe that "graphical methods of spin algebras directly suggest an optimal factorization of the spin diagrams and consequently the most efficient algorithm for matrix element evaluation, which may be difficult to find using algebraic methods based on $U(n)$ representation theory". Furthermore, it can also be used to develop a particle–hole formalism which is essential for methods based on the linked diagram theorem; this will be discussed further in Section 5.11.

The use of the graphical methods of spin algebras in the evaluation of

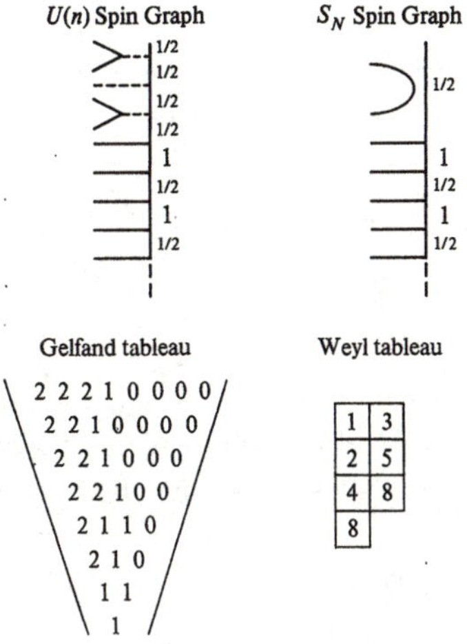

FIG. 5.28.

matrix elements for the unitary group rest on the equivalence (within a phase factor) of the Gelfand–Tsetlin and the Yamanouchi–Kotani states. Paldus has shown that it is useful in using graphical methods for the unitary group to introduce $U(n)$ spin graphs in which zero spins are represented explicitly by dashed lines. A typical $U(n)$ spin graph is shown in Fig. 5.28 together with other graphical and algebraic representations of the same state.

The matrix element for a single generator of the unitary group is equivalent to a cyclic permutation $(i, i+1, \ldots, j)$. Thus, for example, the matrix element for a generator E_{ij} $(i>j)$ is associated with graphs of one of the following forms

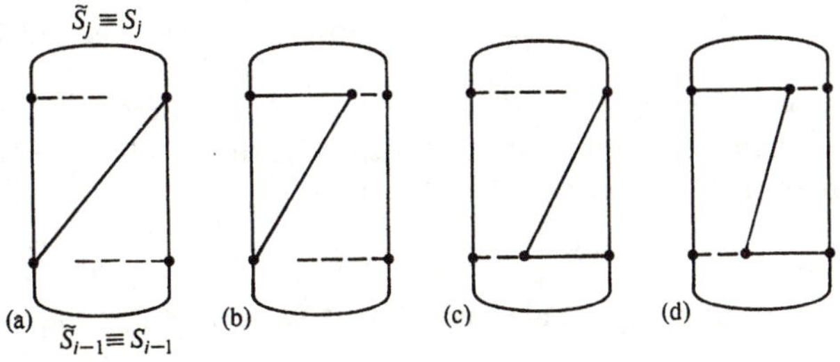

(5.10.11)

which can be shown to be a product of 6-j symbols.

For products of elementary generators, such as $E_{ij}E_{kl}$, the spin graph

will have the forms

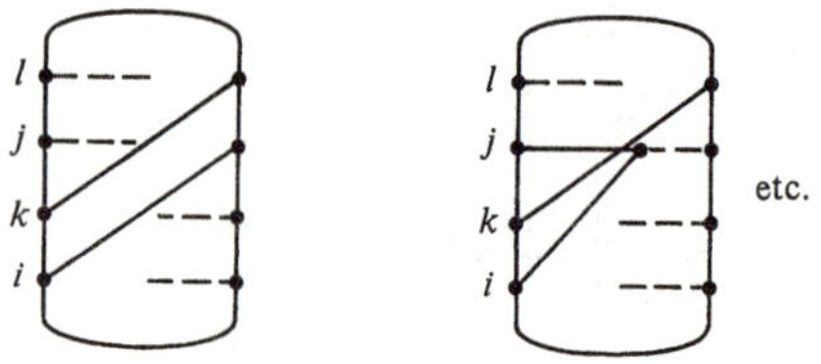

(5.10.12)

which should be compared with (5.6.29). This is a $3n$-j symbol and can be written as a sum of products of n 6-j symbols. It is likely that the graphical methods of spin algebras will prove useful in the determination of matrix elements of products of arbitrary numbers of generators.

5.11 Particle–hole formalism

It was demonstrated in Section 4.4 that the particle–hole formalism is most useful in studies on molecular electronic structure. Indeed, the use of the particle–hole formalism becomes essential in applications to extended systems and in methods based on diagrammatic perturbation theory or cluster expansions.

Flores and Moshinsky (1967) first showed how a particle–hole unitary group approach could be useful for the nuclear shell model problem. More recently, Paldus and Boyle (1980*a*,*b*) have demonstrated that the graphical methods of spin algebras can be used to obtain a powerful particle–hole unitary group formalism for the study of molecular electronic structure.

Consider an N-electron system and $2n$ associated spin orbitals. The relevant unitary group is then $U(2n)$. If the system is described by n_p particle states and n_h hole states then the particle and hole subspaces can be separated using the subduction

$$U(2n) \equiv U(2n_p + 2n_h) \supset U(2n_p) \otimes U(2n_h) \tag{5.11.1}$$

where $\otimes$ denotes the direct product $U(k+l) \supset U(k) \otimes U(l)$. Gelfand–Tsetlin and Yamanouchi–Kotani states can then be separately constructed in each of these two subspaces, using the chain

$$U(2m) \supset U(m) \otimes U(2) \tag{5.11.2}$$

in which $U(m)$ is the orbital unitary group and $U(2)$ the spin group. This leads to the subduction

$$\begin{aligned} U(2n_p) \otimes U(2n_h) &\supset [U(n_p) \otimes U(2)] \otimes [U(n_h) \otimes U(2)] \\ &\approx [U(n_p) \otimes U(n_h)] \otimes [U(2) \otimes U(2)] \end{aligned} \tag{5.11.3}$$

where $\approx$ denotes isomorphism, and the direct product $U(kl) \supset U(k) \otimes U(l)$ is used. Noting that $U(2) \boxtimes U(2) \approx U(2)$, where $\boxtimes$ denotes an inner direct product, allows the following subduction

$$[U(n_p) \otimes U(n_h)] \otimes [U(2) \otimes U(2)] \supset [U(n_p) \otimes U(n_h)] \otimes U(2) \quad (5.11.4)$$

which may be interpreted as the coupling of the total hole spin S_h and the total particle spin S_p to the final spin S.

The irreducible representations of $U(n_p) \otimes U(n_h)$ may be labelled by the highest weights

$$[2^{a_p} 1^{b_p} 0^{c_p} \mid 0^{c_h} -1^{b_h} -2^{a_h}] \equiv \{a_p b_p \mid b_h a_p\} \qquad \begin{array}{l} a_p + b_p + c_p = n_p \\ a_h + b_h + c_h = n_h \end{array} \quad (5.11.5)$$

where the positive numbers correspond to $U(n_p)$ and the negative numbeis to $U(n_h)$. In terms of Young diagrams, the hole component of the label for the irreducible representation is represented by an anti-block (Flores and Moshinsky 1967). A typical Young diagram labelling an irreducible representation of $U(n_p) \otimes U(n_h)$ is shown in Fig. 5.29.

The number of particles N_p and the number of holes N_h are given by

$$\begin{aligned} N_p &= 2a_p + b_p \\ N_h &= 2a_h + b_h \end{aligned} \quad (5.11.6)$$

respectively. For particular values of n_p, n_h, S, and the particle–hole deficiency $N_0 = N_p - N_h$, the irreducible representations for a given excitation level, as determined by N_h, with

$$\max(0, -N_0) \leqslant N_h \leqslant \min(2n_h, 2n_p - N_0) \quad (5.11.7)$$

are labelled, using the notation defined in equation (5.11.5), by

$$\{a_p, N_0 + N_h - 2a_p \mid N_h - a_h, a_h\} \quad (5.11.8)$$

with

$$\max(0, N_h - n_h) \leqslant a_h \leqslant [\tfrac{1}{2} N_h] \quad (5.11.9)$$

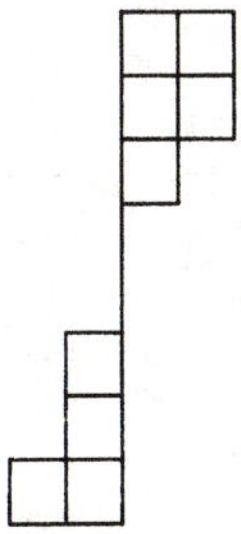

FIG. 5.29.

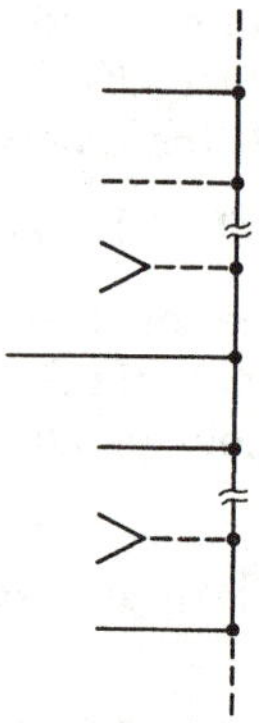

FIG. 5.30.

and

$$\max(0, N_p - n_p, \tfrac{1}{2}N_0 - S + a_h)$$
$$\leq a_p \leq \min([\tfrac{1}{2}N_p], \tfrac{1}{2}N_0 - S + N_h - a_h, \tfrac{1}{2}N_0 + S + a_h) \quad (5.11.10)$$

in which $[M]$ denotes the integral part of M.

The basis, for an irreducible representation of $U(n_p) \otimes U(n_h)$, may be labelled by Weyl tableaux which can be generated from the Young diagram of the type shown in Fig. 5.29 used to label the irreducible representation. The labelling scheme is entirely analogous to that employed in the particle case described in Section 5.8.

The graphical methods of spin algebras proves to be invaluable in the formulation of the particle–hole version of the unitary group approach. If S_p^i denotes the spin obtained after coupling the first i of the particle spin functions and S_h^i the spin obtained after coupling the first i of the hole spin functions and, furthermore, the total particle (hole) spin functions, denoted by S_p (S_h) are coupled to give a total spin S, then a typical graph has the form shown in Fig. 5.30. The particle states have been placed in the lower part of the graph and the hole states in the upper part. Using the rules of the graphical methods of spin algebras presented in Section 5.4, matrix elements of the generators of the unitary group can be evaluated in exactly the same manner as in the particle case and an optimal factorization obtained. The matrix elements of all generators may again be easily expressed as a product of 'segment' contributions. Further details will be found in the work of Paldus and Boyle (1980*a*,*b*).

6
THE ALGEBRAIC APPROXIMATION

6.1 Truncation errors in molecular calculations

In calculations of the electronic structure of atoms, the spherical symmetry of the system allows the problem to be factorized into a radial part and an angular part (see, for example Hartree 1957). The angular part can be treated analytically and the radial part involves a one-dimensional numerical integration. For example, in solving the Hartree–Fock equations for an atom, the radial wave function is obtained by solution of a one-dimensional integro-differential equation which can be handled efficiently by numerical techniques (see, for example Fischer 1977).

Some progress can be made in the numerical treatment of diatomic molecules by making a suitable choice of coordinate system (see, for example, Christiansen and McCullough 1977). However, for an arbitrary polyatomic molecule, it is generally impossible to factorize the molecular field, and the numerical approach does not appear to be tractable.

By expanding molecular orbitals ϕ_i in terms of a finite basis set, i.e.

$$\phi_i = \sum_j \chi_j c_{ij} = \boldsymbol{\chi}\mathbf{c}_i^\dagger \tag{6.1.1}$$

where

$$\boldsymbol{\chi} = (\chi_1\, \chi_2 \ldots) \qquad \mathbf{c}_i = (c_{i1}\, c_{i2} \ldots) \tag{6.1.2}$$

the integro-differential Hartree–Fock equations become a set of algebraic equations for the expansion coefficients c_i. This algebraic approximation is fundamental to most applications of the methods of quantum mechanics to molecular electronic structure. The algebraic approximation is, of course, independent of the Hartree–Fock ansatz and has to be invoked in molecular applications of, for example, geminal models or configuration interaction.

In calculations of molecular correlation effects there are two truncation errors:

(i) That arising from the use of a restricted finite basis set, and
(ii) That resulting from the truncation of the expansion for the correlation energy, correlated wave function, or expectation value.

It is important, particularly in calculations which are not variationally bound, that apparently good results are not obtained by a fortuitous cancellation of these two errors.

Except in the case of very small systems, truncation of the basis set is almost always the largest source of error in molecular calculations and also in most atomic calculations which employ the algebraic approximation. No amount of configuration mixing or summation of high-order terms in perturbation expansions will compensate for an inadequate basis set. The results obtained by performing a full configuration mixing calculation, within the Nth rank direct product space generated by a given basis set, will only approach the complete configuration mixing result as the basis set approaches completeness. The choice of the basis set in studies of atomic and molecular systems ultimately determines the accuracy of the calculation.

It should be noted, however, that in calculations on unstable species within the algebraic approximation the possibility of spurious solutions exists. For example a 'negative ion', which is in fact unstable with respect to the neutral species and a free electron, may appear to be stable in a calculation using a finite basis set. The most important possibility for error when using the algebraic approximation lies in the assumption that the solution exists.

In this chapter, the error in molecular calculations which arises from the use of a restricted basis set is considered in some detail. The errors resulting from the truncation of the expansion for the correlation correction to the energy and other expectation values will be considered in Chapter 7.

The magnitude of the basis set truncation error in typical present-day calculations is described in Section 6.2. In Section 6.3, the nature of the algebraic approximation is discussed. The use of one-centre expansion methods is reviewed in Section 6.4 while the more commonly employed multicentre expansion procedure is considered in Section 6.5. In Section 6.6, the various types of basis functions which are employed in molecular calculations are described, and the advantages and disadvantages of different basis functions are presented. Attention is then turned to the problem of constructing a basis set for energy calculations and molecular property calculations both at the Hartree–Fock level of accuracy and in correlated treatments. The use of even-tempered basis set is reviewed in Section 6.8. Universal basis sets are discussed in Section 6.9; in Section 6.10, the use of systematic sequences of basis sets is described. Finally, in Section 6.11, the convergence of the harmonic expansion in calculations which take account of correlation effects is discussed.

6.2 Basis set truncation errors

Ideally, when performing *ab initio* calculations of atomic and molecular properties it would be preferable to determine rigorous upper and lower bounds (see, for example, Löwdin 1965; Weinhold 1972). The calculation of lower bounds to energy expectation values is particularly difficult and has attracted much attention (cf., for example, Weinstein 1934; Stevenson 1938; Bazley and Fox 1966; Gordon 1968; Blau, Rau, and Spruch 1973). Bunge (1980) summarizes the situation for the calculation of rigorous bounds in atomic studies as follows: 'A few significant calculations have apparently exhausted the present possibilities of current theories .... In general, today's rigorous error bounds are too large to be of much use ...'. The situation with respect to the calculation of rigorous error bounds for molecules is even less satisfactory than for atoms.

The vast majority of contemporary *ab initio* calculations adopt what may be termed a pragmatic approach; that is, no error bounds are determined and the accuracy of the calculation is assessed by comparison with quantities derived from experiment. By means of this procedure, the quality of a particular basis set is established and an empirical range of validity for the basis set thus obtained which can then be used to make educated guesses of the accuracy of a computed molecular property. The art of selecting a basis set is based on previous experience in treating similar systems using basis sets of comparable quality. The problem of selecting a basis set for a particular application will be discussed in some detail in Section 6.7.

There can be little doubt that the error attributable to the truncation of the basis set is the largest source of error in most accurate calculations of electron correlation energies and correlation effects. This is illustrated in Table 6.1 where the results of electron correlation energy calculations for the hydrogen fluoride, nitrogen, and carbon monoxide molecules are

Table 6.1

Relative importance of basis set truncation and higher-order correlation effects[a]

	Correlation energies[b]				
	Basis set A[c]			Basis set B[c]	
Molecule	$E(2)$	$E(3)$	$E(4)$	$E(2)$	$E(3)$
FH	−0.2206	−0.2207	−0.2276	−0.3159	−0.3135
N_2	−0.3291	−0.3193	−0.3486	−0.4470	−0.4447
CO	−0.3076	−0.3028	−0.3312	−0.4274	−0.4294

[a] based on the work of Krishnan, Binkley, Seeger, and Pople 1980, Wilson and Silver 1980, Wilson and Guest 1981.
[b] calculated by diagrammatic many-body perturbation theory; $E(i)$ denotes the correlation energy through ith order.
[c] basis set A, that of Krishnan, Binkley, Seeger, and Pople 1980, is smaller than basis set B, which is that of Wilson and Silver 1980.

Table 6.2

Matrix Hartree–Fock energy and correlation energy of the F^- ion obtained by using different basis sets[a]

Basis set	E_{MHF}	E_{corr}
{13s, 8p, 1d/7s, 4p, 1d}	−99.458 24	−0.251 88
{14s, 9p, 1d/8s, 5p, 1d}	−99.458 44	−0.252 22
{14s, 9p, 2d/8s, 5p, 2d}	−99.458 61	−0.279 61

[a] based on the work of Wilson and Sadlej 1981. The smallest basis set is taken from the work of Kistenmacher, Popkie, and Clementi 1972 and consists of 13s, 8p, and 1d Gaussian-type functions in the contraction 7s, 4p, 1d. The correlation energy was calculated using diagrammatic many-body perturbation theory.

presented. It is demonstrated in Table 6.1 that the use of a larger basis set is often more important than the evaluation of higher-order correlation energy terms. The calculation of electron correlation energies for negative ions and for Rydberg states is particularly sensitive to the quality of the basis set (Simons 1977). This is illustrated in Table 6.2 where correlation energies for the F^- ion using three different basis sets are displayed.

The basis set truncation error is often particularly significant in the calculation of molecular properties other than the energy. The variation of the dipole moment of the water molecule within the matrix Hartree–Fock model with increasing size of basis set is illustrated in Table 6.3. Some properties depend on the quality of the basis set in particular regions of space. For example, calculated polarizabilities can be very sensitive to the choice of basis set as the results given in Table 6.4 for the F^- ion illustrate. Accurate polarizabilities can be computed by including basis functions capable of describing deformations of the electronic charge cloud on applying an electric field. Diffuse basis functions should be included in polarizability calculations to obtain an accurate description

Table 6.3

Dipole moment of the ground state of the water molecule within the matrix Hartree–Fock approximation using different basis sets[a]

Basis set	Dipole moment	Total energy
{5s 4p/3s}	1.0672	−76.0207
{5s 4p/3s 1p}	0.9012	−76.0471
{5s 4p 1d/3s 1p}	0.8576	−76.0559
{5s 4p 2d/3s 2p}	0.7994	−76.0604

[a] the dipole moment and energy are given in atomic units; basis sets consist of contracted Gaussian functions.

Table 6.4
Polarizability of the F^- ion obtained using different basis sets[a]

Basis set	α_{MHF}	α_{corr}
{13s, 8p, 1d/7s, 4p, 1d}	3.25	0.51
{14s, 9p, 1d/8s, 5p, 1d}	4.65	1.29
{14s, 9p, 2d/8s, 5p, 2d}	6.05	1.12

[a] based on the work of Wilson and Sadlej 1981. The smallest basis set is taken from the work of Kistenmacher, Popkie, and Clementi 1972 and consists of 13s, 8p, and 1d Gaussian-type functions in the contraction 7s, 4p, 1d. Polarizabilities were calculated by the finite-field approach and correlation effects were treated by many-body perturbation theory.

of the wave function in regions of space distant from the nuclei. On the other hand, in the calculation of properties such as, for example, spin–orbit coupling constants, it is necessary to include basis functions capable of providing an accurate description of the wave function close to the nucleus (Richards, Trivedi, and Cooper 1981).

An interesting point arises in the calculation of magnetic properties of molecules. The magnetic vector potential **A** can be defined in an infinite number of ways since it can be subjected to a gauge-transformation

$$\mathbf{A}' = \mathbf{A} + \nabla\phi \tag{6.2.1}$$

where ϕ is any scalar function which can be differentiated twice. A change of origin is a gauge transformation and thus **A** is origin dependent. Physical observables are clearly not dependent on the choice of origin; they are gauge-invariant. For a calculation which employed a complete basis set, calculated magnetic constants would not depend on either the choice of the magnetic vector potential **A** or the choice of origin. However, in practice a finite and, therefore, incomplete basis set has to be employed and this yields calculated magnetic properties which are not gauge-invariant. The extent to which calculated magnetic properties are gauge-invariant is an indication of the quality of the calculation and, in particular, the quality of the basis set.

6.3 The algebraic approximation

The determination of the electronic structure of atoms and molecules involves the determination of an appropriate eigenvalue and associated eigenfunction of a semi-bounded, self-adjoint hamiltonian operator $\hat{\mathscr{H}}$ in Hilbert space $\mathscr{h}$. A tractable scheme for solving such equations is the algebraic approximation in which eigenfunctions are parameterized by expansion in a finite set of functions and a set of algebraic equations are obtained for the expansion coefficients.

The algebraic approximation results in the restriction of the domain of the operator $\hat{\mathcal{H}}$ to a finite dimensional subspace S of the Hilbert space $\mathscr{h}$. In most applications of quantum mechanics to molecules, the N-electron wave function is expressed in terms of the Nth rank direct product space V^N generated by a finite dimensional single particle space V^1, i.e.

$$V^N = V^1 \otimes V^1 \otimes \ldots \otimes V^1. \tag{6.3.1}$$

The algebraic approximation may be implemented by defining a suitable orthonormal basis set of M ($>N$) one-electron spin-orbitals and constructing all unique N-electron determinants using the M one-electron functions. The number of unique determinants which can be formed in this way is

$$\eta = \binom{M}{N} \tag{6.3.2}$$

and η is the dimension of the subspace S spanned by the set of determinants. In the algebraic approximation the domain of $\hat{\mathcal{H}}$ is restricted to the η-dimensional subspace S.

Within the algebraic approximation, the electronic Schrödinger equation may, in principle, be solved by the method of configuration mixing. The wave function is then expressed as a superposition of configurations Φ_i with linear coefficients C_i i.e.

$$\Psi = \sum_i \Phi_i C_i \tag{6.3.3}$$

where it is assumed that the Φ_i are associated with a subspace labelled by particular S and M spin quantum numbers. The energy functional is then

$$\mathscr{E}(\mathbf{C}) = \left(\sum_{ij} C_i^* \langle \Phi_i | \hat{\mathcal{H}} | \Phi_j \rangle C_j\right) \Big/ \left(\sum C_i^* \langle \Phi_i \mid \Phi_j \rangle C_j\right) \tag{6.3.4}$$

where $\hat{\mathcal{H}}$ is the molecular hamiltonian. The stationary values of $\mathscr{E}(\mathbf{C})$ with respect to variation of the coefficients C are obtained by solving the secular equations

$$\begin{aligned} &\langle \Phi_1 | \hat{\mathcal{H}} | \Phi_1 \rangle - \mathscr{E} \langle \Phi_1 \mid \Phi_1 \rangle C_1 + \langle \Phi_1 | \hat{\mathcal{H}} | \Phi_2 \rangle C_2 + \ldots = 0 \\ &\langle \Phi_2 | \hat{\mathcal{H}} | \Phi_1 \rangle - \mathscr{E} \langle \Phi_2 \mid \Phi_1 \rangle C_1 + \langle \Phi_2 | \hat{\mathcal{H}} | \Phi_2 \rangle C_2 + \ldots = 0 \\ &\ldots \end{aligned} \tag{6.3.5}$$

If the stationary values of $\mathscr{E}$ are denoted by $E_0 < E_1 < E_2 \ldots$ then, for states of a given symmetry, every E_i is an upper bound to the corresponding exact eigenvalue of $\hat{\mathcal{H}}$ such that

$$E_{i+1}^{\text{exact}} > E_i \geqslant E_i^{\text{exact}}. \tag{6.3.6}$$

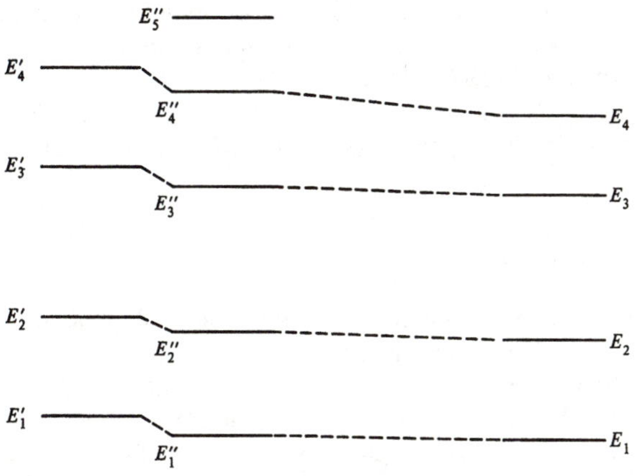

FIG. 6.1. The separation theorem.

This 'separation theorem', due to Hylleraas and Undheim (1930) and MacDonald (1933), is illustrated in Fig. 6.1. As a configuration mixing expansion is extended, the desired energy levels are approached monotonically from above.

In practice, difficulties arise in setting up and solving secular equations of high order and only a small subset of the configurations Φ_i is usually employed in the expansion. The use of the configuration-mixing approach in actual computations will be discussed further in Chapter 7.

The 'separation theorem' (6.3.6) is clearly of theoretical importance in all types of molecular electronic structure calculations. For example, diagrammatic many-body perturbation theory may also be formulated within the algebraic approximation. The domain of the perturbation theory operators is then restricted to the η-dimensional space spanned by the Φ_i and consequently the perturbation theory wave function is generated in terms of an η-dimensional representation. The results of many-body perturbation theory calculations when carried through infinite order are identical to those of full configuration mixing if the same basis set, i.e. subspace S, is employed in both calculations. Full configuration-mixing calculations (see, for example, Bunge 1976; Saxe, Schaefer, and Handy 1980) therefore provide an extremely useful benchmark for assessing other techniques even though the basis set has to be very small in order to keep the computation tractable.

Once the algebraic approximation has been invoked, there is essentially no difference between the atomic problem and the molecular problem except that the multicentre integrals which arise in the latter case are more difficult to evaluate. The evaluation of integrals over typical types of basis functions will be discussed briefly in Section 6.6.

Klahn and Bingel (1977*a,b*) have investigated the problem of establishing the convergence of a complete configuration mixing calculation using a basis set of size N to the exact result as N tends to infinity.

6.4 One-centre expansions

In principle, any molecular wave function can be expanded in terms of a complete set of functions centred at any convenient point in space. For example, the basis functions, $\boldsymbol{\chi}$, in the expansion (6.1.1) of the molecular orbital ϕ_i could all be taken to be centred at the same point in space. This approach is referred to as the one-centre method, the central-field approximation, or the united atom method. This method goes back to the earliest days of molecular physics (Coolidge 1932). Bishop (1967) has reviewed the one-centre method and a typical application is given by Hayes and Parr (1967).

Many more terms have to be included in a one-centre basis set than in the more usual multicentre basis set in order to obtain an adequate representation of the wave function close to off-centre nuclei. However, all of the integrals over the basis functions which arise in the one-centre method can be evaluated analytically. It may be more convenient to recompute the integrals arising in a one-centre calculation each time they are required and thus avoid the necessity of handling large files of integrals.

The basis functions in a one-centre calculation belong to representations of the spherical symmetry group. There is a higher degree of symmetry in the integrals over the basis functions than in the integrals over the molecular orbitals. This can be exploited in the self-consistent-field calculations and in orbital transformations.

For a one-centre basis set, unlike multicentre basis sets problems arising from overcompleteness can usually be controlled if not avoided. The one-centre approach provides control over the convergence of a calculation with respect to the size of the basis set and control over linear dependence. Furthermore, because of the ease with which integrals involving one-centre functions can be handled, the method can be used to explore the use of alternative types of basis functions. The one-centre method is ideally suited to the calculation of energy derivative with respect to the nuclear coordinates.

The one-centre method does, however, usually provide a poor representation of off-centre nuclei. Most applications have been to hydrides with a single heavy atom which is used as an expansion centre.

An interesting development of the one-centre method has been proposed by Ladik and Čížek (1980). They suggest that the expression for the total energy of a molecule which was introduced in Section 3.3, i.e.

$$E = E_{\text{ref}} + E_{\text{corr}} = \langle\Phi|\,\hat{\mathcal{H}}\,|\Phi_{\text{I}}\rangle + \langle\Phi_{\text{II}}|\,\hat{\mathcal{H}}\,|\chi\rangle \qquad (3.3.4)$$

be modified to give

$$E = \langle\Phi_{\text{I}}|\,\hat{\mathcal{H}}\,|\Phi_1\rangle + \langle\phi_{\text{II}}|\,\hat{\mathcal{H}}\,|\chi_{\text{II}}\rangle \qquad (6.4.1)$$

where the subscripts I and II denote that multicentre and one-centre basis sets are used, respectively. Thus, for example, a multicentre basis set is employed in the Hartree–Fock calculation whilst a one-centre expansion is used in the calculation of correlation effects. It is probable that this approach will prove useful for treating valence correlation effects but that atom-centred functions will be required to obtain accurate descriptions of inner shell correlation.

6.5 Multicentre expansions

The vast majority of molecular calculations overcome the convergence problems of the one-centre expansion method, which are associated with the description of off-centre nuclei, by using basis functions located on a number of centres. These centres are often, but not always, chosen to coincide with the nuclei in the molecule. Accurate calculations for molecules containing more than one non-hydrogenic nucleus can be performed by using a multicentre basis set. The use of such basis sets does, however, give rise to some new problems.

The evaluation of integrals, particularly two-electron integrals involving the interelectronic distance, over basis functions on different centres is usually more complicated than the evaluation of one-centre integrals. One-centre integrals can usually be handled analytically in terms of spherical polar coordinates whilst two-centre integrals can be treated in terms of a spheroidal coordinate system. However, three-centre and four-centre, two-electron integrals are difficult particularly when basis sets of exponential-type functions are used (see, for example, Steinborn 1983 and references therein, Saunders 1983 and references therein).

The use of multicentre basis sets often leads to overestimates of interaction energies between molecules. That is mainly attributable to the superposition error or counterpoise error (Clementi 1967; Boys and Bernardi 1970; Johansson, Kollman, and Rotherberg 1973; Liu and McLean 1973; Urban and Hobza 1975; Ostlund and Merrifield 1976; Dacre 1977; Price and Stone 1979). This error is particularly significant when small basis sets are used. For a system XY the description of the subsystem X is improved by the basis functions centred on Y and vice versa. The magnitude of the superposition error can be estimated by performing a calculation on X (Y) using the basis set employed in the study of XY, i.e. the nuclear charges in the fragment Y (X) are set to zero. A superposition correction of this type is made in most accurate calculations of intermolecular energies. However, it should be noted that it is not exact.

6.6 Basis functions

When implementing the algebraic approximation in molecular calculations, there is a considerable degree of freedom in the choice of basis functions used. Any set of functions which can be shown to form a complete set can be employed. The choice of functional form for the basis functions is mainly influenced by two factors:

(i) The rate of convergence of the molecular orbitals when expanded in the basis set.
(ii) The ease with which integrals over the basis functions can be evaluated and, in particular, the ease with which the three-centre and four-centre, two-electron integrals can be computed.

Clearly, these two factors are not unrelated in that if a set of basis functions provides a rapidly convergent representation of a molecular orbital, the number of integrals required is smaller than it would be if the basis functions formed a poorly convergent expansion. The number of two-electron integrals which arise in calculations within the algebraic approximation increases as $\frac{1}{2}m(m+1)$ where $m=\frac{1}{2}n(n+1)$ and n is the number of basis functions. This $\sim n^4$ dependence of the number of integrals is reduced to an n^2 dependence on calculations on large molecules (Ahlrichs 1974) and in calculations using very large basis sets.

The most frequently used basis functions in molecular calculations are exponential-type (or Slater-type functions) and Gaussian-type functions. No one choice of basis function has yet shown overall superiority.

Exponential-type functions are usually taken to have the form

$$\chi_{nlm}(r, \theta, \phi) = N_n r^{n-1} e^{-\zeta r} Y_{lm}(\theta, \phi) \tag{6.6.1}$$

in which the normalization constant is given by

$$N_n = \frac{(2\zeta)^{n+\frac{1}{2}}}{\sqrt{(2n)!}} \tag{6.6.2}$$

and Y_{lm} is a spherical harmonic. These orbitals satisfy the eigenvalue equation

$$(-\tfrac{1}{2}\nabla^2 + V(r))\chi = \epsilon\chi \tag{6.6.3}$$

with the effective potential

$$V(r)=\frac{\zeta n}{r}+\frac{n(n-1)-l(l+1)}{2r^2}. \tag{6.6.4}$$

Exponential-type functions usually provide a rapidly convergent expansion for molecular orbitals. They lead to a good representation of the wave function in regions of space close to the nucleus (they possess a cusp at the nucleus) and in the long-range region. Unfortunately, multicentre integrals over exponential-type basis functions are difficult to compute. Not only is the evaluation of these integrals time-consuming but also the accuracy to which they are usually computed for non-linear, polyatomic systems is not as high as it is when, for example, Gaussian-type orbitals are used.

Two forms of Gaussian-type functions are in common use. Cartesian Gaussian-type functions have the form

$$\chi_{pqr}(x, y, z)=N_{pqr}x^p y^q z^r e^{-\zeta r^2} \tag{6.6.5}$$

with normalization constant

$$N_{pqr}=\left\{\left(\frac{\pi}{2\zeta}\right)^{\frac{3}{2}}\frac{(2p-1)!!\,(2q-1)!!\,(2r-1)!!}{2^{2(p+q+r)}\zeta^{(p+q+r)}}\right\}. \tag{6.6.6}$$

Use of the functions (6.6.5) leads to six components of d-symmetry instead of the true five components. This is equivalent to the addition of a 3s-function which can lead to numerical problems if the s basis set is sufficiently large. The use of spherical Gaussian-type basis functions

$$\chi_{nlm}(r, \theta, \phi)=N_n r^{n-1}e^{-\zeta r^2}Y_{lm}(\theta, \phi) \tag{6.6.7}$$

avoids this problem. In eqn (6.6.6) the normalization constant is given by

$$N_n=\left[\frac{2^{2n}(n-1)!}{(2n-1)!}\sqrt{\left\{\frac{(2\zeta)^{2n+1}}{\pi}\right\}}\right]^{\frac{1}{2}}. \tag{6.6.8}$$

Gaussian-type functions of the form (6.6.7) are eigenfunctions of eqn (6.6.3) with effective potential

$$V(r)=2\zeta^2 r^2+\frac{n(n-1)-l(l+1)}{2r^2}. \tag{6.6.9}$$

The strength of Gaussian-type basis sets lies in the fact that multicentre integrals can be easily evaluated since the product of two Gaussian-type functions on centres A and B, say, is a Gaussian, with appropriate normalization constant, with centre P on the line AB. Explicitly, for s-type functions

$$e^{-\zeta_A r_A^2}e^{-\zeta_B r_B^2}=Ke^{-\zeta_P r_P^2} \tag{6.6.10}$$

where the constant K is given by

$$K = \exp\left(\frac{-\zeta_A\zeta_B}{\zeta_A+\zeta_B}\right)R_{AB}^2 \tag{6.6.11}$$

and

$$R_{AB}^2 = (x_A - x_B)^2 + (y_A - y_B)^2 + (z_A - z_B)^2 \tag{6.6.12}$$

$$\zeta_P = \zeta_A + \zeta_B \tag{6.6.13}$$

$$x_P = \frac{\zeta_A x_A + \zeta_B x_B}{\zeta_A + \zeta_B}$$

$$y_P = \frac{\zeta_A y_A + \zeta_B y_B}{\zeta_A + \zeta_B} \tag{6.6.14}$$

$$z_P = \frac{\zeta_A z_A + \zeta_B z_B}{\xi_A + \xi_B}.$$

Unfortunately, Gaussian-type functions behave incorrectly both in regions close to nuclei (they have no cusp) and in the long-range region. Many more Gaussian-type functions than exponential-type functions are required to approximate a given molecular orbital to a certain accuracy. However, although Gaussian-type functions are not well suited for the description of eigenfunctions of hamiltonians involving Coulomb potentials and therefore relatively large numbers of basis functions are required, this disadvantage is often more than offset by the ease with which the integrals can be computed.

The efficiency of Gaussian basis sets can be increased significantly by employing contracted Gaussian-type functions instead of the primitive Gaussian-type functions defined in eqns (6.6.5–8). Contracted Gaussian-type basis functions are defined by

$$\phi = N \sum_i \chi_i C_i \tag{6.6.15}$$

where χ_i denotes a set of primitive functions of the same symmetry type and centred on the same nucleus, C_i is a set of fixed contraction coefficients, and N is a normalization constant. The contraction coefficient may be determined in a number of ways. They can be chosen so as to provide the best, in the least-squares sense, representation of an exponential-type function. They can be contracted on the basis of atomic matrix Hartree–Fock calculations. The problem of devising suitable contraction schemes has been summarized by van Duijneveldt (1971) as follows:

> In order to reduce the number of integrals that must be stored and handled in setting up the HF-matrix, and in order to reduce the size of this matrix, it is

> useful to contract the primitive basis of GTOs to some extent. Molecular calculations show that contraction of the innermost s- and p-GTOs can be carried through without any loss in accuracy. However, in contracting GTOs in the valence region one should be very careful. In many cases the gain in flexibility by *not* contracting the outer GTOs more than offsets the increase in computation time (which for an efficient program should be small anyway). An additional advantage of loosely contracted basis sets is that the free-atom exponents need no scaling when used in a molecular environment.

In a recent monograph, Čársky and Urban (1980) presented a detailed comparison of the advantages and disadvantages of using exponential-type function and Gaussian-type functions in molecular calculations. They conclude that "Gaussian and Slater basis sets of the same size and quality give comparable energies and other molecular properties". If a basis set is sufficiently large and flexible, different choices of basis function are not reflected in calculated molecular properties.

Some researchers have explored the use of alternative types of basis functions. For example, ellipsoidal functions are useful in the study of diatomic molecules (Ebbing 1963). Hall (1959) has discussed various possible basis functions including hydrogenic functions, Dirichlet functions, and atomic self-consistent-field functions. There has been interest in the use of mixed basis sets in which basis functions of differing types are employed (see, for example, Silver 1971; Le Clerque 1976).

6.7 Selection of basis sets

Contemporary atomic and molecular calculations almost invariably adopt a pragmatic approach to the selection of basis sets. The quality of a particular basis set is determined by means of numerical tests and experiments. The 'art' of quantum chemistry lies in selecting, on the basis of experience gained in previous studies, the type of basis set which is required to calculate the expectation value of a particular property to a certain accuracy.

A great deal of knowledge has been accumulated over the years on the ability of certain basis sets to provide a given accuracy within a given model. For example, within the matrix Hartree–Fock model for closed-shell systems, it is well known (see, for example, Green 1974) that a basis set of Slater or Gaussian functions of 'double-zeta' quality will yield bond lengths within ~0.01 Å of experimental values, at least for molecules containing first-row atoms. However, if a more extensive basis set is used, including functions to describe polarization effects, such bond lengths will differ from experimental values by 0.02–0.04 Å (see, for example, Green 1974). The fact that matrix Hartree–Fock calculations using double-zeta basis sets give good results for bond lengths is an empirical observation based on many calculations. It could be claimed that such results can be explained in terms of a cancellation between polarization effects and correlation effects but fundamentally there is no rigorous theoretical justification. This is not to imply that such empirical observations are not extremely useful. Indeed, the choice of the basis set can be regarded as a part of the concept of a theoretical model chemistry as described in Section 1.3. Once a particular method using a particular basis set has been applied to a range of problems and its accuracy and applicability established, it acquires a certain degree of predictive capability. This type of approach is widely used in applications of quantum mechanical techniques to molecules.

The choice of basis set for a particular application is, on the whole determined by three factors:

(i) The molecular property which is being calculated; for example, as mentioned above, a double-zeta basis set is suitable for the determination on bond lengths within the self-consistent-field approach whereas for calculations of polarizabilities more extended basis sets are required including diffuse functions to describe the distortion of the charge cloud by an applied electric field.
(ii) The accuracy required; for example, in electron correlation energy calculations large basis sets are required if a significant fraction of the correlation energy is to be recovered.
(iii) The cost of a calculation is closely related to the number of basis functions employed.

The evaluation of the integrals over the basis set depends on N^4, where N is the number of basis functions, although this dependence is reduced to N^2 for extended systems (Ahlrichs 1974). The construction of the Fock matrix in self-consistent-field procedures depends on N^4 which is again reduced to N^2 for extended systems. Diagonalization of the Fock matrix depends on N^3. The four-index orbital transformation, i.e. the transformation of the two-electron integrals over basis functions to integrals over molecular orbitals, depends on N^5 and is reduced to N^3 for extended systems.

A number of detailed reviews of basis sets suitable for a variety of applications have been given. Čársky and Urban (1980) review the use of both Slater basis sets and Gaussian basis sets in molecular calculations of many properties. For basis sets of Gaussian functions, particularly informative reviews have been given by Dunning and Hay (1977) and by Ahlrichs and Taylor (1981).

6.8 Even-tempered basis sets

Large basis sets can be efficiently generated by utilizing the concept of an even-tempered basis set. An even-tempered basis set consists of pure exponential or pure Gaussian functions multiplied by a real solid spherical harmonic. (A real solid spherical harmonic is a spherical harmonic $S_l^m(\theta, \phi)$ multiplied by r^l.) A set of even-tempered basis functions is thus defined by

$$\chi(k, l, m) = N_l(\zeta_k)\exp(-\zeta_k r^p) r^l S_l^m(\theta, \phi) \qquad k = 1, 2, \ldots \quad (6.8.1)$$

where $p = 1$ for exponential functions and $p = 2$ for Gaussian functions. Even-tempered atomic orbitals for a given $S_l^m(\theta, \phi)$ do not differ in the power of r, and thus in linear combinations of primitive functions the solid harmonics can be factored. The orbital exponents ζ_k are taken to form a geometric series

$$\zeta_k = \alpha\beta^k; \qquad k = 1, 2, \ldots. \quad (6.8.2)$$

The use of such a series is based on the observation that independent optimization of the exponents with respect to the energy in self-consistent-field calculations yields an almost linear plot of $\ln(\zeta_k)$ against k. The use of orbital exponents which form a geometric progression was originally advocated by Reeves (1963) and the idea was revived and extensively employed by Ruedenberg and his collaborators (Ruedenberg, Raffenetti, and Bardo 1973; Raffenetti 1973*a,b*; Bardo and Ruedenberg 1973; Raffenetti and Ruedenberg 1973; Bardo and Ruedenberg 1974).

A number of advantages accrue to the use of even-tempered basis sets:

(i) Even-tempered basis sets have only two parameters, α and β, which have to be determined for each group of atomic functions belonging to the same symmetry species as opposed to one optimizable orbital exponent per basis function. The determination of orbital exponents by energy minimization is a non-linear optimization problem and there is little possibility of performing a full optimization in polyatomic molecules if all orbital exponents are independent.
(ii) The further restriction of using the same exponents for all values of l, so that there are only two non-linear parameters, α and β, per atom does not produce a very large difference in the calculated energies.
(iii) The proper mixing of basis functions, in terms of principal quantum number, is superfluous since no mixing is used.
(iv) It is evident that the even-tempered basis set approaches a complete set in the limit $\alpha \to 0$, $\beta \to 1$, $\beta^{k_{max}} \to \infty$, and $k_{max} \to \infty$.
(v) An even-tempered basis set cannot become linearly dependent if β remains greater than unity.
(vi) By using an even-tempered basis set, control can be exercised over practical near-linear dependence. As the size of the basis set is increased, the determinant of the overlap matrix decreases and there comes a point where meaningful calculation is impossible with a given numerical word-length.

(vii) Even-tempered basis sets have a unique 'space covering' property. The value of the overlap integral between two normalized 1s-exponential functions on the same centre is

$$\left\{\frac{2\sqrt{(\zeta_i/\zeta_j)}}{1+(\zeta_i/\zeta_j)}\right\}^3 \tag{6.8.3}$$

while the overlap integral between two normalized 1s-Gaussian functions takes the form

$$\left\{\frac{2\sqrt{(\zeta_i/\zeta_j)}}{1+(\zeta_i/\zeta_j)}\right\}^{\frac{3}{2}}. \tag{6.8.4}$$

These integrals are functions of the ratio of the exponents, that is β. Therefore, a set of even-tempered functions of the same symmetry type has an overlap matrix whose elements are constant along diagonal lines.

(viii) Restriction of the basis functions to fewer analytic forms leads to simpler and more efficient integral evaluation procedures.

6.9 Universal basis sets

Computational restrictions have historically made it necessary to limit the number of basis functions used in molecular calculations to a reasonably small number of functions in order to keep the computations tractable. However, in order to achieve high accuracy, particularly in studies of electron correlation effects, moderately large basis sets are ultimately needed. The flexibility of a basis set generally increases as it is extended and, therefore, the need to optimize orbital exponents becomes less important. It is well established that it is almost always more profitable to add another function to a given basis set rather than exhaustively optimize the orbital exponents (see, for example, Cade and Huo 1967). These considerations have led to the concept of a universal basis set (Silver, Wilson, and Nieuwpoort 1978; Silver and Nieuwpoort 1978; Silver and Wilson 1978; Wilson and Silver 1979; Wilson and Silver 1980, 1982). Such a basis set is moderately large and thus has a considerable degree of flexibility. It is, therefore, transferable from system to system with little loss in accuracy even though the orbital exponents are not changed as the nuclear charge varies.

Several advantages accrue to the use of a universal basis set:

(i) Molecular electronic structure studies begin with the evaluation of one-electron and two-electron integrals over the basis functions. For a given set of nuclear positions these integrals can be evaluated once and then used in all subsequent studies without regard to the identity of the constituent atoms. This transferability extends, of course, to all integrals

Table 6.5

Matrix Hartree–Fock ground state energies for first-row atoms obtained using an optimized basis set and using a universal basis set[a]

Atom		Optimized basis set	Universal basis set
B	^2P	−24.529 06	−24.528 92
C	^3P	−37.688 61	−37.688 54
N	^4S	−54.400 92	−54.400 84
O	^3P	−74.809 37	−74.809 33
F	^2P	−99.409 30	−99.409 15
Ne	^1S	−128.547 05	−128.546 81

[a] based on the work of Silver, Wilson, and Nieuwpoort 1978. The basis set which was optimized with respect to the total energy in the matrix Hartree–Fock approximation was taken from the work of Clementi and Roetti 1974; the universal basis set is defined in eqn (6.9.1). The energies are in hartree.

arising in the evaluation of the energy and molecular properties (for example, spin–orbit coupling constants (Cooper and Wilson 1982*a*,*b*,*c*)).
(ii) In the case of diatomic molecules, a universal basis set of ellipsoidal basis functions can be employed. Since the integrals which arise in this case depend on the internuclear distance only as a multiplicative factor (Ebbing 1963), they can be evaluated once and used for different geometries as well as for different nuclear charges.
(iii) A universal basis set is, almost by definition, capable of providing a rather uniform description of a series of atoms and molecules. This uniformity is illustrated in Table 6.5 where matrix Hartree–Fock energies for some first row atoms obtained (Silver, Wilson, and Nieuwpoort 1978) by using a universal even-tempered basis set of exponential functions are given.
(iv) Since a universal basis set is not optimized with respect to the total energy or any other molecular property, it is expected to provide a uniform description of a range of properties. Modification of a universal basis set may, of course, be necessary in order to evaluate properties which are particularly sensitive to the quality of a basis set in one region of space. Polarizability calculations, for example, require the addition of more diffuse functions or alternatively the use of an electric-field-variant (Sadlej 1977*a*,*b*, 1978, 1979) universal basis set.
(v) In order to be flexible, a universal basis set is necessarily moderately large and, therefore, it is capable of yielding high accuracy. The high accuracy which can be attained by using a universal basis set in electron correlation energy studies is illustrated in Table 6.6.

Table 6.6

Empirical correlation energy recovered, at experimental equilibrium geometries, by third-order perturbation theory using an (s, p, d) *universal basis set.*[a] *Comparison with other calculations*

Molecule; and method	R_e	E_{total}	E_{corr}	%
Li_2 (empirical correlation energy = −0.126)				
MBPT; universal basis set	5.0507	−14.9842	−0.1129	89.6
MCSCF-CI[b]		−14.9652		
N_2 (empirical correlation energy = −0.538)				
MBPT; universal basis set	2.0742	−109.4443	−0.4501	83.7
Configuration interaction[c]	2.0674	−109.9692	−0.3140	58.4
MBPT[d]	2.0742	−109.4180	−0.4280	79.5
CO (empirical correlation energy = −0.525)				
MBPT; universal basis set	2.1322	−113.2286	−0.4341	82.7
Configuration interaction[e]	2.132	−113.1456	−0.3645	69.4
MBPT[f]	2.132	−113.1952	−0.4061	77.4
BF (empirical correlation energy = −0.531)				
MBPT; universal basis set	2.3858	−124.5782	−0.4096	77.1
MCSCF[g]	2.391	−124.235		
MBPT[h]	2.391	−124.5029	−0.3461	65.2

[a] based on the work of Wilson and Silver 1980, and Wilson 1981*c*. Energies and bond lengths are in atomic units. The (s, p, d) universal basis set is defined in eqn (6.9.1).
[b] Jonsson, Roos, Taylor, and Siegbahn 1981.
[c] Langhoff and Davidson 1974.
[d] Wilson and Silver 1977*a*.
[e] Siu and Davidson 1970.
[f] Bartlett, Wilson, and Silver 1977.
[g] Bender and Davidson 1969.
[h] Wilson, Silver, and Bartlett 1977.

(vi) A universal basis set can have a higher degree of symmetry than the particular molecule under study. For example, for a homonuclear diatomic molecule, the symmetry of the basis set is the same as that for the molecule, i.e. $D_{\infty h}$, whereas for a heteronuclear diatomic molecule the point symmetry group of the molecule is $C_{\infty v}$ but that of the basis set is $D_{\infty h}$. The higher degree of symmetry associated with a universal basis set can lead to a useful reduction in the length of the two-electron integral list, which in turn can lead to more efficient self-consistent-field iterations and orbital transformations.

Although a universal basis set need not be an even-tempered set, the concept of a universal even-tempered basis set has been shown to be useful (Wilson and Silver 1980, 1982 and references therein) and enable large basis sets to be easily generated. Modern vector-processing and parallel-processing computers are providing the possibility of using large basis sets in routine calculations and thus achieving high precision (Guest and Wilson 1981).

Accurate correlation energies have been obtained for the ground states of some diatomic molecules (Wilson and Silver 1980, 1982) by using a universal even-tempered basis set of exponential functions. The orbital exponents were defined by the geometric series $\alpha\beta^{k-1}$, $k=1,2,\ldots,M$, with the parameters α, β, and M being chosen quite arbitrarily according to the following guidelines:

(i) α must be small enough to ensure a wide range of orbital exponents. The smaller the value of α, the more diffuse will be the first few orbitals of the even-tempered basis set.
(ii) β must be large enough to avoid near-linear dependencies.
(iii) M must be large enough to generate a 'near complete' basis set.

The following values of α, β, and M have been shown to be useful in studies of first-row atoms and diatomic molecules containing them:

$$\begin{array}{llll} \text{1s:} & \alpha=0.5 & \beta=1.55 & M=9 \\ \text{2p:} & \alpha=1.0 & \beta=1.60 & M=6 \\ \text{3d:} & \alpha=1.5 & \beta=1.65 & M=3 \end{array} \tag{6.9.1}$$

Thus the orbital exponents of the 1s-functions range from 0.5 to 16.7 bohr^{-1}, the 2p-exponents range from 1.0 to 10.5 bohr^{-1}, and the 3d-exponents range from 1.5 to 4.1 bohr^{-1}. The energy is generally found to be more sensitive to the particular choice of β than the choice of α.

Universal basis sets of exponential functions have been employed in calculations on first- and second-row atoms within the matrix Hartree–Fock approximation (Silver, Wilson, and Nieuwpoort 1978; Silver and Nieuwpoort 1978) and including electron correlation effects (Silver and Wilson 1978). For molecules such basis sets have been used in matrix Hartree–Fock calculations (Wilson and Silver 1979) and the resulting potential energy curves are compared with calculations using basis sets specifically designed for each molecule in Fig. 6.2 (McLean and Yoshimine 1967; Cade, Sales, and Wahl 1973). Correlation effects have been included in calculations using universal even-tempered basis sets of exponential functions for molecules (Wilson and Silver 1980; Wilson 1981*c*) and some typical results are displayed in Table 6.6 for Li_2, N_2, CO, and BF. At the equilibrium geometries, approximately 77–90 per cent of the correlation energy empirically derived from experimental data is recovered.

Gaussian functions are still widely used in molecular studies, especially in studies of polyatomic systems for which multicentre integrals over exponential functions are difficult to evaluate, and the concept of a

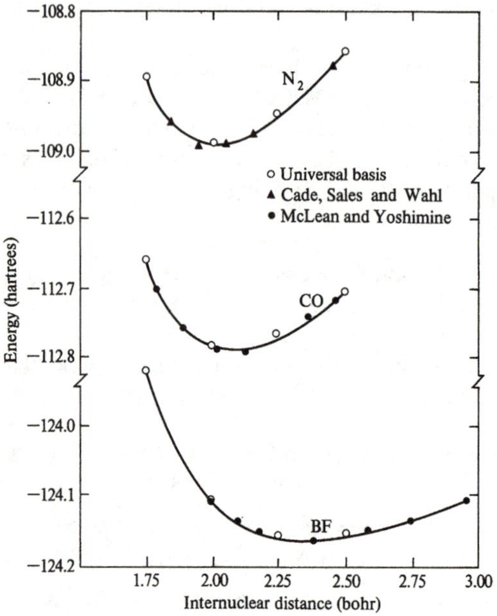

FIG. 6.2. Potential energy curves for the ground states of the N_2, CO, and BF molecules obtained within the matrix Hartree–Fock approximation using a universal even-tempered basis set.

universal even-tempered Gaussian basis set should be briefly considered (cf. Mezey 1979). Clearly, Gaussian basis sets can be transferred from molecule to molecule in precisely the manner demonstrated above for exponential basis sets. However, it is generally accepted that exponential functions do provide a superior basis especially in regions of space close to the nucleus and in the long-range region. Since integrals over universal basis sets are generated once and then stored for future use, it is perhaps more profitable to use exponential functions. Raffenetti (1975) has investigated even-tempered Gaussian functions.

6.10 Systematic sequences of basis sets

In 1963, Schwartz emphasized the need to devise systematic schemes for extending basis sets employed in atomic and molecular calculations:

> "The first essential in talking of convergence rates is to have an orderly plan of procedure. That is, one must choose a set of basis functions to be used and then gradually add more and more of terms to the variational calculation in

some systematic manner. The old habit of picking the 'best' (chosen by art) choice of a fixed number of terms is to be discarded if one wants to see how the problem converges".

Even today, calculations are frequently performed using a single basis set which is constructed in an *ad hoc* fashion using experience gained in previous studies of similar molecules and little effort is made to examine the dependence of calculated molecular properties on the basis set. Ruedenberg and his co-workers (Schmidt and Ruedenberg 1979; Feller and Ruedenberg 1979) and Wilson (1980*a*,*b*) have reiterated the view that the convergence of calculations with respect to the size of the basis set is an important problem to be addressed. As illustrated in Table 6.1, basis set truncation is probably the largest source of error in contemporary molecular calculations.

Ruedenberg and his co-workers (Schmidt and Ruedenberg 1979; Feller and Ruedenberg 1979) devised schemes for systematically extending even-tempered basis sets. Recall from Section 6.8 that such basis sets are restricted to 1s, 2p, 3d, ... functions with exponents given by $\zeta_{lk} = \alpha_l \beta_l^k$, $k = 1, 2, \ldots, M_l$, where the subscript l denotes the angular quantum number. If α_l and β_l are fixed, an even-tempered basis set will not become complete in $L^2(R^3)$ as M_l tends to infinity (see, for example, Klahn and Bingel 1977*a*,*b*, and references therein). In order that an even-tempered basis set tends to a complete set, as M_l tends to infinity, α_l and β_l must be functions of M_l such that

$$\alpha_l \to 0 \qquad \beta_l \to 1 \qquad \beta_l^{M_l} \to \infty. \tag{6.10.1}$$

On the basis of atomic matrix Hartree–Fock calculations using even-tempered Gaussian basis sets of various sizes in which α_l and β_l were optimized with respect to the energy, Ruedenberg and co-workers obtained empirical functional forms for the dependence of α_l and β_l on M_l. In particular, they suggested the empirical functions

$$\ln \ln \beta_l = b_l \ln M_l + b'_l \tag{6.10.2}$$

and

$$\ln \alpha_l = a_l \ln(\beta_l - 1) + a'_l \tag{6.10.3}$$

with

$$-1 < b_l < 0 \tag{6.10.4}$$

and

$$a_l > 0. \tag{6.10.5}$$

These expressions are equivalent to the following recursions (Wilson 1982*a*)

$$\alpha_l(M_l) = \{(\beta_l(M_l) - 1)/(\beta_l(M_l - 1) - 1)\}^{a_l} \alpha_l(M_l - 1) \tag{6.10.6}$$

and

$$\ln \beta_l(M_l) = \{M_l/(M_l - 1)\}^{b_l} \ln \beta_l(M_l - 1) \qquad (6.10.7)$$

which enable a given even-tempered basis set to be increased, or decreased, in size.

Schmidt and Ruedenberg (1979) performed matrix Hartree–Fock calculations on first-row and second-row atoms using even-tempered basis sets of Gaussian functions. For each atom, basis sets of increasing size were used and for each of these the values of α_l and β_l were optimized with respect to the energy expectation value. From these optimal values of α_l and β_l Schmidt and Ruedenberg (1979) determined the parameters a_l, a'_l, b_l, and b'_l in eqns (6.10.2) and (6.10.3) for each atom studied. In Table 6.7, matrix Hartree–Fock energies for some first-row atoms, resulting from the use of even-tempered basis sets of Gaussian functions optimized with respect to the energy, are compared with those obtained

Table 6.7

Matrix Hartree–Fock energies for some first-row atoms using even-tempered basis sets optimized with respect to the energy and basis sets generated by means of eqns (6.10.6–7)[a]

Basis set	Optimal energy	Energy obtained using eqns (6.10.6–7)
Helium		
6s	−2.860 580 068	−2.860 578 549
8s	−2.861 507 099	−2.861 506 747
10s	−2.861 647 441	−2.861 647 263
12s	−2.861 672 957	−2.861 672 945
14s	−2.861 678 287	−2.861 678 285
16s	−2.861 679 529	−2.861 679 529
Carbon		
6s/3p	−37.583 392 9	−37.582 800 6
8s/4p	−37.668 125 9	−37.667 907 3
10s/5p	−37.684 045 2	−37.684 009 3
12s/6p	−37.687 437 1	−37.687 429 9
14s/7p	−37.688 303 5	−37.688 303 5
Neon		
6s/3p	−128.081 587	−128.079 773
8s/4p	−128.439 701	−128.439 172
10s/5p	−128.519 434	−128.519 330
12s/6p	−128.539 378	−128.539 359
14s/7p	−128.544 755	−128.544 745

[a] based on the work of Schmidt and Ruedenberg 1979. All energies are in hartree.

by employing basis sets generated by using eqns (6.10.2–7). It can be seen that there is little loss in accuracy of the calculated energies by using the logarithmic relations to determine the parameters α_l and β_l.

By employing a sequence of basis sets which are determined in some systematic fashion, it is envisaged that reliable extrapolation procedures can be devised to obtain the value of the basis set limit for the property considered. In calculations of the total energy, the Hartree extrapolation (Hartree 1948; Roothaan and Bagus 1963) has been shown to be useful (Schmidt and Ruedenberg 1979; Feller and Ruedenberg 1979; Wilson 1980*a,b*; Cooper and Wilson 1982*a,b,c*). This extrapolation technique is based on the assumption

$$(E[M_2]-E_\infty)=m(E[M_1]-E_\infty) \tag{6.10.8}$$

where $E[M_1]$ and $E[M_2]$ are energy values corresponding to two adjacent basis sets in the systematic sequence, E_∞ is the unknown energy corresponding to the basis set limit, and m is an unknown constant. By considering three successive values of the energy, the constant m can be eliminated to provide an estimate of E_∞ which may be denoted by $E_\infty[M_3]$

$$E_\infty[M_3]=(E[M_3]E[M_1]-E[M_2]^2)/(E[M_3]-2E[M_2]+E[M_1]). \tag{6.10.9}$$

A value of m can also be found by considering three successive energy values and eliminating E_∞ giving

$$m[M_3]=(E[M_3]-E[M_2])/(E[M_2]-E[M_1]). \tag{6.10.10}$$

The value of m provides a useful indication of the convergence properties of the energy values with size of basis set. If $|m|<1$, then the series of energy values is converging, whereas if $|m|>1$, it is diverging. If $m>0$, the series of energies is behaving monotonically, while if $m<0$, it is oscillatory.

Schmidt and Ruedenberg (1979) have shown that the Hartree-extrapolation leads to an empirical upper bound to the true basis set limit, E_∞. This is illustrated for the carbon atom within the Hartree–Fock model in Fig. 6.3. It can further be shown (Schmidt and Ruedenberg 1979) that

$$\hat{E}_\infty[M]=E_\infty[M]-(E[M]-E_\infty[M]) \tag{6.10.11}$$

provides an empirical lower bound to the true basis set limit. The average of the empirical upper bound (6.10.9) and the lower bound (6.10.11), i.e.

$$E_{av}[M]=\tfrac{1}{2}(E_\infty[M]+\hat{E}_\infty[M]) \tag{6.10.12}$$

may be regarded as a 'best' estimate of the basis set limit and the difference

$$D[M]=\tfrac{1}{2}(E_\infty[M]-\hat{E}_\infty[M]) \tag{6.10.13}$$

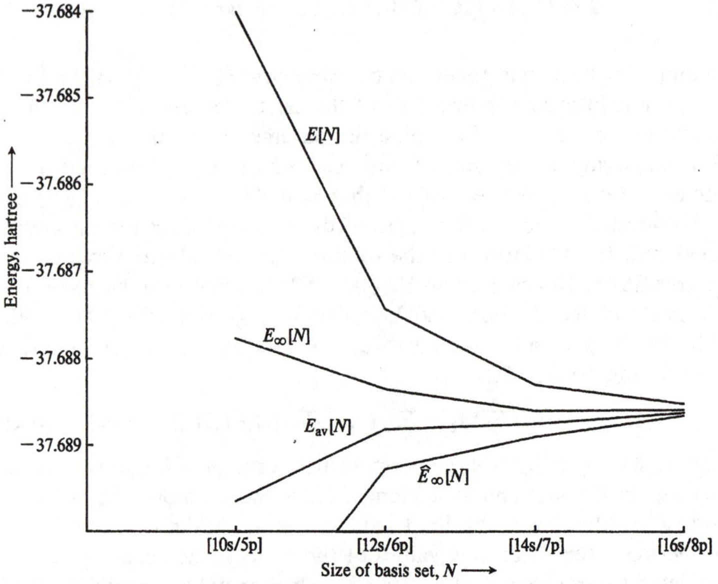

FIG. 6.3. Empirical upper bounds, lower bounds, and estimates of the basis set limit for the matrix Hartree–Fock description of the ground state of the carbon atom obtained by using a systematic sequence of even-tempered basis sets.

Table 6.8

Empirical upper bounds and lower bounds to the basis set limit together with 'best' estimates and accuracy of this limit for matrix Hartree–Fock energies of some first-row atoms[a]

M	$E_\infty[M]$	$\hat{E}_\infty[M]$	$E_{av}[M]$	D[*M*]
Helium				
10s	−2.861 672 3	−2.861 697 4	−2.861 684 9	0.000 012 5
12s	−2.861 678 8	−2.861 684 6	−2.861 681 7	0.000 002 9
14s	−2.861 679 3	−2.861 680 3	−2.861 679 8	0.000 000 5
16s	−2.861 679 7	−2.861 679 8	−2.861 679 8	0.000 000 1
Carbon				
10s/5p	−37.687 766 6	−37.691 524 0	−37.689 645 3	0.001 878 7
12s/6p	−37.688 352 2	−37.689 274 6	−37.688 813 4	0.000 461 2
14s/7p	−37.688 602 2	−37.688 901 1	−37.688 751 7	0.000 149 4
	−37.688 587 0	−37.688 655 9	−37.688 621 5	0.000 034 4
Oxygen				
10s/5p	−74.807 091 9	−74.817 324 7	−74.812 208 3	0.005 116 4
12s/6p	−74.808 833 9	−74.811 643 8	−74.810 238 8	0.001 404 9
14s/7p	−74.809 264 5	−74.810 111 0	−74.809 687 8	0.000 423 3
	−74.809 315 7	−74.809 558 4	−74.809 437 0	0.000 121 3

[a] based on the work of Schmidt and Ruedenberg 1979.

Table 6.9

Diagrammatic many-body perturbation theory calculations of the ground state energy of the neon atom using a systematic sequence of even-tempered basis sets[a]

Basis set M	E_{MHF}	$E[M]$	$E_{\infty}[M]$	$\hat{E}_{\infty}[M]$	$E_{av}[M]$	$D[M]$
6s/3p	−128.079 772 75	−128.221 548	—	—	—	—
8s/4p	−128.439 171 64	−128.606 726	—	—	—	—
10s/5p	−128.519 330 09	−128.699 776	−128.729 415	−128.759 054	−128.744 235	0.014 819
12s/6p	−128.539 358 98	−128.724 439	−128.733 333	−128.742 228	−128.737 781	0.004 447
14s/7p	−128.544 744 84	−128.731 314	−128.733 972	−128.736 629	−128.735 301	0.001 329
16s/8p	−128.546 301 83	−128.733 468	−128.734 451	−128.735 433	−128.734 942	0.000 491
18s/9p	−128.546 816 46	−128.734 247	−128.734 688	−128.735 128	−128.734 908	0.000 220

[a] based on the work of Wilson 1980*f*. Energies are given in hartree. The basis sets are taken from the work of Schmidt and Ruedenberg 1979.

as an estimate of the accuracy which may be assigned to this best estimate.

In Table 6.8, the application of eqns (6.10.9–13) is illustrated for some first-row atoms. The calculations (Schmidt and Ruedenberg 1979) were made within the Hartree–Fock model using basis sets of Gaussian functions.

The Gaussian basis sets of Schmidt and Ruedenberg (1979) have been used in calculations of correlation energies (Wilson 1980*e*). Some typical results for the neon atom are displayed in Table 6.9. The use of eqns (6.10.9–13) for estimating the basis set limit is illustrated in this table for correlation energy calculations. The correlation energy was found to be more sensitive to the degree of completeness of the basis set than the matrix Hartree–Fock energy.

The use of a universal systematic sequence of even-tempered basis sets, that is a single sequence of basis sets which can be used for any atom irrespective of its nuclear charge, has also been proposed (Wilson 1980*f*). Some typical results are presented for beryllium-like ions in Table 6.10,

Table 6.10

Matrix Hartree–Fock energies and correlation energies for radial beryllium-like ions using a universal systematic sequence of even-tempered basis sets[a]

Basis set	E_{MHF}	E_{corr}	E_{MHF}	E_{corr}
	Be		B^+	
6s	−14.534 897	17.140	−24.140 509	17.435
8s	−14.566 442	18.089	−24.222 327	17.250
10s	−14.571 727	18.322	−24.234 213	17.581
12s	−14.572 705	18.384	−24.236 807	17.622
14s	−14.572 951	18.421	−24.237 357	17.639
16s	−14.573 002	18.434	−24.237 519	17.655
18s	−14.573 017	18.440	−24.237 559	17.662
20s	−14.573 021	18.444	−24.237 570	17.666
	C^{2+}		N^{3+}	
6s	−36.230 813	15.687	−50.707 427	16.653
8s	−36.370 037	17.238	−51.013 464	16.579
10s	−36.401 311	17.101	−51.068 088	16.900
12s	−36.406 797	17.207	−51.078 704	16.971
14s	−36.408 060	17.244	−51.081 409	16.980
16s	−36.408 360	17.251	−51.082 079	17.003
18s	−36.408 456	17.259	−51.082 241	17.009
20s	−36.408 484	17.264	−51.082 290	17.013

[a] based on the work of Wilson (1980*f*). The matrix Hartree–Fock energy, E_{MHF}, is given in hartree; the correlation energy, E_{corr}, is given in millihartree with sign reversed.

which demonstrate the usefulness of this approach in calculations made within the Hartree–Fock model and in calculations which take account of electron correlation. The concept of a universal systematic sequence of basis sets of exponential-type functions has been examined (Cooper and Wilson 1982*a*) in calculations of total energies and spin–orbit coupling constants.

It is well known that the components of the total energy of an atom or molecule are often more sensitive to the quality of the basis set than is the total energy itself. Often, therefore, the components of the total energy are used as a rough test of a basis set. For example, in a study of the nitrogen molecule using the matrix Hartree–Fock method, Cade *et al.* (1966) examined the dependence of the total energy E, the kinetic energy T, the potential energy V, and the orbital energies on the quality of the basis set and demonstrated that, although the total energy and the orbital

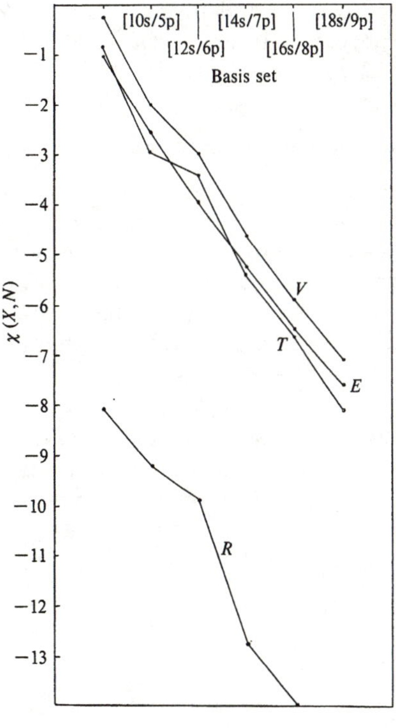

FIG. 6.4. Convergence of the total energy, kinetic energy, potential energy, and virial ratio for the neon atom in the matrix Hartree–Fock model as a function of basis set size for a systematic sequence of even-tempered basis sets.

energies are not very sensitive to the basis set composition, the kinetic energy and the potential energy can behave erratically as the basis set is extended. It is for this reason that the virial ratio

$$R = -V/2T \tag{6.10.14}$$

which in the Hartree–Fock limit should equal unity, is often regarded as a more sensitive test of a basis set than the total energy. In Fig. 6.4, the convergence of the total energy, kinetic energy, potential energy, and virial ratio for the neon atom is displayed as a function of the size of the systematic sequence of basis sets employed (Wilson 1981*a*). In this figure, the convergence is measured in terms of

$$\chi(X, M) = \ln |X[M] - X[M']| \tag{6.10.15}$$

where $X = E$, T, V, or R. It can be seen that $\chi(E, M)$ and $\chi(R, M)$ behave somewhat more smoothly as a function of the basis set size M than do $\chi(T, M)$ and $\chi(V, M)$. Furthermore, it should be noted that $\chi(R, M)$ is considerably smaller than $\chi(E, M)$, $\chi(T, M)$, and $\chi(V, M)$; thus the virial ratio appears to be less sensitive to the basis set than either the total energy or its components.

6.11 Convergence of harmonic expansions

The basis sets employed in molecular calculations are almost always restricted to functions of s-, p-, and often d-symmetry. In orbital models, such as the Hartree–Fock approximation, only functions corresponding to the first few values of the angular quantum number l contribute significantly to the energy and other expectation values. However, in treatments which take account of electron correlation effects the higher harmonics can be important.

Table 6.11

Rough estimate of the maximum value of l required to obtain the energy of $1s^2$ He *to a given accuracy*[a]

Accuracy in atomic units	Maximum value of l
10^{-4}	6
10^{-5}	14
10^{-6}	32
10^{-7}	69
10^{-10}	690
10^{-12}	3204

[a] taken from the work of Carroll, Silverstone, and Metzger 1979.

Table 6.12

Convergence of the harmonic expansion for some electron pair energies in the ground state of the neon atom[a]

l	$(1s2s)\ {}^1S$ Ne	$(1s2s)\ {}^3S$ Ne	$(2p^2)\ {}^1D$ Ne
2	422	73	29 950
3	116	12	7 265
4	43	3	5 900
5	20	1	2 240
6	10		1 030
7	6		545
8	4		306
9	2		185

[a] taken from the work of Jankowski and Malinowski 1980 which employed second-order Rayleigh–Schrödinger perturbation theory. Energies are given in microhartree with signs reversed.

A considerable amount of data is available on the convergence properties of the harmonic expansion for atoms. On the basis of second-order perturbation theory, Schwartz (1962, 1963) established the result

$$E_{2,l} = \frac{-45}{256}(l+\tfrac{1}{2})^{-4}\left\{1 - \frac{19/8}{(l+\tfrac{1}{2})^2} + \mathcal{O}(l^{-4})\right\} \qquad (6.11.1)$$

where $E_{2,l}$ is the component of the second-order energy corresponding to the quantum number l. Carroll *et al.* (1979) have established similar expressions for the convergence of the correlation energy of $1s^2$ He. In Table 6.11, rough estimates of the maximum value of l required to obtain the energy of $1s^2$ He to a given accuracy is presented (Carroll, Silverstone, and Metzger 1979). The convergence of the harmonic expansion

Table 6.13

The importance of basis functions of f-*symmetry in molecular correlation energy calculations. Calculations on the* CO *molecule at various geometries using basis sets of* s, p, *and* d *functions and basis sets of* s, p, d, *and* f *functions*[a]

R	$-E_{MHF}$ (spd)	$-E_{MHF}$ (spdf)	$-e_{corr}$ (spd)	$-e_{corr}$ (spdf)
0.9283	112.640 73	112.646 23	0.273 48	0.298 08
1.0283	112.755 77	112.759 73	0.284 77	0.309 79
1.1283	112.775 51	112.778 36	0.295 51	0.320 62
1.2283	112.745 98	112.748 14	0.305 47	0.330 35
1.3283	112.692 83	112.694 67	0.314 30	0.338 73

[a] taken from the work of Wilson 1982*a*; the nuclear separation R is in Ångstrom units; the matrix Hartree–Fock energies E_{MHF} and correlation energies e_{corr} are in hartree atomic units.

for some of the electron pair energies in the ground state of the neon atom are displayed in Table 6.12 (Jankowski and Malinowski 1980). The convergence of the harmonic expansion for the $2p^2\ ^1D$ electron pair in this system is particularly unfavourable.

The importance of higher terms in the harmonic expansion for molecular systems is illustrated in Table 6.13 where calculations are presented for the carbon monoxide molecule using basis sets of functions with s, p, and d symmetry and basis sets of functions with s, p, d, and f symmetry (Wilson 1982*a*). Matrix Hartree–Fock energies and correlation energies obtained by means of diagrammatic perturbation theory are shown as a function of nuclear geometry.

7

TRUNCATION OF EXPANSIONS FOR EXPECTATION VALUES

7.1 Truncation of expansions

All methods for performing accurate calculations of the electronic structure of atoms and molecules which include a description of correlation effects involve some finite, or infinite, expansion for the wavefunction and corresponding expectation values. This is the case in, for example, the method of configuration mixing, the many-body perturbation theory, the valence bond theory, the group function method, and various cluster expansions. Each method leads to the exact wavefunction and expectation values if all of the terms in the expansion are included. In practice, of course, such expansions have to be truncated in order to keep the calculations tractable and the various methods differ only in the manner in which this truncation is carried out. However, the method of truncation can significantly affect not only the theoretical properties of a particular approach but also, to some extent, its computational feasibility. The fact that one method may include more terms in an expansion than another method does not necessarily imply that it is superior. The terms which are left out of an expansion of the wavefunction or expectation value are, in fact, often just as important as the ones which are actually included. For example, the method of configuration mixing is very often restricted to all single-excitations and double-excitations with respect to a single determinantal reference function; this approach includes many terms, corresponding to unlinked diagrams in the perturbation theoretic analysis of the method, which are exactly cancelled by terms involving higher-order excitations. This will be discussed further in Section 7.2.

This chapter is concerned with the most frequently used and the most promising of the many approaches to the electron correlation problem in molecules which are currently in use. The methods described in this present chapter can be used to calculate correlation energies with respect to the independent electron models discussed in Chapter 2. Qualitative and quantitative analysis of the correlation effects along the lines described in Chapter 3 can be carried out. The various approaches to the correlation problem can be usefully analysed in terms of the diagrammatic perturbation theory which was considered in Chapter 4. Perturbation theory not only provides a very systematic technique for the evaluation of corrections to independent electron models but also provides a powerful method for the analysis of other schemes for handling electron correlation. Group theory can always be employed to simplify the calculation of the electronic structure of molecular systems; the group theoretical apparatus outlined in Chapter 5 has been used, to varying extents, to simplify calculations using the various approaches treated in the present chapter. The use of the algebraic approximation, described in Chapter 6, is common to all of the approaches to the correlation problem in molecules which are considered in this chapter. Indeed, basis set truncation errors are probably the largest source of error in the vast majority of contemporary studies. Hence, the discussion of Chapter 6 should be remembered throughout the whole of the present chapter.

The most commonly used technique for the study of electron correlation in molecules is the method of configuration mixing, configuration interaction, or superposition of configurations. The method of configuration mixing is discussed in Section 7.2. In Section 7.3, schemes for calculating correlation energies using the group function ansatz are considered; particular attention being given to pair function models. Section 7.4 is devoted to the coupled pair approximation and various coupled cluster schemes. Approaches to the correlation problem based on valence bond theory are described in Section 7.5. In Section 7.6, the use of the diagrammatic many-body perturbation theory described in Chapter 4 directly in the calculation of correlation energies and the effects of electron correlation on molecular properties will be described. Finally, in Section 7.7 computational aspects of the correlation problem are very briefly reviewed.

7.2 The method of configuration mixing

The method of configuration mixing, configuration interaction, or superposition of configurations is perhaps the most easily understood and certainly the most widely used approach to the electron correlation problem in atoms and molecules. The total wavefunction is written as a linear combination of N-electron functions, each of which is a product of one-electron functions

$$\Psi = \sum_{\mu} \Phi_{\mu} C_{\mu} \tag{7.2.1}$$

The expansion coefficients C_{μ} are determined by invoking the variation principle which leads to a secular problem of the form

$$\mathbb{H}\mathbf{C} = E\mathbf{C} \tag{7.2.2}$$

where $\mathbb{H}$ is the Hamiltonian matrix

$$\mathbb{H}_{\mu\nu} = \langle \Phi_{\mu} | \hat{\mathcal{H}} | \Phi_{\nu} \rangle. \tag{7.2.3}$$

In general, full configuration mixing for basis sets of an acceptable size is

Table 7.1
Typical values of Weyl's number

Number of basis functions m	Number of electrons N	Total spin S	Weyl's number $D_{m,N,S}$
8	4	0	336
50	4	0	520 625
50	4	1	749 700
10	10	0	19 404
50	10	0	829 515 727 600
75	10	0	53 144 078 124 600

computationally intractable, since the number of terms in the expansion (7.2.1) for the wavefunction is extremely large. The total number of terms for an N-electron system with a spin of S and a basis set of m functions is given by Weyl's formula (Paldus 1976*b*)

$$D_{m,N,S} = \frac{2S+1}{m+1}\binom{m+1}{\frac{1}{2}N-S}\binom{m+1}{m-\frac{1}{2}N-S} \tag{7.2.4}$$

Typical values of Weyl's number are displayed in Table 7.1. From this table it is clear that full configuration-mixing calculations for the ground state of a ten-electron system such as water with a reasonably large basis set of, say, seventy-five basis functions, would remain intractable even if the efficiency of currently available computational machinery were subject to an extraordinary increase!

Applications of the method of configuration mixing to atoms and to molecules almost invariably employ some limited expansion for the wavefunction. One of the most commonly used configuration mixing techniques is to include all single-excitations and double-excitations with respect to a single determinantal reference function. However, this method is not size-consistent, as can easily be seen by considering the simple case of two helium atoms separated by an infinite distance. If the method of single and double excitation configuration mixing is applied to each of the atoms separately, then the exact energy of the system is obtained within the basis set employed. It is clear, however, that it is necessary to include triple excitations and quadruple excitations in a treatment of the supersystem in order to obtain the exact energy. The size-inconsistency of the limited configuration mixing method becomes more and more problematic as the number of electrons in the system being considered increases. Shavitt (1977*b*) sums up the situation as follows:

> "The fact that in a configuration interaction expansion unlinked cluster contributions can only be accounted for by including quadruple- (and higher-order) excitations is one of the principal drawbacks of the method. In

> contrast, such contributions are automatically accounted for without explicitly computing higher-order terms, in some cluster-based methods and in many-body perturbation theory. In this sense the CI expansion is much less compact and less efficient than these approaches and becomes progressively less efficient as the number of electrons increases".

As Shavitt indicates, the method of configuration mixing leads to terms corresponding to unlinked diagrams in expansions for expectation values.

The method of limited configuration mixing with respect to a single determinantal reference function is often extended by employing a multi-determinantal reference function. The reference function is usually chosen to provide a qualitatively correct description of dissociative processes or a number of different electronic states. All configurations which can be generated by single and double excitations from this multi-determinantal reference function are included in the expansion (7.2.1). However, it should be noted that this extended configuration mixing technique, often termed 'multiroot' configuration interaction, is not size consistent; although it is to be expected that the size-consistency error in a calculation using a multi-determinantal reference function will usually be smaller than that associated with a calculation in which a single determinantal reference function is employed.

It is most useful to analyse the configuration mixing expansion in terms of perturbation theory and, in particular, to analyse the various components in terms of the linked diagram theorem of Chapter 4. For simplicity, the configuration mixing expansion with respect to a single determinant will be considered. The order of the perturbation expansion for the energy at which various levels of excitation first arise is illustrated in Fig. 7.1 for reference determinants constructed from three different types of orbitals. In Fig. 7.1(a) it is assumed that the Hartree–Fock orbitals are employed in the construction of the configuration functions. Only double excitations contribute to the hamiltonian matrix through third-order; single-excitations do not contribute until fourth-order. The structure of the configuration mixing matrix when a bare-nucleus model, or a screened bare-nucleus model, is used to generate the orbitals is shown in Fig. 7.1(b). In this case, single-excitations first arise in second-order. Note that the structure of the matrix in the lower right-hand corner is much the same in Figs 7.1(a) and 7.1(b). Indeed, the same structure in the lower right-hand corner of the hamiltonian matrix can be observed in Fig. 7.1(c) where the matrix corresponding to the use of Brueckner orbitals is displayed. Brueckner (1954, 1955*a*) orbitals are not often used in molecular studies. They are defined essentially by the requirement that single-excitations do not contribute to the correlation energy. Unfortunately their determination necessitates a prior knowledge of the exact wavefunction. From Fig. 7.1 it can be seen that the method of single and

	$K^{(0)}$	$K^{(1)}$	$K^{(2)}$	$K^{(3)}$	$K^{(4)}$	$K^{(5)}$	$K^{(6)}$	$K^{(7)}$
$K^{(0)}$	0		2					
$K^{(1)}$		5	4	5				
$K^{(2)}$	2	4	3	4	4			
$K^{(3)}$		5	4	5	5	6		
$K^{(4)}$			4	5	5	6	6	
$K^{(5)}$				6	6	7	7	8
$K^{(6)}$					6	7	7	8
$K^{(7)}$						8	8	9

(a)

	$K^{(0)}$	$K^{(1)}$	$K^{(2)}$	$K^{(3)}$	$K^{(4)}$	$K^{(5)}$	$K^{(6)}$	$K^{(7)}$
$K^{(0)}$	0	2	2					
$K^{(1)}$	2	3	3	4				
$K^{(2)}$	2	3	3	4	4			
$K^{(3)}$		4	4	5	5	6		
$K^{(4)}$			4	5	5	6	6	
$K^{(5)}$				6	6	7	7	8
$K^{(6)}$					6	7	7	8
$K^{(7)}$								

(b)

	$K^{(0)}$	$K^{(1)}$	$K^{(2)}$	$K^{(3)}$	$K^{(4)}$	$K^{(4)}$	$K^{(6)}$	$K^{(7)}$
$K^{(0)}$	0		2					
$K^{(1)}$								
$K^{(2)}$	2		3	4	4			
$K^{(3)}$			4	5	5	6		
$K^{(4)}$			4	5	5	6	6	
$K^{(5)}$				6	6	7	7	8
$K^{(6)}$					6	7	7	8
$K^{(7)}$						8	8	9

(c)

FIG. 7.1. Structure of the Hamiltonian matrix when the single determinantal reference function is constructed from (a) Hartree–Fock orbitals; (b) bare-nucleus orbitals; (c) Brueckner orbitals. The order of perturbation theory in which each block contributes to the correlation energy is indicated.

double excitation configuration mixing is a third-order theory, since some fourth-order components of the correlation energy are neglected.

The unphysical terms corresponding to unlinked diagrams, mentioned above, first arise in fourth-order of the perturbational analysis of the method of single and double excitation configuration mixing. Unlinked diagrams involving doubly excited intermediate states arise which are cancelled by unlinked diagrams involving quadruply excited states. The fourth-order quadruple excitation unlinked diagrams are shown in Fig. 7.2. These diagrams correspond to the term $E_2 \Delta_{11}$, discussed in Section 4.5 in which E_2 is the second-order energy and Δ_{11} is the normalization integral $\langle \Phi_1 | \Phi_1 \rangle$ where Φ_1 is the first-order wavefunction. Davidson first showed (Davidson 1974; see also Langhoff and Davidson 1974; Bartlett and Shavitt 1977*b*; Siegbahn 1978; Wilson and Silver 1978*a,b*, 1979*e*)

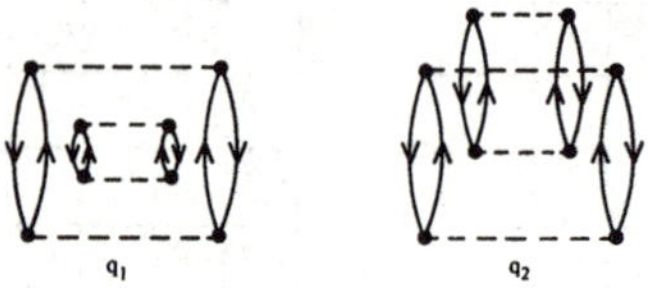

FIG. 7.2. Unlinked fourth-order diagrams involving quadruply excited intermediate states.

that a correction to fourth-order of the perturbation series can be made to double excitation configuration mixing calculations by means of the formula

$$(1-C_0^2)E_D \tag{7.2.5}$$

where E_D is the double excitation configuration mixing energy and C_0 is the coefficient of the reference configuration. The formula (Davidson and Silver 1977; Siegbahn 1978; Wilson and Silver 1978, 1979)

$$\left(\frac{1-C_0^2}{C_0^2}\right)E_D \tag{7.2.6}$$

can be shown to correct the size-consistency error in double excitation configuration mixing calculations through fifth-order.

Useful estimates of the size-consistency error in double excitation configuration mixing calculations can be obtained from perturbation studies (Wilson and Silver 1978*b*). If the first-order wavefunction is written as $\Phi_0+\gamma\Phi_1$, where γ is a parameter, there is a normalization factor of

$$\nu^2=(1+\gamma^2\Delta_{11})^{-1} \tag{7.2.7}$$

and hence

$$1-\nu^2=\frac{\gamma^2\Delta_{11}}{1+\gamma^2\Delta_{11}}=\gamma^2\Delta_{11}+\ldots \tag{7.2.8}$$

which should be compared with eqn (7.2.5). A variational energy result can be obtained from the first-order wavefunction

$$\left(\frac{(2\gamma-\gamma^2)E_2+\gamma^2E_3}{1+\gamma^2\Delta_{11}}\right)_{\text{optimum}}=\gamma_{\text{opt}}E_2. \tag{7.2.9}$$

Equations (7.2.8) and (7.2.9) can now be combined to give

$$\gamma_{\text{opt}}^3E_2\Delta_{11} \tag{7.2.10}$$

as a correction to variational calculations.

Table 7.2

Variation of the fourth-order unlinked diagram component of the ground state energy of the hydrogen fluoride molecule with nuclear separation[a]

R	$-E_2$	$-E_2\,\Delta_{11}$	$-\gamma^3_{\text{opt}}E_2\,\Delta_{11}$
1.0	0.291 81	0.011 36	0.010 49
1.5	0.300 24	0.013 94	0.012 20
2.0	0.311 81	0.017 54	0.014 47
2.5	0.323 94	0.022 14	0.017 26
3.0	0.336 33	0.028 06	0.020 70
3.5	0.349 26	0.035 98	0.024 97

[a] taken from the work of Wilson and Silver (1978). Atomic units are used throughout.

In Table 7.2, values of $E_2\,\Delta_{11}$ and $\gamma^3_{\text{opt}}E_2\,\Delta_{11}$ are given as a function of internuclear separation R for the hydrogen fluoride molecule. Between R-values of 1.0 and 3.5 bohr, the unlinked diagram energy changes considerably, both as a percentage of the correlation energy ($\sim E_2$) and in absolute value. Since the fourth-order unlinked diagram energy component represents the lowest order correction to a configuration mixing calculation including all doubly excited configurations, it is evident that an uncorrected potential energy curve for the FH molecule would contain an error that varies strongly (10–20 millihartree) over a small range of R.

The effectiveness of configuration mixing in calculations for large molecules has been discussed by Sasaki (1977) using a model system which consists of N equivalent, non-interacting electron pairs. The exact ground state wavefunction for one of these pairs is written as

$$\phi = C_0\phi_0 + C_1\phi_1 \tag{7.2.11}$$

where ϕ_0 is the reference wavefunction and ϕ_1 is its orthogonal complement. The correlation correction for the pair may be expressed in terms of the hamiltonian matrix elements as

$$\begin{aligned}\Delta\epsilon &= \langle\phi|\,\hat{\mathcal{H}}\,|\phi\rangle - \langle\phi_0|\,\hat{\mathcal{H}}\,|\phi_0\rangle \\ &= b-(b^2+a^2)^{\frac{1}{2}}\end{aligned} \tag{7.2.12}$$

where

$$2b = \langle\phi_1|\,\hat{\mathcal{H}}\,|\phi_1\rangle - \langle\phi_0|\,\hat{\mathcal{H}}\,|\phi_0\rangle \tag{7.2.13a}$$

$$a = \langle\phi_1|\,\hat{\mathcal{H}}\,|\phi_0\rangle. \tag{7.2.13b}$$

The configuration mixing expansion for the whole system of N pairs may be written

$$\Psi = d_0\Phi_0 + d_1\sum_{k=1}^{N}\Phi^k + d_2\sum_{k>l=1}^{N}\Phi^{kl} + \dots \tag{7.2.14}$$

where Φ_0 is the product of the ϕ_0, Φ^k is a doubly excited configuration function, Φ^{kl} is a quadruply excited configuration function, ... etc. The exact energy for the whole system is, of course, $N\,\Delta\epsilon$. Now let

$$\Delta E = E - \langle \Phi_0 | \hat{\mathcal{H}} | \Phi_0 \rangle \tag{7.2.15}$$

and, following Sasaki (1977), define

$$\gamma = \Delta E / N\,\Delta\epsilon \tag{7.2.16}$$

and

$$\beta = C_1^2 / C_0^2. \tag{7.2.17}$$

The secular equations can be put in the form

$$\begin{aligned} 0 &= \langle \Phi_0 | \hat{\mathcal{H}} - E | \Psi \rangle \\ &= -\Delta E d_0 + N a d_1 \\ 0 &= \langle \Phi^k | \hat{\mathcal{H}} - E | \Psi \rangle \\ &= a d_0 + (2b - \Delta E) d_1 + (N-1) a d_2 \\ 0 &= \langle \Phi^{kl} | \hat{\mathcal{H}} - E | \Psi \rangle \\ &= 2 a d_1 + (4b - \Delta E) d_2 + (N-2) a d_3 \ldots \end{aligned} \tag{7.2.18}$$

and the secular determinant is

$$\begin{vmatrix} \gamma N & N & 0 & 0 & 0 & \cdots \\ \dfrac{1}{\beta} & \dfrac{1}{\beta} - 1 + \gamma N & N-1 & 0 & 0 & \cdots \\ 0 & \dfrac{2}{\beta} & 2\left(\dfrac{1}{\beta} - 1\right) + \gamma N & N-2 & 0 & \cdots \\ 0 & 0 & \dfrac{3}{\beta} & 3\left(\dfrac{1}{\beta} - 1\right) + \gamma N & N-3 & \cdots \\ \vdots & \vdots & \vdots & \vdots & \vdots & \end{vmatrix} = 0 \tag{7.2.19}$$

Clearly, the double-excitation configuration mixing method yields the secular determinant

$$\begin{vmatrix} \gamma N & N \\ \dfrac{1}{\beta} & \dfrac{1}{\beta} - 1 + \gamma N \end{vmatrix} = 0 \tag{7.2.20}$$

or

$$\frac{1}{\gamma} - 1 = \beta(\gamma N - 1) \tag{7.2.21}$$

and for double and quadruple excitation configuration mixing one obtains

$$\begin{vmatrix} \gamma N & N & 0 \\ \frac{1}{\beta} & \frac{1}{\beta}-1+\gamma N & N-1 \\ 0 & \frac{2}{\beta} & 2\left(\frac{1}{\beta}-1\right)+\gamma N \end{vmatrix} = 0 \tag{7.2.22}$$

or

$$\frac{1}{\gamma}-1=\frac{\beta(\gamma N-1)(\gamma N-2)}{\left(3\gamma N+\frac{2}{\beta}-2\right)}. \tag{7.2.23}$$

For large values of N it is apparent that γ is proportional to $(\beta N)^{-\frac{1}{2}}$. In particular, Sasaki (1977) showed that for large N

$$\begin{aligned} &\gamma=(\beta N)^{-\frac{1}{2}} \quad \text{for double-excitation configuration mixing} \\ &\gamma=3^{\frac{1}{2}}(\beta N)^{-\frac{1}{2}} \quad \text{for double and quadruple excitation configuration mixing} \\ &\gamma=(3+6^{\frac{1}{2}})^{\frac{1}{2}}(\beta N)^{-\frac{1}{2}} \quad \text{for up to and including sixfold excitations} \\ &\gamma=(5+10^{\frac{1}{2}})^{\frac{1}{2}}(\beta N)^{-\frac{1}{2}} \quad \text{for up to and including eightfold excitations} \end{aligned} \tag{7.2.24}$$

Using the above results, Sasaki estimated that full single and double excitation configuration mixing will recover about 80 per cent of the total correlation energy for, say pyrrole, and about 50–55 per cent for porphine. Full single, double, triple, and quadruple excitation configuration mixing will recover about 95 per cent of the correlation energy for pyrrole and 75 per cent for porphine. Sasaki concludes "It is shown that even SDTQ CI is quite unsatisfactory for large molecules such as porphine".

Meunier, Levy, and Berthier (1976) have also examined the spurious terms which arise in the single and double excitation configuration mixing technique. They conclude that "a truncated CI calculation . . . is theoretically and numerically hazardous". In a later publication (Kutzelnigg, Meunier, Levy, and Berthier 1977) they discuss the relation between the terms which depend on N^2 in the perturbation series and the $\sqrt{N}$ dependence of limited configuration mixing. This relation is essentially that between $E_2\,\Delta_{11}$ and eqn (7.2.10).

In spite of the size-inconsistency of the configuration mixing technique and the slow convergence of the expansion (7.2.1), this approach is widely employed and there are a number of excellent reviews of the method (see, for example, Davidson 1974; Buenker and Peyerimhoff 1975;

Meyer 1977; Shavitt 1977*b*; Roos and Siegbahn 1977; Buenker, Peyerimhoff, and Butscher 1978). A wide variety of tricks and devices are employed in order to render the method computationally tractable. The direct configuration interaction method (Roos 1972; Roos and Siegbahn 1977; Siegbahn 1983) avoids the explicit construction of the hamiltonian matrix (7.2.3). In the self-consistent electron pair method (Meyer 1976; Dykstra, Schaefer, and Meyer 1976) an explicit transformation of integrals over atomic orbitals to integrals over molecular orbitals is avoided. Techniques have been developed for the selection of configurations using perturbational considerations (see, for example, Raffenetti, Hsu, and Shavitt 1977) and schemes for performing energy extrapolations have been constructed (Buenker and Peyerimhoff 1975). In the pair-natural-orbital configuration interaction method (Meyer 1977; Taylor 1981), the long configuration mixing expansion is approximated by a shorter expansion based on non-orthogonal orbitals, that is, the pair natural orbitals for each pair of occupied spin-orbitals. However, as Čársky and Urban (1980) point out, most of these developments are of a "more or less technical nature and do not remove the inherent drawback of the ordinary CI expansion".

Full configuration-mixing calculations do not suffer from the problem described above for limited configuration mixing. In two important cases, full configuration mixing expectation values, or very accurate approximations to them, can be obtained. Because of their high symmetry, large configuration mixing calculations can be performed for small atomic

Table 7.3

Full and limited configuration mixing calculations for the ground state of the water molecule using a 'double zeta' basis set

Method	Number of configurations	Total energy	Correlation energy	C_0
Matrix Hartree–Fock	1	−76.009 838	0.0	1.0
Limited configuration mixing including all:				
Double excitations	342	−76.149 178	−0.139 340	0.979 38
Single and double excitations	361	−76.150 015	−0.140 177	0.978 74
Single, double, and triple excitations	3 203	−76.151 156	−0.141 318	0.978 19
Double and quadruple excitations	14 817	−76.155 697	−0.145 859	0.976 68
Single, double, triple, and quadruple excitations	17 678	−76.157 603	−0.147 765	0.975 43
Full configuration mixing	256 473	−76.157 866	−0.148 028	0.975 28

[a] taken from the work of Saxe *et al.* (1981); all energies are in hartree; C_0 is the coefficient of the reference configuration.

systems (e.g. Bunge 1976, 1980). For molecules, Handy and his co-workers (Handy 1980; Saxe, Schaefer, and Handy 1981) have developed techniques for performing full configuration mixing calculation using small basis sets. Their method assumes that all two-electron integrals can be stored in the computer's central core memory simultaneously. Although the basis set truncation errors in such studies are extremely large, they do provide invaluable benchmark results with which expectation values obtained by means of other techniques, using the same basis set, may be compared. In Table 7.3 the results of a full configuration mixing calculation for the ground state of the water molecule are compared with various limited configuration mixing expansions; all calculations being performed with a basis set of 'double zeta' quality.

The group theoretical aspects of the electron correlation problem, which were discussed in Chapter 5, have seen wide usage in applications of the method of configuration mixing. The theory of the symmetric group and that of the unitary group have been employed in the evaluation of matrix elements of the hamiltonian. Such matrix elements may be written in the form

$$\int \Phi_i^* \hat{\mathscr{H}} \Phi_j \, d\tau = \sum_{IJ} a_{IJ}^{ij} f_{IJ} + \sum_{IJKL} b_{IJKL}^{ij} g_{IJKL} \tag{7.2.25}$$

where f_{IJ} is a one-electron integral and g_{IJKL} is a two-electron integral. The coefficients a_{IJ}^{ij} and b_{IJKL}^{ij} depend on the nature of the configurations Φ_i^* and Φ_j and on the space–spin coupling schemes defining them. The use of the theory of the symmetric group, and of the unitary group, leads to efficient computational schemes for the evaluation of the coefficients a_{IJ}^{ij} and b_{IJKL}^{ij}.

Exploitation of the theory of the symmetric group in configuration mixing studies has been investigated by Gallup and Norbeck (1976), by Karwowski and Duch (Karwowski 1973*a*,*b*; Duch and Karwowski 1979; Duch 1980) Karwowski and Duch (1977), and by Rettrup and Sharma (1977). Duch (1980) has devised an algorithm for performing direct configuration mixing calculations using a general multiconfigurational reference function.

The use of the unitary group in configuration mixing has been investigated by Shavitt (1977*a*, 1978), by Schafer and his co-workers (Brooks and Schaefer 1979; Brooks, Laudig, Saxe, Goddard, and Schaefer 1981), and by Siegbahn (1979, 1980). In reviewing the impact of the unitary group approach on matrix element evaluation, Shavitt (1979*b*) writes:

> ... the total computational effort required for formula determination in multireference direct CI calculations becomes negligible in comparison with the rate-determining steps, the integral transformation and the eigenvector iterations procedure.

Wormer and Paldus (1979, 1980), Wormer (1981), and Paldus and Wormer (1979) have investigated the relation between the symmetric group approach and the unitary group approach to the evaluation of matrix elements (7.2.25) and the more traditional methods such as the method of bonded functions (McWeeny and Cooper 1966; Sutcliffe 1966).

7.3 The group function model

One of the simplest and physically more appealing methods for the description of electron correlation effects in molecules which avoids, to some extent, the spurious dependence on the number of electrons observed in the method of limited configuration mixing, is the group function model. In the group function model the quantum mechanical description of molecules is simplified by emphasizing the partition of the system into weakly interacting groups of electrons. The philosophy of this approach has been succinctly expressed by Parr (1963), who commented that we should 'destroy the fallacy that large molecules are forever inaccessible to accurate treatment... in a molecule there are never more than a few electrons in one region of space'. The pair function model, or geminal model, initialized in the pioneering work of Fock (1950) and of Hurley, Lennard-Jones, and Pople (1953) is the simplest example of the group function approach. Since the concept of the electron pair is fundamental to the theory of chemical bonding, the pair function or geminal model has been widely studied (e.g. McWeeny and Ohno 1960; Csizmadia, Sutcliffe, and Barnett 1965; Klessinger and McWeeny 1965; Miller and Ruedenberg 1968*a*,*b*; Ahlrichs and Kutzelnigg 1968; Silver 1969; Silver, Mehler, and Ruedenberg 1970; Wilson and Gerratt 1975). Often, however, there may be more than a single pair of electrons in one particular region of space. It is then desirable to describe these strongly interacting electrons by a group function (Parr, Ellison, and Lykos 1956; McWeeny 1959; McWeeny 1960; McWeeny and Sutcliffe 1963; McWeeny and Sutcliffe 1976).

In the group function model the spatial wavefunction is approximated by

$$\Phi = \prod_{\mu=1}^{\mu=n} \Phi_\mu \qquad (7.3.1)$$

where the group function Φ_μ describes the μth group of electrons. The division of an N-electron system into weakly interacting subsystems is, of course, somewhat arbitrary and the success of the model will depend on our ability to achieve this partition in a reasonable fashion. In order to obtain a tractable formalism, group functions are usually required to satisfy some orthogonality conditions which will be discussed further below.

Let us first consider the simplest of the group function models, the pair function model. In general, the wavefunction employed in the antisymmetrized product of geminals approach may be written in the form

$$\Psi_{S,M} = \mathscr{A}(\Lambda_1(\mathbf{r}_1, \mathbf{r}_2)\, \Lambda_2(\mathbf{r}_3, \mathbf{r}_4) \ldots \Lambda_{\frac{1}{2}N}(\mathbf{r}_{N-1}, \mathbf{r}_N)\Theta_{S,M}) \qquad (7.3.2)$$

where $\mathscr{A}$ is the idempotent antisymmetrizing operator, and $\Theta_{S,M}$ is the total spin function for the system. The spatial wavefunction is a product of $\frac{1}{2}N$ two-electron functions, $\Lambda_{\frac{1}{2}i}(\mathbf{r}_{i-1}, \mathbf{r}_i)$, otherwise termed pair functions or geminals. The geminals are required to be normalized

$$\int \Lambda_i^*(\mathbf{r}_1, \mathbf{r}_2)\Lambda_i(\mathbf{r}_1, \mathbf{r}_2)\, d\mathbf{r}_1\, d\mathbf{r}_2 = 1 \qquad (7.3.3)$$

and, in the most commonly used antisymmetrized product of strongly orthogonal geminals method, they are required to satisfy the condition

$$\int \Lambda_i^*(\mathbf{r}_1, \mathbf{r}_2)\Lambda_j(\mathbf{r}_1, \mathbf{r}_2)\, d\mathbf{r}_1 = 0, \qquad i \neq j \qquad (7.3.4)$$

which is the strong orthogonality condition which will be discussed further below.

Geminal models of molecular electronic structure are appealing chemically, since they constitute a quantum mechanical representation of the classical theory of bonding advocated, in the early part of this century by Lewis (1916) and Langmuir (1919). Hurley, Lennard-Jones, and Pople (1953) introduced the geminal model as an extension of the molecular orbital theory to include a description of the electrostatic interactions between electrons associated with the same molecular orbital. Hurley *et al.* observe that for a molecular orbital wavefunction 'electrons of the same spin may . . . be said to be correlated in their spatial arrangement by the property of exclusion. Electrostatic forces play only a secondary role in causing electrons of the same spin to avoid each other'. However, 'the position of one electron relative to another of opposite spin is influenced only by the electrostatic force between them'. Hurley *et al.* also regarded the geminal model as an extension of the valence bond approach. If the geminals are restricted to have the form of a product of two non-orthogonal orbitals

$$\Lambda_i(\mathbf{r}_1, \mathbf{r}_2) = \phi_{ia}(\mathbf{r}_1)\phi_{ib}(\mathbf{r}_2) \qquad (7.3.5)$$

each electron pair is described by a Coulson–Fischer type wavefunction, which was discussed in Chapter 2.

Let us now examine the form of the energy expression obtained with the antisymmetrized product of strongly orthogonal geminals approach.

The expectation value of the molecular hamiltonian for a wavefunction of the form (7.3.2) is given by

$$E_{Sk}=(\Delta_{Skk})^{-1}\sum_{P\in S_N} U^{NS}_{kk}(P)\,\langle\Lambda_1\,\Lambda_2\ldots\Lambda_{\frac{1}{2}N}|$$
$$\times\left(\sum_i h_i+\sum_{i>j} g_{ij}\right)|P^r\Lambda_1\,\Lambda_2\ldots\Lambda_{\frac{1}{2}N}\rangle \qquad (7.3.6)$$

where h_i is the one-electron hamiltonian for the ith electron and $g_{ij}=1/r_{ij}$. $U^{NS}_{kk}(P)$ is an element of the representation matrices for an appropriate spin permutation group. The only permutations which lead to non-vanishing contributions to (7.3.6), when the geminals are required to be strongly orthogonal are those which permute the coordinates of the two electrons associated with a given geminal and the transposition P^r_{ij} which corresponds to a term $1/r_{ij}$. The normalization integral is given by

$$\Delta_{Skk}=d_1^{Sk}\,d_2^{Sk}\ldots d_{\frac{1}{2}N}^{Sk} \qquad (7.3.7)$$

where, if we employ the Serber coupling scheme (see Section 5.6),

$$d_i^{Sk}=\langle\Lambda_i|\,(I+(-1)^{S_i}P^r_{2i-1,2i})\,|\Lambda_i\rangle. \qquad (7.3.8)$$

in which S_i is the total spin for the ith geminal.

The expectation value of the total hamiltonian can be written as

$$E_{Sk}=\sum_{i=1}\{H_i^{Sk}+G_i^{Sk}\}+\sum_{i>j=1}\{J_{ij}^{Skk}+U^{NS}_{kk}(P_{2i,2j})K_{ij}^{Skk}\} \qquad (7.3.9)$$

in which the one-electron term is defined by

$$H_i^{Sk}=(d_i^{Sk})^{-1}\,\langle\Lambda_i|\,(h_1+h_2)(I+(-1)^{S_k}P^r_{12})\,|\Lambda_i\rangle \qquad (7.3.10)$$

and is associated entirely with intrageminal effects. The intrageminal two-electron Coulomb and exchange terms are given by

$$G_i^{Sk}=(d_i^{Sk})^{-1}\,\langle\Lambda_i(\mathbf{r}_1,\mathbf{r}_2)|\,g_{12}(I+(-1)^{S_i}P^r_{12})\,|\Lambda_i(\mathbf{r}_1,\mathbf{r}_2)\rangle. \qquad (7.3.11)$$

The intergeminal Coulomb interaction is given by

$$J_{ij}^{Skk}=(d_i^{Sk}d_j^{Sk})^{-1}\,\langle\Lambda_i(\mathbf{r}_1,\mathbf{r}_2)\Lambda_j(\mathbf{r}_3,\mathbf{r}_4)|\,(g_{13}+g_{14}+g_{23}+g_{24})$$
$$\times\{1+(-1)^{S_i}P^r_{12}\}\{1+(-1)^{S_i}P^r_{34}\}\,|\Lambda_i(\mathbf{r}_1,\mathbf{r}_2)\Lambda_j(\mathbf{r}_3,\mathbf{r}_4)\rangle \qquad (7.3.12)$$

and the intergeminal exchange interaction by

$$K_{ij}^{Skk}=(d_i^{Sk}d_j^{Sk})^{-1}\,\langle\Lambda_i(\mathbf{r}_1,\mathbf{r}_2)\Lambda_j(\mathbf{r}_3,\mathbf{r}_4)|\,(g_{13}P^r_{13}+g_{14}P^r_{14}+g_{23}P^r_{23}+g_{24}P^r_{24})$$
$$\times\{1+(-1)^{S_i}P^r_{12}\}\{1+(-1)^{S_j}P^r_{34}\}\,|\Lambda_i(\mathbf{r}_1,\mathbf{r}_2)\Lambda_j(\mathbf{r}_3,\mathbf{r}_4)\rangle. \qquad (7.3.13)$$

Matrix elements of the hamiltonian between geminal product functions which differ in the spin coupling index k are given by

$$\langle\Psi_{S,M;l}|\,\hat{\mathcal{H}}\,|\Psi_{S,M;k}\rangle=\sum_{i>j} U^{NS}_{lk}(P_{2i,2j})K_{ij}^{Slk} \qquad (7.3.14)$$

the only surviving contribution being that due to the intergeminal exchange interaction

$$
\begin{aligned}
K_{ij}^{Slk} = (d_i^{Sk} d_j^{Sk} d_i^{Sl} d_j^{Sl})^{-\frac{1}{2}} \langle \Lambda_i(\mathbf{r}_1, \mathbf{r}_2) \Lambda_j(\mathbf{r}_3, \mathbf{r}_4)| \\
\times \{(-1)^{S_i^l + S_j^l + S_i^k + S_j^k} g_{13} P_{13}^r + (-1)^{S_i^l + S_i^k} g_{14} P_{14}^r \\
+ (-1)^{S_j^l + S_j^k} g_{23} P_{23}^r + g_{24} P_{24}^r\} \{I + (-1)^{S_i^k} P_{12}^r\} \\
\times \{I + (-1)^{S_j^k} P_{34}\} |\Lambda_i(\mathbf{r}_1, \mathbf{r}_2) \Lambda_j(\mathbf{r}_3, \mathbf{r}_4)\rangle.
\end{aligned} \tag{7.3.15}
$$

in which S_i^k is the spin of the ith geminal in $\Psi_{SM;k}$.

In general, a geminal may be written as a bilinear expansion of the form

$$\Lambda_i(\mathbf{r}_1, \mathbf{r}_2) = \sum_{k,l} a_{kl}^{(i)} \varphi_k^{(i)}(\mathbf{r}_1) \varphi_l^{(i)}(\mathbf{r}_2). \tag{7.3.16}$$

By means of a linear transformation, this can be rewritten in natural orbital, or pure quadratic, form

$$\Lambda_i(\mathbf{r}_1, \mathbf{r}_2) = \sum_k C_k^{(i)} \phi_k^{(i)}(\mathbf{r}_1) \phi_k^{(i)}(\mathbf{r}_2). \tag{7.3.17}$$

It is clear that the strong orthogonality condition (7.3.4) will be satisfied if the orbitals, $\varphi_l^{(i)}$ or $\phi_k^{(i)}$, employed in these parameterizations of the geminals can be divided into mutually orthogonal subsets, that is

$$\langle \phi_k^{(i)} | \phi_l^{(j)} \rangle = 0 \quad \text{if} \quad i \neq j. \tag{7.3.18}$$

The converse statement, that the strong orthogonality condition implies the partition of the orbitals employed in a bilinear expansion into mutually orthogonal sets, has been proved by Arai (1960) and by Löwdin (1961) and will be shown below.

For a system of $M = \frac{1}{2}N$ well-separated two-electron systems, for example, helium atoms, the antisymmetrized product of strongly orthogonal geminals ansatz leads to the exact non-relativistic ground state wave function. For an array of M helium atoms both the Hartree–Fock theory and the limited single and double excitation configuration mixing method lead to unphysical results, as N becomes large (see the discussion of Section 7.2). Limited configuration mixing yields a vanishingly small correlation energy as M tends to infinity and it is, therefore, instructive to examine the relation between the antisymmetrized product of strongly orthogonal geminals approach and the configuration mixing approach in more detail. Consider a N-electron primitive spatial wavefunction

$$\Phi = \Lambda_1(\mathbf{r}_1, \mathbf{r}_2)\, \Lambda_2(\mathbf{r}_3, \mathbf{r}_4) \ldots \Lambda_{\frac{1}{2}N}(\mathbf{r}_{N-1}, \mathbf{r}_N) \tag{7.3.19}$$

in which each geminal is expanded in natural orbital form. This spatial function may be rewritten as

$$\begin{aligned}\Phi = C_1^{(1)}C_1^{(2)}\dots C^{(\frac{1}{2}N)}\Big[&\phi_1^{(1)}\phi_1^{(1)}\phi_1^{(2)}\phi_1^{(2)}\dots\phi_1^{(\frac{1}{2}N)}\phi_{(1)}^{\frac{1}{2}N}\\ &+\sum_{k=2}^{\infty}\left\{\frac{C_k^{(1)}}{C_1^{(1)}}\phi_k^{(1)}\phi_k^{(1)}\phi_1^{(2)}\phi_1^{(2)}\dots+\frac{C_k^{(2)}}{C_1^{(2)}}\phi_1^{(1)}\phi_1^{(1)}\phi_k^{(2)}\phi_k^{(2)}\dots\right\}\\ &+\sum_{k=2}^{\infty}\sum_{l=2}^{\infty}\left\{\frac{C_k^{(1)}C_l^{(2)}}{C_1^{(1)}C_1^{(2)}}\phi_k^{(1)}\phi_k^{(1)}\phi_l^{(2)}\phi_l^{(2)}\dots\right\}+\dots\Big].\end{aligned} \tag{7.3.20}$$

The first term in (7.3.20) is constructed from the principal natural orbitals and closely resembles the Hartree–Fock function. The first summation represents all configurations which are doubly excited with respect to the principal natural orbital configuration. The second, third, ... summations are over configurations which are quadruply excited, sextuply excited, ..., respectively. The coefficients of these quadruply excited, sextuply excited, ... states are merely products of the double-excitation coefficients. The terms in (7.3.20) involving quadruple- and higher-order excitations are said to be of unlinked cluster type. An extremely large number of terms would have to be included in a configuration mixing expansion in order to allow an exact description of M well-separated electron pairs. The antisymmetrized product of strongly orthogonal geminals ansatz provides a much more efficient and economical description of the correlation effects which arise in systems consisting of well separated pairs of electrons than does the limited configuration mixing approach.

The strongly orthogonal geminal model provides an exact description of intrageminal correlation effects if a sufficiently large number of terms are included in the bilinear expansion (7.3.16) or (7.3.17). However, the method provides no description of intergeminal correlation effects. Clearly, the number of intrageminal interactions in an N electron system increases as $\frac{1}{2}N$, whereas the number of intergeminal interactions increases as $\frac{1}{2}N(N-2)$. Thus, as pointed out by Sinanoğlu and Skutnik (1968), even though in a system of well-separated electron pairs the intergeminal correlation effects are small, the total number of intergeminal terms increases as $\sim N^2$, as compared with N for the intrageminal terms, and thus the former will eventually dominate as systems of increasing size are considered. Sinanoğlu and Skutnik (1968) demonstrated that in a description of the methane molecule in the geminal model the components of the correlation energy corresponding to intergeminal effects have a total magnitude which is roughly twice the total correlation energy arising from intrageminal effects. In Table 7.4, the decreasing percentage of the total correlation energy recovered by the

Table 7.4

Percentage of the correlation energy recovered by the antisymmetrized product of strongly orthogonal geminals ansatz for some simple systems

System	Number of electrons	Percentage of correlation energy recovered	Reference
Be	4	91%	a
LiH	4	89%	b
BH	6	74%	b
NH	8	20%	c
NH_3	10	48%	d

[a] Miller and Ruedenberg (1968*a*,*b*).
[b] Mehler, Ruedenberg, and Silver (1970).
[c] Silver, Mehler, and Ruedenberg (1970).
[d] Saunders and Guest (1974).

antisymmetrized product of strongly orthogonal geminals ansatz is illustrated. The beryllium atom and the lithium hydride molecule are systems which contain two pairs of well-separated electrons and the strongly orthogonal geminal model provides an accurate description of such systems, yielding approximately 90 per cent of the total correlation energy. However, for ten-electron systems, such as NH_3, only 50 per cent of the correlation energy is recovered by the strongly orthogonal geminal approach, in spite of the fact that in the qualitative picture of the electronic structure of such molecules afforded by the theories of Lewis (1916), Sidgwick and Powell (1940), and Gillespie and Nyholm (1957), the electron pairs are well defined and believed to be well separated.

The strongly orthogonal geminal description of electronic structure can be improved in two ways. Firstly, the strong orthogonality restriction can be removed. Geminals can be required to be weakly orthogonal, that is

$$\int \Lambda_i^*(\mathbf{r}_1, \mathbf{r}_2)\Lambda_j(\mathbf{r}_1, \mathbf{r}_2)\, d\mathbf{r}_1\, d\mathbf{r}_2 = 0; \qquad i \neq j \tag{7.3.21}$$

or they can be allowed to be non-orthogonal. The strongly orthogonal geminal ansatz does not describe any intergeminal correlation; the non-orthogonal geminal ansatz can be said to describe intergeminal static correlation effects. The second method by means of which the strongly orthogonal geminal model can be improved, is by admitting intergroup mixing. To obtain the exact wavefunction, in fact, all partitions of the electrons into pairs must be included. However, in discussing this more general strongly orthogonal group function model, Parks and Parr (1958) suggest that 'it is much more profitable to improve the wavefunction within each group than to refine the whole wavefunction by admitting intergroup mixing'. This philosophy retains the physical interpretability of the model. If intergroup mixing is of importance, then the groups are not well separated and group functions describing larger numbers of electrons should be assumed.

In order to obtain a tractable formalism, geminals, or more generally, group functions, are usually required to satisfy some orthogonality condition. In most studies, group functions have been required to be strongly orthogonal; a condition which corresponds to the assumption that physically only the exchange of single electrons between the groups is important (Wilson 1976, 1978*a*). The strong orthogonality condition does not allow for the description of any intergroup correlation effects.

Let us consider the strong orthogonality condition in more detail. Consider two group functions Φ_μ and Φ_ν. If the groups of electrons which Φ_μ and Φ_ν describe are separated by an infinite distance, then $\phi_{\mu i} \in \mathscr{E}_\mu$ and $\phi_{\nu i} \in \mathscr{E}_\nu$ where $\phi_{\mu i}$ denotes the set of orbitals used to construct the group function Φ_μ and $\mathscr{E}_\mu$ and $\mathscr{E}_\nu$ are mutually orthogonal subspaces. It will now be shown that the strong orthogonality condition, which for group functions takes the form

$$\int \Phi_\mu^*(\mathbf{r}_1, \mathbf{r}_2, \ldots, \mathbf{r}_{N_\mu})\Phi_\nu(\mathbf{r}_1, \mathbf{r}_2, \ldots, \mathbf{r}_{N_\nu})\, \mathrm{d}r_1 = 0 \qquad \mu \neq \nu \quad (7.3.22)$$

is a necessary and sufficient condition that $\phi_{\mu i} \in \mathscr{E}_\mu$ and $\phi_{\nu i} \in \mathscr{E}_\nu$ where $\mathscr{E}_\mu$ and $\mathscr{E}_\nu$ are again orthogonal subspaces even though the two groups of electrons are no longer an infinite distance apart.

Let $\{\phi_i\}$ denote a complete set of one-electron functions and let

$$\Phi_\mu = \sum_{\mu_1, \mu_2, \ldots} p_{\mu_1\mu_2\ldots}\phi_{\mu_1}(\mathbf{r}_1)\phi_{\mu_2}(\mathbf{r}_2)\ldots \quad (7.3.23)$$

and

$$\Phi_\nu = \sum_{\nu_1, \nu_2, \ldots} q_{\nu_1\nu_2\ldots}\phi_{\nu_1}(\mathbf{r}_1)\phi_{\nu_2}(\mathbf{r}_2)\ldots \quad (7.3.24)$$

with

$$\langle \phi_{\mu i} \mid \phi_{\nu j}\rangle = 0; \qquad \forall i, \forall_j, \mu \neq \nu. \quad (7.3.25)$$

The sets $\{\phi_{\mu i}\}$ and $\{\phi_{\nu j}\}$ span mutually orthogonal subspaces. Now eqn (7.3.23) may be written

$$\Phi_\mu = \sum_{\mu_1} p_{\mu_1}\phi_{\mu_1}(\mathbf{r}_1) \quad (7.3.26)$$

where

$$p_{\mu_1} = \sum_{\mu_2\mu_3\ldots} p_{\mu_2\mu_3\ldots}\phi_{\mu_2}(\mathbf{r}_2)\phi_{\mu_3}(\mathbf{r}_3)\ldots. \quad (7.3.27)$$

Similarly,

$$\Phi_\nu = \sum_{\nu_1} q_{\nu_1}\phi_{\nu_1}(\mathbf{r}_1). \quad (7.3.28)$$

Substituting eqns (7.3.26) and (7.3.28) into the left-hand side of eqn (7.3.22) and using eqn (7.3.25) gives

$$\sum_{\mu_1,\nu_1} p_{\mu_1}q_{\nu_1} \int \phi^*_{\mu_1}(\mathbf{r}_1)\phi_{\nu_1}(\mathbf{r}_1)\,\mathrm{d}\mathbf{r}_1 = 0 \tag{7.3.29}$$

and thus proves that the strong orthogonality condition is a necessary condition for the functions $\{\phi_{\mu i}\}$ and $\{\phi_{\nu i}\}$ to span mutually orthogonal subspaces. Now, following Arai (1960) and Löwdin (1961), sufficiency will be proved.

Let $\boldsymbol{\phi}$ denote a complete set of one-electron function and put

$$\Phi_\mu = \boldsymbol{\phi}\mathbf{a} = \sum_i \phi_i a_i$$

and (7.3.30)

$$\Phi_\nu = \boldsymbol{\phi}\mathbf{b} = \sum_i \phi_i b_i$$

where $\mathbf{a}^\dagger = (a_1, a_2, \ldots)$ and $\mathbf{b}^\dagger = (b_1, b_2, \ldots)$ are coefficients. Substituting eqns (7.3.30) into the condition (7.3.22) leads to

$$\sum_i \sum_j a_i^* b_j \int \phi_i^*(\mathbf{r}_1)\phi_j(\mathbf{r}_1)\,\mathrm{d}\mathbf{r}_1 = 0$$

$$\Rightarrow \sum_i a_i^* b_i = 0 \tag{7.3.31}$$

or

$$\mathbf{a}^\dagger\mathbf{b} = 0. \tag{7.3.32}$$

Multiplying from the left by $\mathbf{a}$

$$\mathbf{a}\mathbf{a}^\dagger\mathbf{b} = \mathbf{0} \tag{7.3.33}$$

and introducing the unitary matrix $\mathbf{U}$ yields

$$(\mathbf{U}^\dagger\mathbf{a}\mathbf{a}^\dagger\mathbf{U})\mathbf{U}^\dagger\mathbf{b} = \mathbf{0}. \tag{7.3.34}$$

Choosing $\mathbf{U}$ so that the term in parenthesis in eqn (7.3.34) is the diagonal matrix $\mathbf{D}$ and putting

$$\boldsymbol{\alpha} = \mathbf{U}^\dagger\mathbf{a} \qquad \boldsymbol{\beta} = \mathbf{U}^\dagger\mathbf{b} \tag{7.3.35}$$

leads to

$$\boldsymbol{\alpha}\boldsymbol{\alpha}^\dagger\boldsymbol{\beta} = \mathbf{D}\boldsymbol{\beta} = \mathbf{0} \tag{7.3.36}$$

which implies that

$$D_{ii}\beta_i = 0. \tag{7.3.37}$$

Since $\boldsymbol{\alpha}\boldsymbol{\alpha}^\dagger$ is an hermitian matrix

$$D_{ii} = \alpha_i\alpha_i^* = |\alpha_i|^2 \geqslant 0 \tag{7.3.38}$$

and

$$D_{ii} = 0 \Rightarrow \alpha_i = 0. \tag{7.3.39}$$

Hence from eqn (7.3.37)

$$\begin{aligned} \alpha_i \neq 0 &\Rightarrow \beta_i = 0 \\ \beta_i \neq 0 &\Rightarrow \alpha_i = 0 \end{aligned} \tag{7.3.40}$$

where $\boldsymbol{\alpha}$ and $\boldsymbol{\beta}$ are expansion coefficients in the set $\boldsymbol{\psi} = \boldsymbol{\phi}\mathbf{U}$. Hence the sufficient conditions has been obtained; if two functions are strongly orthogonal, they are constructed from orbitals spanning mutually orthogonal subspaces.

For the group function model, the strong orthogonality condition (7.3.22) is just one of a set of constraints which the group functions may be required to obey. If two group functions obey the condition

$$\begin{aligned} \int \Phi_\mu^*(P_\mu\{\mathbf{r}_1, \mathbf{r}_2, \ldots, \mathbf{r}_p, \mathbf{r}'_{p+1}, \ldots, \mathbf{r}_{N_\mu}\}) \\ \times \Phi_\nu(P_\nu\{\mathbf{r}_1, \mathbf{r}_2, \ldots, \mathbf{r}_p, \mathbf{r}''_{p+1}, \ldots, \mathbf{r}_{N_\nu}\}) \, d\mathbf{r}_1 \, d\mathbf{r}_2 \ldots d\mathbf{r}_p = 0 \\ \mu \neq \nu; \qquad \forall P_\mu \in S_{N_\mu}; \qquad \forall P_\nu \in S_{N_\nu} \end{aligned} \tag{7.3.41}$$

they are said to be p-orthogonal (Wilson 1976; Wilson 1978). The imposition of this constraint on the group functions Φ_μ and Φ_ν ensures that the contributions to the electronic energy arising from the interchange of more than p electrons between these two groups are identically zero. The most restrictive condition which can be placed on the group functions is the 1-orthogonality, or strong orthogonality, condition. If the group functions Φ_μ and Φ_ν obey the p-orthogonality condition, then the N_μ-electron and N_ν-electron configurations in which these functions are expanded may have no more than $(p-1)$ of the orbitals in common. For the simplest case of two group functions, the orbitals are divided into three sets S_μ, S_ν, and $S_{\mu\nu}$. To maintain the p-orthogonality condition, Φ_μ is expanded in N_μ-electron configurations built from the orbitals of the direct sum $S_\mu \oplus S_{\mu\nu}$. Φ_ν, on the other hand, is expanded in N_ν-electron configurations constructed from the orbitals of $S_\nu \oplus S_{\mu\nu}$. However, no configuration may have more than $(p-1)$ orbitals from $S_{\mu\nu}$. In the most restrictive case, that is strong, or 1-orthogonality, $S_{\mu\nu}$ is a null set and the orbitals are divided into mutually disjoint sets.

For an arbitrary spin-independent operator, $\hat{\Lambda}$, which is symmetric in the spatial coordinates, the expectation value in the group function model may be written as

$$\Lambda = \sum_{P \in S_N} \mathbf{U}^{NS}(P) \left\langle P\left\{\prod_{\mu=1} \Phi_\mu\right\}\right| \hat{\Lambda} \left|\left\{\prod_{\mu=1} \Phi_\mu\right\}\right\rangle. \tag{7.3.42}$$

Now any permutation $P \in S_N$ may be expressed in the form

$$P = QR$$
$$R = \prod_{\mu} P_{\mu} \qquad P_{\mu} \in S_{N_{\mu}}. \tag{7.3.43}$$

The permutation Q interchanges electrons between the group functions. For a molecular system which has been divided into several subsystems, it is very useful to represent these interchanges by arrow diagrams (Herring 1966; Matsen and Klein 1971; Wilson 1978*a*). In these diagrams, each dot represents a group function and each arrow the transfer of a single electron between the group functions. Diagrams containing up to six arrows are displayed in Fig. 7.3. It is assumed that the number of electrons in each group remains constant and charge-transfer effects will be negligible if the electrons are partitioned in a reasonable fashion. Now from (7.3.43) it should be recognized that the representation matrices for the symmetric group satisfy

$$\mathbf{U}^{NS}(P) = \mathbf{U}^{NS}(Q) \prod_{\mu=1} \mathbf{U}^{NS}(P_{\mu}) \tag{7.3.44}$$

which, together with (7.3.42), leads to

$$\mathbf{\Lambda} = \sum_{Q} \mathbf{U}^{NS}(Q) \sum_{R} \left\{ \prod_{\mu} \mathbf{U}^{NS}(P_{\mu}) \right\} \left\langle Q \prod_{\mu} \{P_{\mu} \Phi_{\mu}\} \right| \hat{\Lambda} \left| \prod_{\mu} \Phi_{\mu} \right\rangle \tag{7.3.45}$$

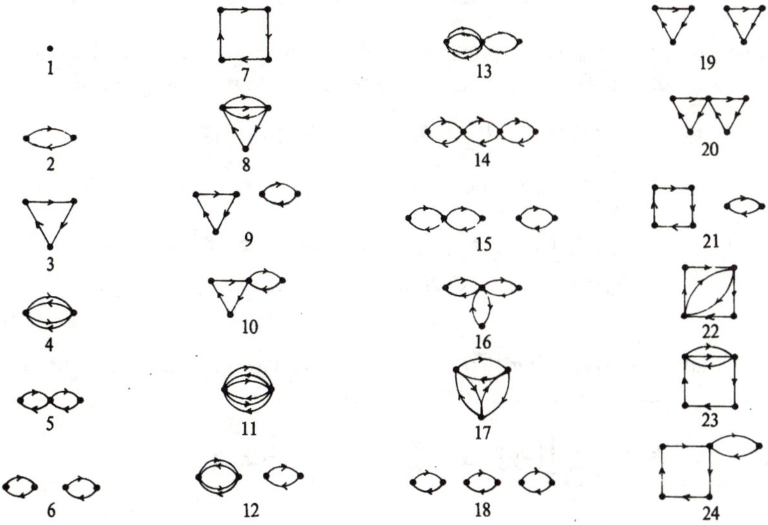

FIG. 7.3. Arrow diagrams representing interchanges between group functions.

and thus the expectation value of $\hat{\Lambda}$ may be written in the form

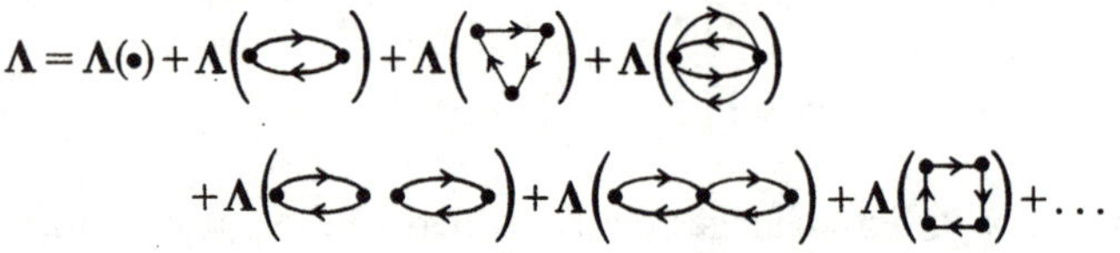

(7.3.46)

that is, an expansion over arrow diagrams. In the first term, there are no interchanges of electrons between the groups and the description of intergroup effects is purely classical.

To make further progress, it is necessary to be more specific about the nature of the operator $\hat{\Lambda}$. Here the overlap matrix ($\hat{\Lambda}=\hat{I}$)

$$\mathbf{\Delta}_{S,M}=\sum_{P\in S_N}\mathbf{U}^{NS}(P)\left\langle P^r\left\{\prod_\mu \Phi_\mu\right\}\middle|\left\{\prod_\mu \Phi_\mu\right\}\right\rangle \tag{7.3.47}$$

will be considered; one-electron and two-electron components of a spin-independent operator are considered in detail elsewhere (Wilson 1978). The overlap term in which no electrons are interchanged between the group functions is a simple product

$$\mathbf{\Delta}_{S,M}(\cdot)=\prod_\mu \mathbf{\Delta}_\mu \tag{7.3.48}$$

where each factor is associated with a single group function

$$\mathbf{\Delta}_\mu=\sum_{P\in S_{N_\mu}}\mathbf{U}^{NS}(P_\mu)\langle P_\mu\Phi_\mu\,|\,\Phi_\mu\rangle. \tag{7.3.49}$$

The other terms in the expansion for the overlap matrix can be expressed in terms of the m-electron spin-independent group density matrices

$$\begin{aligned}\boldsymbol{\rho}_\mu(i,\ldots,m;i',\ldots,m')&=\boldsymbol{\rho}_\mu(\mathbf{r}_i,\ldots,\mathbf{r}_m;\mathbf{r}'_i,\ldots,\mathbf{r}'_m)\\ &=\sum_{P_\mu\in S_{N_\mu}}\mathbf{U}^{N,S}(P)\int d\mathbf{r}_1\ldots d\mathbf{r}_{i-1}\,d\mathbf{r}_{i+1}\ldots d\mathbf{r}_{m-1}\,d\mathbf{r}_{m+1}\ldots d\mathbf{r}_{N_\mu}\\ &\quad\times\Phi^*_\mu(P_\mu\{\mathbf{r}_1,\ldots,\mathbf{r}_{i-1},\mathbf{r}_i,\mathbf{r}_{i+1},\ldots,\mathbf{r}_{m-1},\mathbf{r}_m,\mathbf{r}_{m+1},\ldots,\mathbf{r}_{N_\mu}\})\\ &\quad\times\Phi_\mu(\mathbf{r}_1,\ldots,\mathbf{r}_{i-1},\mathbf{r}'_i,\mathbf{r}_{i+1},\ldots,\mathbf{r}_{m-1},\mathbf{r}'_m,\mathbf{r}_{m+1},\ldots,\mathbf{r}_{N_\mu}).\end{aligned} \tag{7.3.50}$$

For diagrams containing two arrows, the overlap term may be written in terms of the one-electron group density matrix

$$\Leftrightarrow\frac{1}{2!}\sum_{\mu,\nu}{}'\left[\left(\prod_{\gamma\neq\mu,\nu}\mathbf{\Delta}_\gamma\right)\left(\sum_{\substack{Q_{ij}\\ i\in\mu\\ j\in\nu}}\left\{\mathbf{U}^{N,S}(Q_{ij})\int d\mathbf{r}_i\,d\mathbf{r}_j\times\boldsymbol{\rho}_\mu(\mathbf{r}_j;\mathbf{r}_i)\boldsymbol{\rho}_\nu(\mathbf{r}_i;\mathbf{r}_j)\right\}\right)\right] \tag{7.3.51}$$

while for diagrams containing three arrows

$$\Leftrightarrow \frac{1}{3!}\sum_{\mu,\nu,\sigma}' \left[\left(\prod_{\gamma\neq\mu,\nu,\sigma} \Delta_\gamma\right)\left(\sum_{\substack{Q_{ijk}\\ i\in\mu\\ j\in\nu\\ k\in\sigma}} \left\{\mathbf{U}^{N,S}(Q_{ijk})\int d\mathbf{r}_i\, d\mathbf{r}_j\, d\mathbf{r}_k \boldsymbol{\rho}_\mu(\mathbf{r}_k;\mathbf{r}_i) \times \boldsymbol{\rho}_\nu(\mathbf{r}_i;\mathbf{r}_j)\boldsymbol{\rho}_\sigma(\mathbf{r}_j;\mathbf{r}_k)\right\}\right)\right] \quad (7.3.52)$$

and for diagrams containing four arrows

$$\Leftrightarrow \frac{1}{2!}\sum_{\mu,\nu}' \left[\left(\prod_{\gamma\neq\mu,\nu} \Delta_\gamma\right)\left(\sum_{\substack{Q_{ijkl}\\ i,j\in\mu\\ k,l\in\nu}} \left\{\mathbf{U}^{N,S}(Q_{ijkl})\int d\mathbf{r}_i\, d\mathbf{r}_j\, d\mathbf{r}_k\, d\mathbf{r}_l \times \boldsymbol{\rho}_\mu(\mathbf{r}_k,\mathbf{r}_l;\mathbf{r}_i,\mathbf{r}_j)\boldsymbol{\rho}_\nu(\mathbf{r}_i,\mathbf{r}_j;\mathbf{r}_k,\mathbf{r}_l)\right\}\right)\right] \quad (7.3.53)$$

$$\Leftrightarrow \frac{1}{3!}\sum_{\mu,\nu,\sigma}' \left[\left(\prod_{\gamma\neq\mu,\nu,\sigma} \Delta_\gamma\right)\left(\sum_{\substack{Q_{ijkl}\\ i\in\mu; j,k\in\nu\\ k\in\sigma}} \left\{\mathbf{U}^{N,S}(Q_{ijkl})\int d\mathbf{r}_i\, d\mathbf{r}_j\, d\mathbf{r}_k\, d\mathbf{r}_l \times \boldsymbol{\rho}_\mu(\mathbf{r}_j;\mathbf{r}_i)\boldsymbol{\rho}_\nu(\mathbf{r}_i,\mathbf{r}_j;\mathbf{r}_l,\mathbf{r}_k)\boldsymbol{\rho}_\sigma(\mathbf{r}_k;\mathbf{r}_l)\right\}\right)\right] \quad (7.3.54)$$

$$\Leftrightarrow \frac{1}{4!}\sum_{\mu,\nu,\sigma,\tau}' \left[\left(\prod_{\gamma\neq\mu,\nu,\sigma,\tau} \Delta_\gamma\right) \sum_{\substack{Q_{ijkl}\\ i\in\mu; j\in\nu\\ k\in\sigma; l\in\tau}} \left\{\mathbf{U}^{N,S}(Q_{ijkl})\int d\mathbf{r}_i\, d\mathbf{r}_j\, d\mathbf{r}_k\, d\mathbf{r}_l \times \boldsymbol{\rho}_\mu(\mathbf{r}_l;\mathbf{r}_i)\boldsymbol{\rho}_\nu(\mathbf{r}_i;\mathbf{r}_j)\boldsymbol{\rho}_\sigma(\mathbf{r}_j;\mathbf{r}_k)\boldsymbol{\rho}_\tau(\mathbf{r}_k;\mathbf{r}_l)\right\}\right)\right] \quad (7.3.55)$$

$$\Leftrightarrow \left(\frac{1}{2!}\right)^2 \sum_{\substack{\mu,\nu,\sigma,\tau\\ (\mu,\nu)>(\sigma,\tau)}} \left[\left(\prod_{\substack{\gamma\neq\mu,\\ \nu,\sigma,\tau}} \Delta_\gamma\right)\left(\sum_{\substack{Q_{ijkl}\\ i\in\mu; j\in\nu;\\ k\in\sigma; l\in\tau}} \left\{\mathbf{U}^{N,S}(Q_{ijkl}) \times \int d\mathbf{r}_i\, d\mathbf{r}_j\, d\mathbf{r}_k\, d\mathbf{r}_l\, \boldsymbol{\rho}_\mu(\mathbf{r}_j;\mathbf{r}_i)\boldsymbol{\rho}_\nu(\mathbf{r}_i;\mathbf{r}_j)\boldsymbol{\rho}_\sigma(\mathbf{r}_l;\mathbf{r}_k)\boldsymbol{\rho}_\tau(\mathbf{r}_k;\mathbf{r}_l)\right\}\right)\right]. \quad (7.3.56)$$

The series terminates when the number of arrows connected to any dot exceeds the number of electrons in that group.

In this section, the extent to which intergroup correlation effects are described by various group functions and geminal models has been considered in some detail. Let us conclude by commenting on the description of intragroup correlation effects by the group function model. There is a large degree of freedom in the choice of the functional form of the individual group functions and any of the ansatz for describing electron correlation effects which are discussed in this chapter may be employed to represent the individual group functions—a limited configuration mixing approach, a cluster-type expansion, or diagrammatic perturbation theory. Indeed, different methods may be used to describe correlation effects in different group functions. The group function model provides a size-consistent description of intergroup correlation; the dependence of intragroup correlation energies on the number of electrons will depend on the form adopted for the group functions.

7.4 Cluster expansions

In the group function model, which is discussed in Section 7.3, the electrons in a molecule are divided into disjoint subsystems and electron correlation effects are described to differing degrees of accuracy depending upon their origin—that is, whether they arise from intragroup interactions of intergroup effects. This ansatz clearly leads to difficulties if the electrons do not fall into well separated groups. However, it is possible to decompose an N-electron system into smaller subsystems without dividing it into disjoint subsystems. This may be achieved by means of the cluster expansion approach (Sinanoğlu 1961*b*, 1962, 1964; Čížek 1966; see also Coester 1958; Coester and Kummel 1960). Each electron is then correlated with a small cluster of other electrons but is not restricted to belong to just one such cluster function.

Let us write the exact non-relativistic wavefunction in the form

$$\Psi = \Phi_0 + \chi \tag{7.4.1}$$

where Φ_0 is a reference function and χ is the correlation correction. This correlation correction may be written as an expansion of the form

$$\chi = \sum_{i} \chi^{(i)} + \sum_{i>j} \chi^{(ij)} + \sum_{i>j>k} \chi^{(ijk)} + \sum_{i>j>k>l} \chi^{(ijkl)} + \cdots \tag{7.4.2}$$

where $\chi^{(i)}$ is a one-electron cluster, $\chi^{(ij)}$ is a two-electron cluster, $\chi^{(ijk)}$ is a three-electron cluster, etc. $\chi^{(ijkl)}$ describes the correlation of four electrons and consists of four types of effects: (i) the simultaneous correlation of four electrons, (ii) the correlation of three electrons and a one-electron cluster function, (iii) the interaction of two pairs of electrons, and (iv) four one-electron cluster functions. Case (i) leads to what are termed linked clusters or connected clusters while cases (ii), (iii), and (iv) lead to unlinked or disconnected clusters. Any cluster can be described by a sum of linked and unlinked clusters. If the linked clusters are denoted by

$$\omega^{(i)}\Phi_0 \tag{7.4.3a}$$

$$\omega^{(ij)}\Phi_0 \ldots \text{etc.} \tag{7.4.3b}$$

where the operator $\omega^{(i_1 \ldots i_n)}$ results in an n-fold excitation in the reference determinant, then

$$\chi^{(i)} = \omega^{(i)}\Phi_0 \tag{7.4.4a}$$

$$\chi^{(ij)} = \omega^{(ij)}\Phi_0 + \omega^{(i)}\omega^{(j)}\Phi_0 \tag{7.4.4b}$$

$$\begin{aligned}\chi^{(ijk)} = {} & \omega^{(ijk)}\Phi_0 + \omega^{(i)}\omega^{(jk)}\Phi_0 + \omega^{(j)}\omega^{(ik)}\Phi_0 \\ & + \omega^{(k)}\omega^{(ij)}\Phi_0 + \omega^{(i)}\omega^{(j)}\omega^{(k)}\Phi_0\end{aligned} \tag{7.4.4c}$$

$$\begin{aligned}\chi^{(ijkl)} = {} & \omega^{(ijkl)}\Phi_0 + \omega^{(i)}\omega^{(jkl)}\Phi_0 + \omega^{(ij)}\omega^{(kl)}\Phi_0 \\ & + \omega^{(i)}\omega^{(j)}\omega^{(k)}\omega^{(l)}\Phi_0 \ldots \text{etc.}\end{aligned} \tag{7.4.4d}$$

Now let us define

$$\hat{\Omega}_1 = \sum_i \omega^{(i)} \tag{7.4.5a}$$

$$\hat{\Omega}_2 = \sum_{i>j} \omega^{(ij)} \tag{7.4.5b}$$

$$\hat{\Omega}_3 = \sum_{i>j>l} \omega^{(ijl)} \tag{7.4.5c}$$

$$\hat{\Omega}_4 = \sum_{i>j>k>l} \omega^{(ijkl)} \tag{7.4.5d}$$

$$\ldots \text{etc.}$$

and

$$\hat{\Omega} = \hat{\Omega}_1 + \hat{\Omega}_2 + \ldots + \hat{\Omega}_N \tag{7.4.6}$$

for an N-electron system. The wavefunction (7.4.1) may then be written in the compact form

$$\begin{aligned}\Psi & = \exp(\hat{\Omega})\Phi_0 \\ & = \left(1 + \hat{\Omega} + \frac{1}{2!}\hat{\Omega}^2 + \ldots\right)\Phi_0.\end{aligned} \tag{7.4.7}$$

The exact Schrödinger equation

$$\mathcal{H}\Psi = E\Psi \tag{7.4.8}$$

may be written as

$$\mathcal{H}e^{\hat{\Omega}}\Phi_0 = Ee^{\hat{\Omega}}\Phi_0 \tag{7.4.9}$$

The expression (7.4.7) is exact and difficult to handle completely. If the reference determinant is a Hartree–Fock function, then we can use perturbation theory to analyse the importance of the various components of the cluster expansion. This analysis is presented in Table 7.5 for a closed-shell system and in Table 7.6 for an open-shell system. In the following we shall restrict our attention to closed-shell systems for the

Table 7.5

Order of the perturbation expansion for the wavefunction in which various linked and unlinked clusters first arise for closed-shell systems†

	Order of perturbation
Linked clusters	
ω_1	2
ω_2	1
ω_3	2
ω_4	3
ω_5	4
ω_6	5
Unlinked clusters	
ω_1^2	4
$\omega_1\omega_2$	3
$\omega_1\omega_2^2$	4
ω_2^2	2
$\omega_1\omega_3$	4
$\omega_2\omega_3$	3
$\omega_2^2\omega_3$	4
ω_3^2	4
$\omega_1\omega_4$	5
$\omega_2\omega_4$	4
$\omega_3\omega_4$	5
ω_4^2	6
ω_2^3	3
ω_2^4	4

† $\omega_n \equiv \omega^{(i_1 \ldots i_n)}$

sake of simplicity. The terms in the expansion for the wavefunction may be represented diagrammatically using the conventions given in Chapter 4.

It is clear from Table 7.5 that the most important clusters are those corresponding to pairs of electrons. The approximation

$$\hat{\Omega} \approx \hat{\Omega}_2 \tag{7.4.10}$$

is the basis of the coupled pair many-electron theory (CPMET). If the coupled electron pair energy is denoted by $\mathscr{E}$ we have

$$\hat{\mathscr{H}}\left(1+\hat{\Omega}_2+\frac{1}{2!}\hat{\Omega}_2^2+\ldots\right)\Phi_0=\mathscr{E}\left(1+\hat{\Omega}_2+\frac{1}{2!}\hat{\Omega}_2+\ldots\right)\Phi_0. \tag{7.4.11}$$

Projection from the left with Φ_0 and $\Phi_i^{(2)}$ where the latter is a state obtained from Φ_0 by the excitation of two electrons, gives, respectively

$$\langle\Phi_0|\,\hat{\mathscr{H}}\hat{\Omega}_2\,|\Phi_0\rangle=\mathscr{E} \tag{7.4.12}$$

Table 7.6

Order of the perturbation expansion for the wavefunction in which various linked and unlinked clusters first arise for open-shell systems†

	Order of perturbation
Linked clusters	
ω_1'	0
ω_2'	1
ω_3'	2
ω_4'	3
ω_5'	4
Unlinked clusters	
$\omega_1'\omega_2$	1
$\omega_1'\omega_1$; $\omega_1'\omega_2^2$; $\omega_1'\omega_3$; $\omega_2'\omega_2$	2
$\omega_1'\omega_4$; $\omega_1'\omega_1\omega_2$; $\omega_1'\omega_2\omega_3$; $\omega_1'\omega_2^3$; $\omega_1'\omega_4$; $\omega_2'\omega_1$; $\omega_2'\omega_3$; $\omega_2'\omega_2^2$; $\omega_3'\omega_2$	3
$\omega_1'\omega_1\omega_3$; $\omega_1'\omega_2\omega_4$; $\omega_1'\omega_1^2$; $\omega_1'\omega_5$; $\omega_1'\omega_3^2$; $\omega_1'\omega_2^4$; $\omega_1'\omega_1\omega_2^2$; $\omega_1'\omega_3\omega_2^2$; $\omega_2'\omega_1\omega_2$; $\omega_2'\omega_2\omega_3$; $\omega_2'\omega_2^3$; $\omega_3'\omega_1$; $\omega_3'\omega_3$; $\omega_3'\omega_2^2$; $\omega_4'\omega_2$	4

† $\omega_n \equiv \omega^{(i_1 \cdots i_n)}$; ω_i' is associated with open-shell orbitals.

and

$$\langle \Phi_i^{(2)} | \hat{\mathcal{H}} \left(1 + \hat{\Omega}_2 + \frac{1}{2!} \hat{\Omega}_2^2\right) | \Phi_0 \rangle = \mathcal{E} \langle \Phi_i^{(2)} | \hat{\Omega}_2 | \Phi_0 \rangle. \tag{7.4.13}$$

Cubic and higher-order terms in Ω_2 do not appear in (7.4.13) because $\hat{\mathcal{H}}$ does not contain more than two-electron interactions. Equations (7.4.12) and (7.4.13) are represented diagrammatically in Fig. 7.4 using Hugenholtz diagrams and considering only the two-electron interactions. The disconnected parts appearing on the left- and right-hand sides of the second of these two equations cancel and thus

$$\langle \Phi_i^{(2)} | \left\{ \hat{\mathcal{H}} \left(1 + \hat{\Omega}_2 + \frac{1}{2!} \hat{\Omega}_2^2\right) \right\}_{\mathrm{C}} | \Phi_0 \rangle = 0 \tag{7.4.14}$$

where the subscript C denotes the fact that only connected terms appear. The cluster equations (7.4.12) and (7.4.14) are then solved iteratively.

The coupled-pair many-electron theory does not lead to a variational upper bound. However, Čársky and Urban (1980) comment that 'the advantage of having an upper bound to the energy is probably not so important when the method is accurate enough to give the correlation energy with an accuracy of a few per cent'.

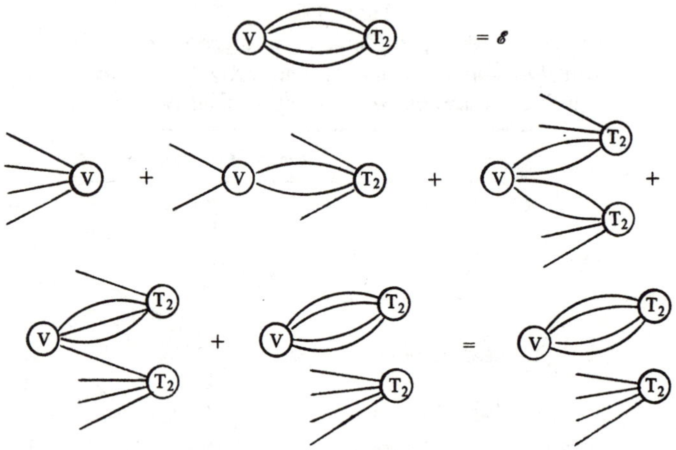

FIG. 7.4. Diagrammatic representation of eqns (7.4.12) and (7.4.13) using Hugenholtz diagrams.

Table 7.7

Application of coupled pair theories to the H_2 dimer[a]

	Calculated total energies							
α	F-CI	D-CI DQ-CI	CPMET	L-CPMET ACP-D123	ACP-B ACP-D45	ACP-A ACP-C	D-CI + Davidson's correction	E_4 (MBPT)
0.500	53.690	52.515 53.511	53.572	56.260 56.086	53.276 53.756	54.884 54.054	55.759	53.861
0.200	57.260	55.861 57.163	57.233	60.553 60.374	56.836 57.417	58.933 57.844	59.815	56.059
0.100	65.321	63.340 65.227	65.318	71.017 70.566	64.824 65.691	68.273 66.448	69.252	62.187
0.050	76.429	73.785 76.401	76.635	91.829 90.990	76.113 77.507	83.125 79.240	84.268	69.517
0.020	92.148	88.781 92.124	93.034	320.410 NC	92.265 94.195	111.728 99.937	112.667	75.768
0.015	96.711	93.220 96.686	97.900	−517.543 NC	96.872 98.763	121.474 106.305	122.538	76.908
0.005	109.196	105.552 109.188	111.355	−8.129 NC	109.062 110.369	148.182 123.484	152.316	79.242

[a] taken from the work of Jankowski and Paldus (1980) and Wilson, Jankowski, and Paldus (1983).

The H atoms are taken to be in a planar configuration with the H_1–H_2, H_2–H_3, and H_3–H_4 bond distances being equal to 2 bohr and the H_1–H_2–H_3 and H_2–H_3–H_4 bond angles being given by $\alpha\pi$. (This is called the H4 model in the above publications.)

All energies are in millihartree with signs reversed. NC indicates no convergence.

F-CI: fuel configuration interaction; D-CI: double-excitation configuration interaction; DQ-CI: double- and quadruple-excitation configuration interaction; other acronyms are defined in Table 7.8 and Fig. 7.6.

The practical implementation of the coupled pair many-electron theory is illustrated in Table 7.7 where results are presented for the H_2 dimer. These model calculations were performed by Jankowski and Paldus (1980) in order to investigate the convergence of the method in the presence of quasi-degeneracy effects. Full configuration mixing calculations were performed in the same basis set for comparative purposes. The results of full fourth-order diagrammatic perturbation theory calculations for the same system are also given in Table 7.7.

It should be noted that Paldus and his co-workers (see, for example, Adams and Paldus 1979) have extended the coupled pair many-electron theory to account for linked three-electron clusters. This is termed the extended coupled pair many-electron theory. As we shall show in Section 7.6, the component of the correlation energy corresponding to triply excited states can often be quite important. This clearly suggests that it is much more productive to evaluate all terms which arise in the fourth-order perturbational analysis of the coupled pair many-electron theory than to sum through all orders, terms arising from two-electron clusters and products of two-electron clusters.

In a discussion of the coupled pair many-electron theory, Jankowski and Paldus (1980) express the view that "it seems quite unlikely that the . . . technique can be routinely used in the near future for larger than 10–20-electron systems or even for smaller problems requiring very extensive orbital basis sets". Jankowski and Paldus stress the need to develop reliable approximate coupled pair theories (ACP). Perturbation theory provides what is perhaps the most balanced and systematic approach for obtaining such approximate schemes; however, we shall delay discussion of perturbational schemes until Section 7.6. There are a whole range of approximate coupled pair theories from the linear theory of Čížek (1966), which is equivalent to the summation of all double-excitation, linked diagrams in the perturbation series for the correlation energy, through infinite order as we illustrate in Fig. 7.5, to various coupled electron pair

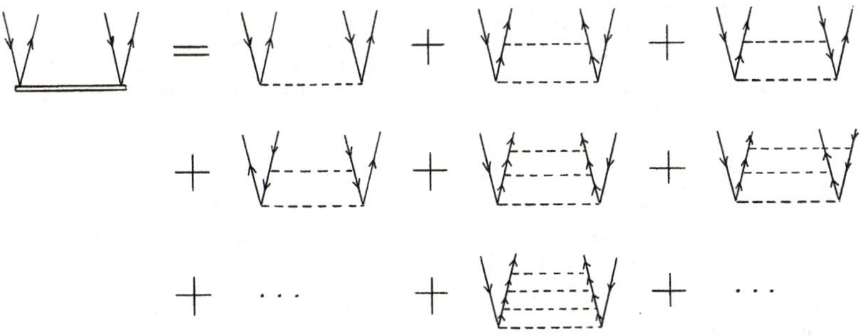

FIG. 7.5. Diagrammatic representation of Čížek linear approximation.

Table 7.8

Approximation to the coupled pair many-electron theory†

Approximation	Acronym	Reference
Complete neglect of all non-linear terms	L-CPMET	Čížek 1966
Only diagrams in Fig. 7.5 considered; All diagrams in Fig. 7.6 neglected	CEPA (0)	Kutzelnigg 1977
		Ahlrichs 1979
	DE-MBPT	Bartlett and Silver 1976*a*
Evaluation of components corresponding to diagrams (i), (ii), and (iii) in Fig. 7.6	ACP-D123	Jankowski and Paldus 1980
Evaluation of components correspond to diagrams (iv) and (v) in Fig. 7.6	ACP-D45	Jankowski and Paldus 1980
Evaluation of diagrams (iv) and (v) in Fig. 7.6 with $b_1=c_1$ and $b_2=c_2$ or $b_1=c_2$ and $b_2=c_1$	ACP-A	Jankowski and Paldus 1980
	CEPA (2)	Kutzelnigg 1977
	CPA′	Hurley 1976*b*
Addition of terms associated with (iv) in which $b_1 \neq c_1, c_2$ to above	ACP-B	Jankowski and Paldus 1980
	CEPA (3)	Kutzelnigg 1977
	CPA″	Hurley 1976*b*
	'Kelly's' CEPA	Kelly 1964*a*
		Kelly and Sessler 1963
Addition of $\frac{1}{2}$(ACP-A+ACP-B)	ACP-C	Jankowski and Paldus 1980
	CEPA (1)	Kutzelnigg 1977

† L-CPMET: linear coupled pair many-electron theory; CEPA: coupled electron pair approximation; DE-MBPT: double excitation many-body perturbation theory; ACP: approximate coupled pair; CPA: coupled pair approximation.

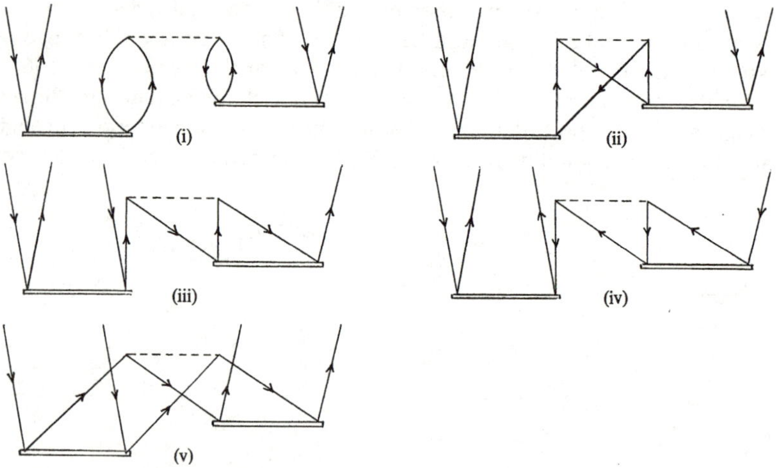

FIG. 7.6. Some of the diagrams which are included in various coupled electron pair approximations.

Table 7.9
Some calculated energies for various small molecules obtained by application of the coupled electron pair approximation[a]

System	E_{MHF}	E_{corr}	E_{total}
BeH_2	−15.770 51	−75.28	−15.845 79
BH	−25.128 08	−97.49	−25.225 57
BH_3	−26.398 81	−130.99	−26.529 80
CH_4	−40.211 47	−213.75	−40.425 22
NH_3 (D_{3h})	−56.207 19	−241.68	−56.448 88
NH_3 (C_{3v})	−56.215 44	−242.16	−56.457 60
H_2O	−76.052 93	−260.04	−76.312 97
FH	−100.051 95	−267.64	−100.319 60
Ne	−128.527 34	−267.74	−100.795 09
CH_3^- (D_{3h})	−39.516 67	−206.44	−39.723 11
CH_3^- (C_{3v})	−39.520 01	−204.54	−39.724 55
H_3O^+ (D_{3h})	−76.327 51	−225.35	−76.552 86
H_3O^+ (C_{3v})	−76.328 72	−226.54	−76.555 26

[a] taken from the work of Ahlrichs *et al.* (1975*a*,*b*); the matrix Hartree–Fock energy E_{MHF} and the total energy E_{total} are given in hartree units; the correlation energy E_{corr} is given in millihartree.

approximations (Meyer 1971, 1974; Meyer and Rosmus 1975; Ahlrichs, Keil, Lischka, Staemmler, and Kutzelnigg 1975; Taylor, Bacskay, Hush, and Hurley 1976, 1978; Hurley 1976*b*; Kutzelnigg 1977; Meyer 1977; Ahlrichs 1979). In the linear coupled pair theory all non-linear terms are neglected while in other approximations various methods for estimating the non-linear part of eqns (7.4.12) and (7.4.14) are employed. The various coupled electron pair approximations have been reviewed recently by Ahlrichs (1979, 1983). In Table 7.8, a summary is given of the various approximation schemes which are currently in use together with the acronyms which have become all too common in this field. The interrelation of the various schemes is also indicated.

In order to illustrate the accuracy which may be achieved by the coupled electron pair approximation, we present in Table 7.9 typical results for a selection of small molecules. These results are taken from the work of Ahlrichs *et al.* (1975*a*,*b*) and compare favourably with calculations based on other theories. The dominant error in these results is certainly attributable to basis set truncation.

7.5 Valence bond approaches

Multistructure valence bond approaches to the electronic structure of molecules have traditionally employed a wavefunction which consists of a linear combination of structure functions each of which is a product of atomic functions (see, for example, Raimondi 1975, 1977). As Doggett (1980) observes 'large numbers of structures are needed in order to achieve a satisfactory description of the change in electronic structure on molecule formation. Unfortunately, the resulting wavefunctions are just as unwieldy as their molecular orbital—configuration interaction counterparts'.

By employing improved orbitals of the type discussed in Section 2.6, it is envisaged that a more rapidly convergent expansion for the wavefunction might be obtained. The situation is somewhat similar to the use of multiconfiguration self-consistent-field reference function in configuration interaction expansions. However, an important difference between the two approaches is that the extended valence bond model does not abandon the independent electron picture in zero-order.

Gerratt and Raimondi (1980) write their valence bond wavefunction in the form

$$\Psi_{S,M_S} = |\{\phi_1 \phi_2 \ldots \phi_N\}SM_S\rangle + \sum_{\mu_1\mu_2\ldots\mu_N} C\begin{pmatrix}\mu_1 & \mu_2 & \cdots & \mu_N \\ 1 & 2 & & N\end{pmatrix} |\{\phi_{\mu_1} \phi_{\mu_2} \ldots \phi_{\mu_N}\}SM_S\rangle. \tag{7.5.1}$$

The first term here is the reference configuration and consists of the optimal description of the system in terms of a product of non-orthogonal orbitals. Each of the orbitals, $\phi_1, \phi_2 \ldots \phi_N$, in the reference configuration, is an eigenfunction of a different effective hamiltonian (Gerratt 1971, 1974)

$$F(i)\phi_{\mu i} = \epsilon_{\mu i}\phi_{\mu i} \qquad \mu = 0, 1, \ldots \qquad \phi_i = \phi_{0i} \tag{7.5.2}$$

The additional configurations are generated by replacing ϕ_i by $\phi_{\mu i}$ where the $\phi_{\mu i}$ are eigenfunctions of the same operator as ϕ_i. Gerratt and Raimondi (1980) demonstrate that this leads to a rapidly convergent series, although they note that linear dependence problems will arise if many highly excited functions are included.

In Table 7.10, the effectiveness of the extended valence bond approach of Gerratt and Raimondi (1980) is illustrated by presenting the results of some calculations for the BeH molecule. A comparison is made with previous molecular orbital–configuration interaction studies. The results of a 71 structure valence bond calculation are seen to compare quite favourably with a 1039 configuration interaction expansion (Bender and Davidson 1969).

There are two major reasons for the superiority of the valence bond model in calculations of this type:

(i) The reference model describes molecular dissociation correctly.

(ii) The low-lying unoccupied orbitals are spatially more contracted than they are when orthogonal orbitals are employed.

Table 7.10

Spectroscopic constants for the BeH molecule resulting from spin-coupled valence bond calculations

Calculation	W_{min} a.u.	D_e eV	ω_e cm^{-1}	$\omega_e X_e$ cm^{-1}	B_e cm^{-1}	R_e Å
Reference configuration[a]	−15.178 997	2.04[a]	1974.6	36.6	9.9858	1.3658
Valence bond calculation with						
80 structures[a]	−15.197 642	1.87	1985.5	39.3	10.0596	1.3610
71 structures[a]	−15.196 877	1.84	1974.6	39.6	9.9994	1.3650
53 structures[a]	−15.193 965	1.88	1985.4	42.3	10.0422	1.3620
Molecular orbital-	[b] −15.196 35	2.115	2059.4	38.03	10.279	1.345
Configuration interaction	[c] −15.220 7	1.48				
	−15.232 4					
experimental		2.16	2060.8	36.3	10.3164	1.3426

[a] taken from Gerratt and Raimondi (1980); [b] Bagus, Maser, Goethals, and Verhaegen (1973); [c] Bender and Davidson (1969).

However, as seen in Section 2.6, the non-orthogonality problem leads to severe problems in the valence bond approach when larger numbers of electrons are being considered. It is then necessary to impose orthogonality conditions on the orbitals in order to keep the computation tractable. van Lenthe and Balint-Kurti (1980) and Walch, Dunning, Raffenetti, and Bobrowicz (1980) have both presented schemes for performing valence bond studies which invoke orthogonality constraints in order to make the calculation possible for larger systems.

Just as the limited configuration mixing method has a size-consistency error, so does the limited multistructure valence bond approach. One means of overcoming this problem is to use the valence bond reference function in a perturbational expansion for the correction arising from electron correlation effects. A scheme for performing such calculations has been given in Section 4.11. Kirtman and Cole (1978) have reported valence bond perturbation calculation for some two-electron systems. They reported that accurate results can be obtained in second order of the perturbation series.

7.6 Diagrammatic perturbation theory

In addition to providing a powerful approach to the analysis of the various methods available for attacking the electron correlation problem in atoms and molecules, the linked diagram theorem of many-body perturbation theory forms the basis of very efficient computational schemes for calculations. Perturbation theory provides a clearly defined order parameter in the expansion for the wavefunction and corresponding expectation values, giving a 'least biased' (Brandow 1975) indication of the relative importance of the various components and an unambiguous criterion for truncation of the series. In discussing various approaches to the electron correlation problem, Brandow (1975) comments:

> The structure of each formalism tends to suggest that certain types of approximation are most reasonable or 'natural', regardless of the actual quantitative characteristics of the physical system. The many-body literature is full of papers where people have tried to make the physics fit into their preconceived approximation schemes, instead of vice versa. . . . perturbation theory . . . seems to be the 'least biased' of all the many-body techniques.

Brandow's comment is particularly poignant when one considers the methods which are employed to determine the most important configurations in, say, a limited configuration mixing calculation or a coupled cluster expansion. In, for example, the method of limited configuration mixing, this question is answered by appealing to perturbation theory and, as is shown in Fig. 7.1, it is observed that double excitations contribute in lower order than any other configurations. It is, therefore, usual to include doubly excited configurations in the method of limited configuration mixing, but this component has then been summed to all orders of the perturbation series, whereas strictly perturbation theory indicates that double excitation configurations are the most important through third-order; in fourth-order single, triple, and quadruple excitation configurations are just as significant.

The pioneering work on the application of the diagrammatic many-body perturbation theory to atomic and molecular systems was performed by Kelly (Kelly 1963, 1964, 1966, 1968, 1969*a,b*, 1970*a,b*). He applied the method to atoms using numerical solutions of the Hartree–Fock equations. Many other calculations on atomic systems using this approach were subsequently reported (e.g. Pu and Chang 1966; Chang *et al.* 1968; Dutta *et al.* 1969; Lee *et al.* 1970*a,b,c*; Lee *et al.* 1971). The first molecular calculations using diagrammatic perturbation theory employed a single-centre expansion and were limited to simple hydrides where it is possible to treat the hydrogen nuclei as additional perturbations (e.g. Kelly 1969*b*; Miller and Kelly 1971; Lee *et al.* 1970*a,b,c*; Lee and Das 1972; Dutta *et al.* 1970). More recently, the theory has been applied to arbitrary molecules by employing the algebraic approximation which, as shown in Chapter 6, is fundamental to most molecular calculations (Wilson and Silver 1976; see also Schulman and Kaufman 1970, 1972; Robb 1973, 1974; Kaldor 1973*a,b*, 1975*a,b*; Bartlett and Silver 1975*a,b*; Freeman and Karplus 1976; Urban, Kellö, and Hubač 1977; Prime and Robb 1976; Pople, Binkley, and Seeger 1976; Wilson, Silver, and Bartlett 1977; Wilson and Silver 1977*a,b*; Wilson 1977*a,b*, 1978*a,c*; for a recent review see Wilson 1981, 1983).

Diagrammatic many-body perturbation theory forms the basis of a highly systematic approach to the calculation of correlation energies and the effects of correlation on molecular properties. In this section, various aspects of the truncation of the perturbation expansion are discussed and a review of applications is given. The use of Padé approximants in perturbation theory is outlined and the special invariance properties of the $[N+1/N]$ Padé approximants in diagrammatic perturbation theory are emphasized. Scaling of the zero-order hamiltonian is considered and the construction of upper bounds to the energy described. Fourth-order and higher-order terms are discussed in some detail. The effects of quasi-degeneracy on the convergence properties of perturbation series is examined. A selective survey of calculations which illustrate the application of the diagrammatic perturbation theory to the calculation of molecular electronic structure is given.

The usual application of perturbation theory leads to a power series, or $[N/0]$ Padé approximant, for the energy or other expectation value. Other Padé approximants provide useful representations of the perturbation series. Some Padé approximants provide a convergent sequence of approximations to the correlation energy when the usual power series, the $[N/0]$ Padé approximant, diverges. Unlike the power series, some Padé approximants can handle a class of functions with various types of singularities still providing correct uniform convergence.

The $[P/Q]$ Padé approximant to a Taylor series $T(\lambda)$ of order M is defined by (Padé 1892; Baker 1965, 1975)

$$[P/Q](\lambda) = P(\lambda)/Q(\lambda) \tag{7.6.1}$$

where $P(\lambda)$ is a polynomial of order P and $Q(\lambda)$ is a polynomial of order Q such that

$$P(\lambda) - Q(\lambda)[P/Q](\lambda) = \mathcal{O}(\lambda^{M+1}) \tag{7.6.2}$$

in which $\mathcal{O}(\lambda^{M+1})$ denotes terms of order $M+1$ and higher.

In the Lennard-Jones–Brillouin–Wigner perturbation theory, which was discussed in Section 4.2, Padé approximants can be used to obtain upper and lower bounds to the exact energy (Goldhammer and Feenberg 1956; Goscinski 1967). On the other hand, in the Rayleigh–Schrödinger perturbation theory, from which the diagrammatic perturbation theory is derived, the $[N+1/N]$ Padé approximants to the energy are special in that, when the expansion parameter λ is set equal to unity (the only physically meaningful value), the numerical value of the Padé approximant is invariant to two simple modifications of the zero-order hamiltonian; namely a change of scale and a shift of origin in the reference spectrum (Wilson, Silver, and Farrell 1977).

Invariants in Lennard-Jones–Brillouin–Wigner perturbation theory and in Rayleigh–Schrödinger perturbation theory were first considered in the pioneering work of Feenberg and his co-workers (Goldhammer and Feenberg 1956; Feenberg 1958; see also Kumar 1963). Let us consider two modifications of a given zero-order operator; a uniform displacement of the zero-order energy spectrum and a uniform change of scale of the zero-order energy levels. Consider the zero-order operator

$$\hat{\mathcal{H}}_0^{\mu,\nu} = \mu\hat{\mathcal{H}}_0 + \nu\hat{I} \tag{7.6.3}$$

where $\hat{I}$ is the identity operator and μ and ν are scalars. The perturbing operator corresponding to the modified reference hamiltonian (7.6.3) has the form

$$\hat{\mathcal{H}}_1^{\mu,\nu} = \hat{\mathcal{H}}_1 + (1-\mu)\hat{\mathcal{H}}_0 - \nu\hat{I}. \tag{7.6.4}$$

Thus when the modified zero-order hamiltonian and the perturbation are added, the total hamiltonian is recovered. The energy coefficients, $E_i^{\mu,\nu}$, where i denotes the order of perturbation, may be related to the values of $E_i^{1,0}$ by the expressions

$$E_0^{\mu,\nu} = \mu E_0^{1,0} + \nu \tag{7.6.5}$$

$$E_1^{\mu,\nu} = E_1^{1,0} + (1-\mu)E_0^{1,0} - \nu \tag{7.6.6}$$

and

$$E_n^{\mu,\nu} = \frac{1}{\mu^{n-1}} \sum_{k=2}^{k=n} \binom{n-2}{k-2} (\mu-1)^{n-k} E_k^{1,0} \qquad n = 2, 3, \ldots \tag{7.6.7}$$

where $\binom{p}{q}$ is a binomial coefficient.

It can be shown (Wilson, Silver, and Farrell 1977) that, when the perturbation parameter λ is set equal to unity all $[N/M]$ Padé approximants for which $N > M$ are invariant with respect to the choice of the scalar ν. Furthermore, uniquely among all Padé approximants of order $2N+1$, the $[N+1/N]$ Padé approximants to the energy expansion are invariant to both the choice of μ and ν.

From third-order diagrammatic perturbation theory calculations, the [2/1] Padé approximants can be formed and several additional reasons can be advanced for their use:

(i) There is a large amount of evidence demonstrating that their use leads to improved results (see, for example, Wilson 1979).
(ii) This may be regarded as the choice of scaling parameter which makes the third-order energy vanish (Amos 1970, 1972, 1978).
(iii) It may also be regarded as the energy expression for which the third-order energy is a minimum with respect to the choice of the scaling parameter (Tuan 1970), although it should be remembered that there is no variation principle here.

It should be noted that for one-electron properties, such as dipole moments, the $[N/N]$ Padé approximants are invariant to changes of scale and shift of origin in the reference spectrum (Wilson 1978*c*), whereas for second-order properties, such as polarizabilities, the $[N/N+1]$ Padé approximants are to be preferred. Indeed, for polarizabilities, the use of the form

$$\alpha_0\left(1-\frac{\alpha_1}{\alpha_0}\right)^{-1} \tag{7.6.8}$$

has been advocated by a number of authors (Wilson 1978*c*; Howat, Trsic, and Goscinski 1977; Wilson 1980*b*; Wilson and Sadlej 1981; Sadlej and Wilson 1981).

Full fourth-order correlation energy calculations lead to the need for fourth-order forms which share the invariance properties of the [2/1] Padé approximants in third-order calculations. Following the early work of Feenberg (1958), the expression

$$E[2/1]+\left(\frac{E[2/1]}{E_2}\right)^3\left(E_4-\frac{E_3^2}{E_2}\right) \quad \text{where} \quad E[2/1]=E_2/(1-E_3/E_2) \tag{7.6.9}$$

has been shown to be useful in this respect (Wilson 1980*c*; Wilson and Guest 1981). Recent work has examined the use of Padé approximants in the evaluation of derivatives of the energy, with respect to the nuclear geometry (Wilson 1980*a*) and in the calculation of molecular interactions (Silver and Wilson 1982).

Although full fifth-order calculations are not being performed routinely at present, it should be noted that the [3/2] Padé approximants can be constructed from the energy coefficients obtained in such a calculation. Explicitly, the [3/2] Padé approximants may be written as

$$E[3/2]=E_0+E_1\lambda+\mu_1(\lambda)E_2\lambda^2+\mu_2(\lambda)E_3\lambda^3 \tag{7.6.10}$$

with

$$\mu_1(\lambda)=(1+b_1\lambda)(1+b_1\lambda+b_2\lambda^2)^{-1} \tag{7.6.11a}$$

and

$$\mu_2(\lambda)=(1+b_1\lambda+b_2\lambda^2)^{-1} \tag{7.6.11b}$$

in which

$$b_1=d(E_2E_5-E_3E_4) \tag{7.6.11c}$$

$$b_2=d(E_4^2-E_3E_5) \tag{7.6.11d}$$

$$d=(E_3^2-E_2E_4)^{-1}. \tag{7.6.11e}$$

In performing a perturbation calculation to $2N+1$ order for the energy, one also implicitly calculates the wavefunction to Nth order (Hirschfelder, Byers Brown, and Epstein 1964). This Nth order wavefunction may be substituted in the Rayleigh quotient to obtain an upper bound to the exact energy. Let the wavefunction through first-order be written as

$$\phi_0+\gamma\phi_1 \tag{7.6.12}$$

where the parameter γ has been introduced. The upper bound obtained by substitution of (7.6.10) in the Rayleigh quotient has the form (Wilson and Silver 1976)

$$E_{\text{var}}(\gamma)=E_0+E_1+\frac{(2\gamma-\gamma^2)E_2+\gamma^2E_3}{1+\gamma^2\,\Delta_{11}} \tag{7.6.13}$$

where the normalization integrals $\Delta_{11}=\langle\phi_1\,|\,\phi_1\rangle$. γ may be regarded as a variational parameter and its optimal parameter obtained by invoking the variation principle, giving

$$\gamma_{\text{opt}}=\Delta_{11}^{-1}\{\surd(\mathcal{S}^2+\Delta_{11})-\mathcal{S}\} \tag{7.6.14a}$$

with

$$\mathcal{S}=\tfrac{1}{2}(1-E_3/E_2). \tag{7.6.14b}$$

Unfortunately, by forming an upper bound, one introduces terms which have an unphysical dependence on the number of electrons and the quality of this upper bound deteriorates as the number of electrons in the system being studied increases. This is illustrated in Table 7.11 for an array of well-separated N_2 molecules. It could be argued that no upper bound is useful, unless it is accompanied by a lower bound of comparable

Table 7.11

Upper bounds to the correlation energy of an array of well-separated N_2 *molecules obtained by using eqns* (7.6.11) *and* (7.6.12)[a]

Number of molecules n	γ^n_{opt}	Variational upper bound $\gamma^n_{\text{opt}}E^n_2$	Perturbation theory $E^n_2+E^n_3$
1	0.8882	−0.2903	−0.3177
2	0.8264	−0.2701	−0.6354
3	0.7781	−0.2544	−0.9531
4	0.7388	−0.2415	−1.2707
5	0.7059	−0.2307	−1.5884
10	0.5944	−0.1943	−3.1769
20	0.4798	−0.1558	−6.3537

[a] based on the results of Guest and Wilson (1980).

quality, however, energies obtained by using (7.6.13) are of interest in comparisons with single and double excitation configuration mixing.

It is possible to modify the diagrammatic many-body perturbation series by employing alternative potentials in the zero-order hamiltonian. For example, the modified potential

$$V^{N-1}(\mathbf{r}_1)=\sum_{k=1}^{N-1}\int \phi_k^*(r_2)r_{12}^{-1}(1-P_{12})\phi_k(\mathbf{r}_2)\,d\mathbf{r}_2 \qquad (7.6.15)$$

has been widely used in atomic studies (Kelly 1969*a*). Molecular calculations, within the algebraic approximation, have demonstrated (Silver and Bartlett 1976; Silver, Wilson, and Bartlett 1977) that there is little difference between results obtained by using the Hartree–Fock potential and the modified potential (7.6.15) when *all* terms are included through third-order. For the FH molecule, using the model perturbation series, 80.2% of the empirical correlation energy was recovered in second-order when the Hartree–Fock potential was employed, whereas 101.2% of the correlation energy was recovered when the V^{N-1} modified potential was used. The [2/1] Padé approximant to the perturbation series using the Hartree–Fock potential gave 79.6% of the correlation energy, which is close to the 77.8% given by the [2/1] Padé approximant to the series based on the V^{N-1} potential. The infinite order result should, of course, be independent of the choice of zero-order hamiltonian.

In Table 7.12 a comparison of full third-order many-body perturbation theory calculations using both the Hartree–Fock model zero-order hamiltonian and the shifted denominator expansion is made with a full single- and double-excitation configuration mixing calculation for the neon atom in its ground state using three basis sets of Slater functions taken from the work of Clementi (1964), Clementi, Roothaan, and Yoshimine (1962), and Barr and Davidson (1970). The percentage of the configuration mixing energy recovered is given in parentheses, although it should be emphasized that, because of the size-consistency error in limited configuration mixing, exact agreement is not expected. The two-body energy corrections can be seen to overestimate the configuration mixing values by 14–23 per cent; the three-body and four-body effects in third-order accounting for most of this difference. The perturbation series based on the Hartree–Fock model zero-order operator, which corresponds to the Moeller–Plesset perturbation expansion (Moeller and Plesset 1934) which has been advocated by Pople and his co-workers (Binkley and Pople 1975; Pople, Binkley, and Seeger 1976) leads to closer agreement with the configuration mixing result than the shifted denominator scheme. The [2/1] Padé approximants to the energy expansion are within 0.9% of the configuration mixing results. The use of this approximant leads to a significant change in the results obtained by using the shifted denominator

Table 7.12

Comparison of full single- and double-excitation configuration mixing energy corrections for the ground state of the neon atom with many-body perturbation theory values, [2/1] Padé approximants, and variational upper bounds derived from the perturbation theory calculations[a]

Basis set	A		B		C	
$\hat{\mathcal{H}}_0$	$\hat{\mathcal{H}}_{model}$	$\hat{\mathcal{H}}_{shifted}$	$\hat{\mathcal{H}}_{model}$	$\hat{\mathcal{H}}_{shifted}$	$\hat{\mathcal{H}}_{model}$	$\hat{\mathcal{H}}_{shifted}$
Configuration interaction[b] (single and double excitations)						
(CI)	−0.133 56		−0.174 65		−0.247 60	
Restricted perturbation (two-body)						
$E_2+E_3^2$	−0.159 38	−0.164 09	−0.207 65	−0.214 39	−0.281 36	−0.286 69
(% CI)	(119.3)	(122.9)	(118.9)	(122.8)	(113.6)	(115.8)
Full many-body perturbation (two-, three-, and four-body)						
E_2+E_3	−0.134 08	−0.126 27	−0.174 28	−0.166 04	−0.249 79	−0.240 89
(% CI)	(100.4)	(94.5)	(99.8)	(95.1)	(100.9)	(97.3)
Padé approximant for many-body perturbation						
$E[2/1]$	−0.134 15	−0.133 67	−0.174 41	−0.173 00	−0.249 79	−0.248 14
(% CI)	(100.4)	(100.1)	(99.9)	(99.1)	(100.9)	(100.2)
Many-body perturbative upper bound						
E_{var} ($\gamma=1$)	−0.132 07	−0.123 45	−0.170 31	−0.160 60	−0.244 61	−0.233 74
(% CI)	(98.9)	(92.4)	(97.5)	(92.0)	(98.8)	(94.4)
Optimized many-body perturbative upper bound						
γ optimal	0.964 38	0.797 77	0.952 82	0.814 21	0.976 94	0.835 77
E_{var} ($\gamma=$opt)	−0.132 25	−0.131 72	−0.170 72	−0.169 11	−0.244 74	−0.242 84
(% CI)	(99.0)	(98.6)	(97.7)	(96.8)	(98.8)	(98.1)

[a] Energies are in hartrees. Based on Wilson and Silver (1976).
[b] Barr and Davidson 1970.

expansion, which has the effect of bringing the model and shifted schemes into closer agreement. This agreement suggests that the result is independent of the particular splitting of $\hat{\mathcal{H}}$ chosen for the calculations and is a qualitative measure of the rapid convergence of the perturbation series. Finally, in Table 7.12, two upper bounds to the exact non-relativistic energy obtained by using (7.6.11); the first corresponds to taking $\gamma=1$ and the second to the optimum value of γ given by (7.6.12).

As an example of the application of full third-order diagrammatic perturbation theory to molecules, in Tables 7.13, 7.14, and 7.15 results are presented for the $X^1\Sigma^+$ states of first-row and second-row diatomic hydrides using basis sets given by Cade and Huo (1967) as extended by Bartlett and Silver (1975*a*, 1976*b*). In Table 7.13, the energies associated with the various second-order and third-order diagrams are given. The importance of three- and four-body interpair interactions should be

Table 7.13

Components of the perturbation expansion for LiH, BH, FH, NaH, AlH, *and* HCl[a]

$\hat{\mathscr{H}}_0$	LiH		BH		HF	
	$\hat{\mathscr{H}}_{\text{model}}$	$\hat{\mathscr{H}}_{\text{shifted}}$	$\hat{\mathscr{H}}_{\text{model}}$	$\hat{\mathscr{H}}_{\text{shifted}}$	$\hat{\mathscr{H}}_{\text{model}}$	$\hat{\mathscr{H}}_{\text{shifted}}$
Intrapair energies: diagonal terms (two-body)						
E_2^2	−0.065 39	−0.076 69	−0.111 03	−0.145 10	−0.305 49	−0.353 55
$E_3^2(pp)$	+0.006 18	0.0	+0.012 97	0.0	+0.023 04	0.0
$E_3^2(hp)$	−0.026 48	0.0	−0.054 20	0.0	−0.104 57	0.0
$E_3^2(hh)$	+0.011 28	0.0	+0.019 22	0.0	+0.041 06	0.0
Total	−0.074 41	−0.076 69	−0.133 04	−0.145 10	−0.345 96	−0.353 55
Intrapair interactions: non-diagonal terms (two-body)						
$E_3^2(pp)$	+0.009 52	+0.013 85	+0.016 07	+0.033 00	+0.023 88	+0.032 58
$E_3^2(hp)$	−0.010 50	−0.015 26	−0.015 00	−0.031 77	−0.030 56	−0.042 77
Total	−0.000 98	−0.001 41	+0.001 07	+0.001 23	−0.006 68	−0.010 19
Interpair interactions between pairs having a common hole state (three-body)						
$E_3^3(hp)$	+0.000 11	+0.000 19	−0.000 55	−0.001 25	+0.046 35	+0.067 49
$E_3^3(hh)$	+0.000 07	+0.000 09	+0.000 76	+0.001 97	−0.000 03	−0.000 08
Total	+0.000 18	+0.000 28	+0.000 21	+0.000 72	+0.046 32	+0.067 41
Interpair interactions between pairs having no common hole state (four-body)						
$E_3^4(hh)$	+0.000 01	+0.000 01	+0.001 17	+0.003 29	+0.002 97	+0.004 01
Overlap Δ_{11}	+0.017 58	+0.028 49	+0.037 17	+0.102 87	+0.050 73	+0.073 15

$\hat{\mathscr{H}}_0$	NaH		AlH		HCl	
	$\hat{\mathscr{H}}_{\text{model}}$	$\hat{\mathscr{H}}_{\text{shifted}}$	$\hat{\mathscr{H}}_{\text{model}}$	$\hat{\mathscr{H}}_{\text{shifted}}$	$\hat{\mathscr{H}}_{\text{model}}$	$\hat{\mathscr{H}}_{\text{shifted}}$
Intrapair energies: diagonal terms (two-body)						
E_2^2	−0.291 44	−0.331 33	−0.332 11	−0.387 11	−0.457 81	−0.536 37
$E_3^2(pp)$	+0.018 56	0.0	+0.022 42	0.0	+0.031 26	0.0
$E_3^2(hp)$	−0.086 52	0.0	−0.096 43	0.0	−0.135 79	0.0
$E_3^2(hh)$	+0.034 04	0.0	+0.032 88	0.0	+0.043 89	0.0
Total	−0.362 48	−0.331 33	−0.373 24	−0.387 11	−0.518 45	−0.536 37
Intrapair interactions: non-diagonal terms (two-body)						
$E_3^2(pp)$	+0.018 09	+0.024 91	+0.019 62	+0.034 61	+0.022 57	+0.036 98
$E_3^2(hp)$	−0.025 09	−0.033 68	−0.020 13	−0.031 42	−0.023 99	−0.039 21
Total	+0.007 00	−0.008 77	−0.000 51	+0.003 19	−0.001 42	−0.002 23
Interpair interactions between pairs having a common hole state (three-body)						
$E_3^3(hp)$	+0.030 71	+0.039 61	+0.020 91	+0.023 60	+0.042 28	+0.069 15
$E_3^3(hh)$	+0.000 27	+0.000 35	+0.001 09	+0.001 82	+0.000 92	+0.001 08
Total	+0.030 98	+0.039 96	+0.022 00	+0.025 42	+0.043 20	+0.070 23
Interpair interactions between pairs having no common hole state (four-body)						
$E_3^4(hh)$	+0.001 93	+0.002 41	+0.003 01	+0.005 68	+0.004 45	+0.007 01
Overlap, Δ_{11}	+0.040 13	+0.056 80	+0.053 25	+0.132 59	+0.071 68	+0.134 09

[a] Energies are in hartrees. E_i^j denotes the ith order j body term. Third order energy components are distinguished by the nature of the central interaction line in the corresponding diagram (See Fig. 4.16). Based on the work of Wilson and Silver (1977*b*).

Table 7.14
Convergence of the perturbation series for LiH, BH, FH, NaH, AlH, *and* HCl[a]

	LiH		BH		HF	
$\hat{\mathscr{H}}_0$	$\hat{\mathscr{H}}_{\text{model}}$	$\hat{\mathscr{H}}_{\text{shifted}}$	$\hat{\mathscr{H}}_{\text{model}}$	$\hat{\mathscr{H}}_{\text{shifted}}$	$\hat{\mathscr{H}}_{\text{model}}$	$\hat{\mathscr{H}}_{\text{shifted}}$
Convergence with increasing number of interacting bodies						
E^0	+0.9950	+0.9950	+2.1404	+2.1404	+5.1939	+5.1939
E^1	−5.4932	−8.9823	−17.3652	−27.2718	−59.9289	−105.2632
E^2	−3.5645	−0.0781	−10.0386	−0.1439	−45.6869	−0.3637
E^3	+0.0002	+0.0003	+0.0002	+0.0007	+0.0463	+0.0674
E^4	+0.0000	+0.0000	+0.0012	+0.0033	+0.0030	+0.0040
Convergence with increasing order						
E_0	−4.4982	−7.9873	−15.2248	−25.1314	−54.7350	−100.0693
E_1	−3.4891	0.0	−9.9066	0.0	−45.3343	0.0
E_2	−0.0654	−0.0767	−0.1110	−0.1451	−0.3055	−0.3536
E_3	−0.0098	−0.0011	−0.0196	+0.0052	+0.0021	+0.0612

	NaH		AlH		HCl	
$\hat{\mathscr{H}}_0$	$\hat{\mathscr{H}}_{\text{model}}$	$\hat{\mathscr{H}}_{\text{shifted}}$	$\hat{\mathscr{H}}_{\text{model}}$	$\hat{\mathscr{H}}_{\text{shifted}}$	$\hat{\mathscr{H}}_{\text{model}}$	$\hat{\mathscr{H}}_{\text{shifted}}$
Convergence with increasing number of interacting bodies						
E^0	+3.0847	+3.0847	+4.1747	+4.1747	+7.0577	+7.0577
E^1	−96.1119	−165.4776	−147.6139	−246.6386	−284.4868	−467.1687
E^2	−69.6981	−0.3401	−99.3985	−0.3839	−183.2017	−0.5386
E^3	+0.0310	+0.0400	+0.0220	+0.0254	+0.0432	+0.0702
E^4	+0.0019	+0.0024	+0.0030	+0.0057	+0.0045	+0.0070
Convergence with increasing order						
E_0	−93.0272	−162.3929	−143.4392	−242.4639	−277.4291	−460.1109
E_1	−69.3657	0.0	−99.0248	0.0	−182.6818	0.0
E_2	−0.2914	−0.3313	−0.3321	−0.3871	−0.4578	−0.5364
E_3	−0.0080	+0.0336	−0.0166	+0.0343	−0.0144	+0.0750

[a] The energy E^m is a sum of all m-body components through third order; E_n is the nth order term. Energies are in hartrees. Taken from Wilson and Silver (1977*b*).

emphasized. The convergence of the perturbation series both with the number of interacting bodies, denoted by superscripts, and with the order of perturbation, denoted by subscript, is displayed in Table 7.14. The zero-body term E^0 is the nuclear repulsion energy. As was the case for the neon atom, convergence by order is seen to be more rapid for the model perturbation expansion for the FH, NaH, AlH, and HCl molecules, whereas for LiH and BH, the shifted denominator expansion appears to converge more rapidly. The empirical correlation energies for the $X^1\Sigma^+$ states of first-row and second-row diatomic hydrides are compared with those resulting from perturbation theory studies in Table 7.15.

Table 7.15

Comparison of perturbative results, Padé approximants, and upper bounds with the empirical correlation energy for LiH, BH, FH, NaH, AlH, *and* HCl[a]

$\hat{\mathscr{H}}_0$	LiH $\hat{\mathscr{H}}_{\text{model}}$	LiH $\hat{\mathscr{H}}_{\text{shifted}}$	BH $\hat{\mathscr{H}}_{\text{model}}$	BH $\hat{\mathscr{H}}_{\text{shifted}}$	HF $\hat{\mathscr{H}}_{\text{model}}$	HF $\hat{\mathscr{H}}_{\text{shifted}}$
$E_{\text{empirical correlation}}$[b]	−0.083		−0.155		−0.381	
$E_{\text{empirical relativistic}}$[c]	−0.0006		−0.0060		−0.0829	
Second-order, E_2	−0.065 39	−0.076 69	−0.111 03	−0.145 10	−0.305 49	−0.353 55
Two-body, $E_2+E_3^2$	−0.075 41	−0.078 11	−0.131 97	−0.143 87	−0.352 63	−0.363 74
Many-body, E_2+E_3	−0.075 22	−0.077 82	−0.130 59	−0.139 87	−0.303 34	−0.292 31
[2/1] Padé $E_2/(1-E_3/E_2)$	−0.076 96	−0.077 84	−0.134 78	−0.140 05	−0.303 36	−0.301 35
E_{var} ($\gamma=1$)	−0.073 92	−0.075 66	−0.125 91	−0.126 82	−0.288 70	−0.272 39
γ optimal	+1.149 57	+0.986 77	+1.153 84	+0.887 08	+0.947 77	+0.811 32
E_{var} (γ optimal)	−0.075 17	−0.075 68	−0.128 11	−0.128 71	−0.289 53	−0.286 84

$\hat{\mathscr{H}}_0$	NaH $\hat{\mathscr{H}}_{\text{model}}$	NaH $\hat{\mathscr{H}}_{\text{shifted}}$	AlH $\hat{\mathscr{H}}_{\text{model}}$	AlH $\hat{\mathscr{H}}_{\text{shifted}}$	HCl $\hat{\mathscr{H}}_{\text{model}}$	HCl $\hat{\mathscr{H}}_{\text{shifted}}$
$E_{\text{empirical correlation}}$[b]	−0.432		−0.478		−0.699	
$E_{\text{empirical relativistic}}$[c]	−0.2002		−0.4206		−1.3717	
Second-order, E_2	−0.291 44	−0.331 33	−0.332 11	−0.387 11	−0.457 81	−0.536 37
Two-body, $E_2+E_3^2$	−0.332 36	−0.340 10	−0.373 74	−0.383 92	−0.519 88	−0.538 60
Many-body, E_2+E_3	−0.299 46	−0.297 74	−0.348 74	−0.352 81	−0.472 23	−0.461 36
[2/1] Padé, $E_2/(1-E_3/E_2)$	−0.299 68	−0.300 83	−0.349 61	−0.355 60	−0.472 70	−0.470 56
E_{var} ($\gamma=1$)	−0.287 90	−0.281 74	−0.331 11	−0.311 51	−0.440 64	−0.406 81
γ optimal	+0.988 00	+0.869 01	+0.996 98	+0.833 92	+0.963 78	+0.801 70
E_{var} (γ optimal)	−0.287 94	−0.287 93	−0.331 11	−0.322 82	−0.441 23	−0.430 01

[a] energies are in hartrees. Based on Wilson and Silver (1977*b*).
[b] from Cade and Huo 1967.
[c] relativistic energy of heavy atom from Veillard and Clementi 1968.

The empirical relativistic energy is also given in this table and it should be noted that for HCl this is roughly twice as large as the correlation energy. Comparison of the energies displayed in Table 7.15 with results obtained by configuration mixing (Wilson and Silver 1977*b*; Bender and Davidson 1966, 1969) has demonstrated that third-order diagrammatic many-body perturbation theory can provide an accuracy comparable with that afforded by configuration mixing, whilst avoiding size-consistency problems.

Many-body perturbation theory forms the basis of a very efficient computational scheme for the determination of electron correlation energies (Silver 1978*a*,*b*; Wilson 1978*b*). The method has been used, for example, to describe the correlation effects involving all twenty-two electrons in the carbon monosulphide molecule in its $X^1\Sigma^+$ electronic state (Wilson 1977*b*). A comparison of calculated correlation energies

Table 7.16

Comparison of calculations on the ground state of the carbon monosulphide molecule

Method	Reference	$-E_{ref}$	$-E_{corr}$	$-E_{total}$
Molecular orbital calculation	(a)	435.329 69	...	...
Configuration interaction calculation	(b)	435.330 37	0.075	435.405 37
Configuration interaction calculation	(c)	435.309 4	0.146 8	435.456 2
Configuration interaction calculation	(d)	435.331 7	0.259 1	435.590 8
Diagrammatic perturbation theory, E_{var}	(e)	435.346 27	0.417 76	435.764 03
Diagrammatic perturbation theory, $E[2/1]$	(e)	435.346 27	0.465 35	435.811 62

[a] Richards 1967.
[b] Robbe and Schamps 1976, $r=2.8$ bohr.
[c] Bruna, Kramer, and Vasudevan 1975, $r=2.9$ bohr.
[d] Green 1971, $r=2.899\,64$ bohr.
[e] Wilson 1977*b*.

obtained using diagrammatic perturbation theory and configuration mixing is presented in Table 7.16. The estimated non-relativistic energy of the carbon monosulphide molecule at its equilibrium nuclear geometry is −436.226 62 hartree. This is 0.415 hartree below the $E[2/1]$ value in Table 7.16 and 0.636 hartree below the lowest reported value obtained by configuration mixing. There are two deficiencies in this calculation on the CS molecule and in the calculations on diatomic hydrides discussed above. The first, and almost certainly the major effect, is that resulting from the truncation of the basis set. The second deficiency results from the truncation of the perturbation expansion for the energy at third-order. These truncation errors will be discussed in detail below.

Potential energy curves have been calculated for a number of diatomic molecules using diagrammatic perturbation theory; including FH (Wilson 1978*a*), CO (Wilson 1977*a*), CS (Wilson 1977*b*), N_2 (Urban and Kellö 1979; Wilson and Silver 1980), BF (Wilson and Silver 1980), F_2 (Urban and Kellö 1979), Be_2 (Bartlett and Purvis 1978*a*; Blomberg and Siegbahn 1977; Blomberg, Siegbahn, and Roos 1980; Robb and Wilson 1980), Mg_2 (Bartlett and Purvis 1978*b*), and CH^+ (Wilson 1979*d*). Portions of the potential energy surface for the ground state of the water molecule have been computed (Silver and Wilson 1977; Bartlett, Shavitt, and Purvis 1979). Here the results for the FH and CO molecules will be

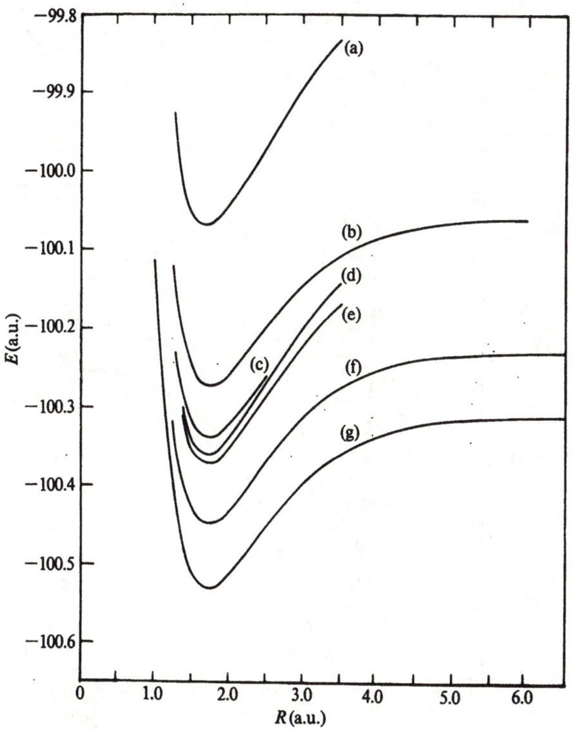

FIG. 7.7

discussed; the CH^+ potential energy curve will be considered below where quasi-degeneracy effects are analysed.

Potential energy curves for the FH molecule are shown in Fig. 7.7. Curves (d) and (e) in this figure were determined by full third-order diagrammatic perturbation theory calculation (Wilson 1978). Curve (a) is the potential energy curve obtained by means of the self-consistent-field approximation. Curves (b) and (c) were determined by Dunning (1976) using the generalized valence bond method and by Meyer and Rosmus (1975) using the coupled electron pair approximation. Curve (d) corresponds to the upper bound determined from a third-order perturbation calculation using the model perturbation series and curve (e) to the [2/1] Padé approximant constructed from this series. Morse curves derived from experimental data are labelled (f) and (g). Curve (f) includes a correction for relativistic effects and represents an estimate of the non-relativistic potential energy curve.

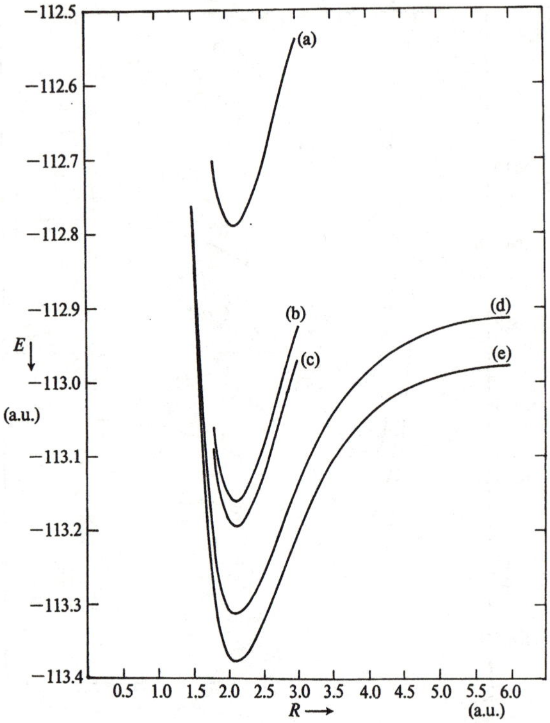

FIG. 7.8

Potential energy curves for the ground state of the CO molecule are displayed in Fig. 7.8. Curve (a) is the self-consistent-field potential energy function; curves (b) and (c) are derived from diagrammatic perturbation theory calculations (Wilson 1977*a*), the former being an upper bound to the true non-relativistic energy and the latter corresponding to the [2/1] Padé approximant to the perturbation series. Curves (d) and (e) are derived from experimental data; the latter is a Morse function determined from the experimentally determined values of r_e, D_e, and ω_e, while the former includes an empirical correction for the relativistic energy.

The results of diagrammatic perturbation theory studies of a variety of small atoms and molecules are presented in Tables 7.17 and 7.18. In all of these calculations, all components of the valence correlation energy through third-order were evaluated. The nuclei in molecules were taken to be at their experimental equilibrium geometry. Basis sets of moderate size were employed (Guest and Wilson 1980). In Table 7.17, the self-consistent-field energy is given together with the second-order and third-order

Table 7.17

Application of third-order diagrammatic perturbation theory to various small systems[a]

System	E_{scf}	E_2	E_3	$E[2/1]$
Ne	−128.540 45	−210.784	+0.753	−210.034
Ar	−526.801 94	−149.200	−11.657	−161.846
BH	−25.128 90	−63.700	−16.915	−86.731
FH	−100.060 07	−223.633	+1.212	−222.427
AlH	−242.455 53	−52.514	−14.605	−72.745
ClH	−460.094 56	−153.719	−15.311	−170.724
BeH_2	−15.770 24	−51.998	−12.597	−68.623
OH_2	−76.055 58	−220.304	−3.108	−223.457
MgH_2	−200.720 28	−45.934	−12.498	−63.105
SH_2	−398.700 77	−150.402	−19.784	−173.182
BH_3	−26.398 34	−97.988	−18.943	−121.471
CH_3^+	−39.245 33	−112.112	−18.635	−134.462
NH_3 (D_{3h})	−56.209 57	−197.500	−10.641	−208.747
NH_3 (C_{3v})	−56.216 35	−198.60	−11.003	−210.245
OH_3^+ (D_{3h})	−76.332 76	−215.249	−6.161	−221.592
OH_3^+ (C_{3v})	−76.334 75	−216.907	−6.224	−223.315
AlH_{33}	−243.637 70	−79.302	−19.460	−105.090
PH_3 (D_{3h})	−342.418 87	−141.017	−23.001	−168.516
PH_3 (C_{3v})	−342.477 17	−138.965	−23.798	−167.680
N_2	−108.976 84	−326.887	+9.201	−317.938
CO	−112.775 51	−300.509	+5.083	−295.510
BF	−124.156 42	−252.307	−2.390	−254.719
SiO	−363.827 90	−284.326	+9.308	−275.312
SiS	−686.484 88	−206.813	−16.878	−225.190

[a] Based on the work of Guest and Wilson (1980).

components of the valence correlation energy corresponding to the use of the model zero-order hamiltonian and the [2/1] Padé approximant which can be constructed from them. For systems such as the neon atom, the FH molecule, or water, the total third-order energy is a small percentage (0.3–1.4%) of the total second-order energy, implying that the perturbation expansion is rapidly converging. For other systems, such as BeH_2 or MgH_2, the third-order energy represents a larger percentage (24–27%) of the second-order component. In Table 7.18, [3/0] and [2/1] Padé approximants are given for all of the atoms and molecules considered in Table 7.17 for the perturbation series based on the Hartree–Fock model zero-order hamiltonian and that based on the shifted denominator series. It should be noted that for systems such as Ne, FH, and H_2O, the [3/0] and [2/1] Padé approximants for the model scheme are in close agreement, whereas the corresponding approximants for the shifted denominator scheme differ somewhat. Furthermore, the [2/1] Padé approximants for the two expansions are in closer agreement than the [3/0] Padé approximants. On the other hand, for BeH_2 and MgH_2, the [3/0] and [2/1] Padé

Table 7.18

Comparison of the model perturbation expansion and the shifted denominator perturbation expansion for various small systems[a]

	Model perturbation expansion		Shifted denominator perturbation expansion	
System	$E[3/0]$	$E[2/1]$	$E[3/0]$	$E[2/1]$
Ne	−210.031	−210.034	−201.253	−209.472
Ar	−160.858	−161.846	−152.283	−161.454
BH	−80.615	−86.731	−89.893	−89.893
FH	−222.421	−222.427	−210.262	−221.208
AlH	−67.119	−72.745	−74.334	−74.349
ClH	−169.030	−170.724	−162.270	−169.975
BeH_2	−64.596	−68.623	−69.181	−69.187
OH_2	−223.413	−223.457	−210.657	−221.857
MgH_2	−58.433	−63.105	−63.506	−63.693
SH_2	−170.186	−173.182	−167.182	−172.549
BH_3	−116.931	−121.471	−122.542	−122.593
CH_3^+	−130.748	−134.462	−135.462	−135.451
NH_3 (D_{3h})	−208.141	−208.746	−200.974	−208.077
NH_3 (C_{3v})	−209.600	−210.245	−202.576	−209.543
OH_3^+ (D_{3h})	−221.411	−221.592	−213.513	−220.229
OH_3^+ (C_{3v})	−223.131	−223.315	−216.247	−222.409
AlH_3	−98.762	−105.091	−106.862	−107.067
PH_3 (D_{3h})	−164.029	−168.516	−165.865	−168.229
PH_3 (C_{3v})	−162.763	−167.680	−166.193	−167.891
N_2	−317.686	−317.938	−257.443	−310.457
CO	−295.697	−295.510	−269.302	−293.641
BF	−254.697	−254.719	−247.456	−256.778
SiO	−275.017	−275.312	−244.702	−273.423
SiS	−223.691	−225.190	−205.403	−222.582

[a] Based on the work of Guest and Wilson (1980).

approximants for the shifted denominator expansion are in closer agreement than the corresponding approximants for the model scheme and the [2/1] Padé approximants for the two schemes are in closer agreement than the [3/0] Padé approximants. Quasi-degeneracy effects explain the behaviour of the perturbation series for systems such as BeH_2 and MgH_2. These effects will be discussed in more detail below, but first let us summarize the results of studies of correlation effects using third-order diagrammatic perturbation theory.

It has been found that:

(i) the use of the Hartree–Fock model operator as a zero-order hamiltonian (the Moeller–Plesset expansion) yields good results if no near degeneracy is present (Wilson 1978*c*; Wilson and Silver 1976, 1977*b*);
(ii) the use of shifted denominators (the Epstein–Nesbet expansion) is useful when quasi-degeneracy is present, especially if [2/1] Padé approx-

imants are constructed (Wilson and Silver 1977*b*; Silver, Wilson, and Bunge 1979; Wilson 1979);
(iii) $[N+1/N]$ Padé approximants have special invariance properties for Rayleigh–Schrödinger perturbation theory (Wilson, Silver, and Farrell 1977) and [2/1] Padé approximants appear to be very useful in third-order studies (Wilson and Silver 1976; Wilson and Silver 1977*b*; Bartlett and Shavitt 1977*a*, 1978; Urban and Kellö 1979);
(iv) there appears to be little or no advantage in employing modified potentials in calculations of correlation energies if all terms are included through third-order (Silver, Wilson, and Bartlett 1977);
(v) there is a danger in making a partial evaluation of energy components in any order, e.g. neglecting third-order many-body effects in FH over-estimates the correlation energy available within a given basis set by twenty-five per cent (Wilson, Silver, and Bartlett 1977);
(vi) the method is computationally very efficient (Silver 1978; Wilson 1978*b*, 1983).

These observations are true not only for closed-shell systems but also for open-shell functions for which a restricted Hartree–Fock reference function can be used.

It was noted above that the convergence properties of the non-degenerate many-body perturbation theory deteriorate when quasi-degeneracy effects are present. However, in view of its simplicity, there is considerable interest in exploring the range of applicability of the non-degenerate formalism.

The low-lying 2p-state in the Be atom makes the calculation of the ground state correlation energy for this system a severe test of the non-degenerate perturbation series. Full configuration mixing and third-order diagrammatic perturbation theory calculations have been reported (Silver, Wilson, and Bunge 1979) for the ground state of the Be atom within the same basis set. For a basis set limited to functions of s-symmetry there are no near degeneracy problems and the agreement between the two methods is excellent. For a basis set consisting of functions with s-, p-, and d-symmetry, only 93% of the correlation correction given by the configuration mixing calculation was obtained on forming the [2/1] Padé approximant to the perturbation series corresponding to the Hartree–Fock model hamiltonian, whereas 99.5% is obtained by forming the [2/1] Padé approximant to the shifted denominator perturbation series.

Third-order non-degenerate diagrammatic perturbation theory studies of the CH^+ ion have been reported (Wilson 1979*a*). Quasi-degeneracy effects in this system become increasingly important as the internuclear separation is increased. Calculations were performed within the same basis set used by Green *et al.* (1972) in their two-root configuration mixing study. For this system it was again shown that the [2/1] Padé approximant to the shifted denominator perturbation series is useful when quasi-degeneracy is present.

The generalized zero-order hamiltonian,

$$\hat{\mathscr{H}}_0 = (1-\mu)\hat{\mathscr{H}}_0^{\text{model}} + \mu\hat{\mathscr{H}}_0^{\text{shifted}} \tag{7.6.16}$$

where μ is an arbitrary scalar, has been investigated (Wilson 1979*b*) in order to eliminate the arbitrary use of the model and the shifted denominator perturbation series depending on the degree of quasi-degeneracy in the problem. Calculations have been reported for both the Be atom, in which quasi-degeneracy effects are important, and for the Ne atom, in which no near degeneracy is present. For Be the [2/1] Padé approximant to the energy series has a minimum corresponding to ~0.9, while for Ne a minimum at ~0.2 is observed. It should be noted, however, that there is no variational theorem here.

It was noted above that one of the truncation errors in contemporary calculations based on the diagrammatic perturbation theory is that arising from basis set truncation. Indeed, this is almost certainly the dominant error in the majority of cases.

For first-row atoms in their ground states, Eggarter and Eggarter (1978*a,b,c,d*) and Jankowski and Malinowski (1978, 1979, 1980*a,b*) have shown that, provided no quasi-degeneracy effects are present, accurate results can be obtained in second-order perturbation theory if a large basis set containing functions with high l quantum number is employed. For example, about ninety-eight per cent of the empirical correlation energy of the neon atom in its ground state was recovered in second-order by including functions with l quantum number up to 6 in the basis set. The percentage of the correlation energy recovered in second-order calculations on other first-row atoms is shown in Fig. 7.9.

In order to achieve high accuracy in applications to diatomic molecules,

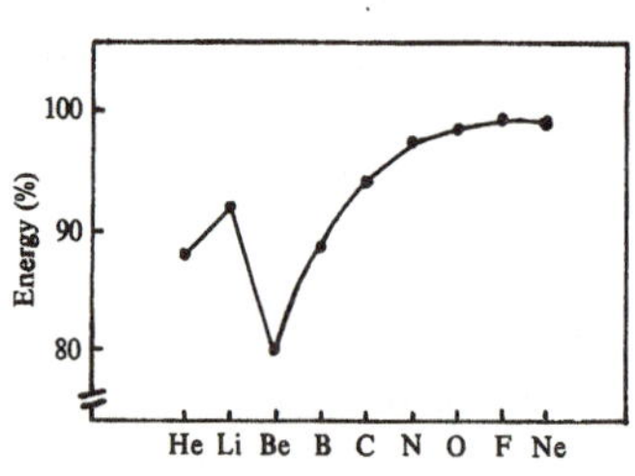

FIG. 7.9. Percentage of empirical correlation energy obtained by second-order perturbation theory for first-row atoms.

Table 7.19

A comparison of diagrammatic perturbation theory calculations using universal basis sets for the LiH *and* FH *molecules with other studies*

Method		E_{total}	% $E_{correlation}$
LiH molecule (R_e = 3.015 bohr)			
Bender and Davidson (1969)	CI	−8.0606	88.3
Meyer and Rosmus (1975)	PNO-CI	−8.0647	93.3
	CEPA	−8.0660	94.8
Wilson and Silver (1977*b*)	DPT $E[2/1]$	−8.0643	92.8
	$\tilde{E}[2/1]$	−8.0652	93.9
Diagrammatic perturbation	$E[2/1]$	−8.0653	94.0
theory/universal basis set	$\tilde{E}[2/1]$	−8.0661	94.9
FH molecule (R_e = 1.738 bohr)			
Bender and Davidson (1969)	CI	−100.3564	75.1
Meyer and Rosmus (1975)	PNO-CI	−100.3274	67.5
	CEPA	−100.3392	70.6
Wilson and Silver (1977*b*)	$E[2/1]$	−100.3727	79.4
	$\tilde{E}[2/1]$	−100.3707	78.9
Diagrammatic perturbation	$E[2/1]$	−100.3837	82.3
theory/universal basis set	$\tilde{E}[2/1]$	−100.3770	80.6

moderately large basis sets are ultimately required. Universal even-tempered basis sets, which were discussed in Section 6.8, have been employed in a number of studies using diagrammatic many-body perturbation theory. (Wilson and Silver 1980; Wilson 1981*c*; Wilson and Silver 1982). Some of the results are summarized in Table 7.19, where a comparison is given with results obtained using other methods for handling correlation effects and different basis sets. Diagrammatic perturbation theory using universal basis sets leads to the most accurate calculated correlation energies of the ground states of the molecules considered in Table 7.19.

Systematic sequences of even-tempered basis sets and of universal basis sets (Wilson 1980*e,f*) have also been reported. In Fig. 7.10, the convergence of the matrix Hartree–Fock energy is compared with the convergence of the correlation energy for the beryllium atom. It can be seen that the correlation energy converges less rapidly than the matrix Hartree–Fock energy with increasing size of basis set. By using the systematic techniques for extending basis sets described in Section 6.10, reliable extrapolation procedures can be used to obtain the basis set limit for the matrix Hartree–Fock energy and the correlation correction.

The second source of error in the calculations described earlier in the present section, is that associated with the neglect of fourth-order and higher-order terms in the perturbation series. Most theories currently employed in the study of electron correlation effects in atoms and

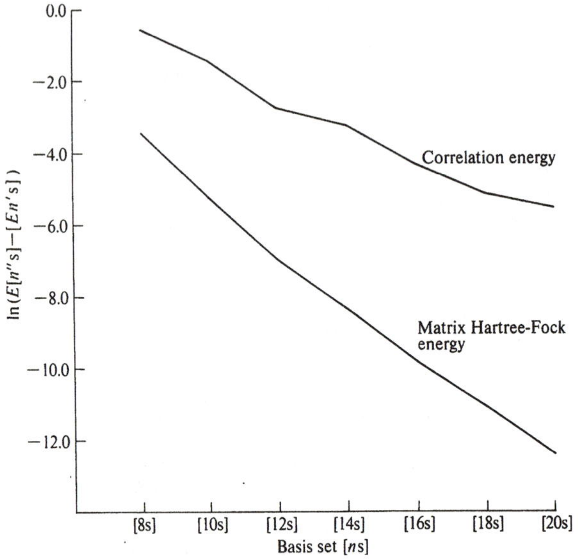

FIG. 7.10. Convergence of the matrix Hartree–Fock energy and the correlation energy for the beryllium atom with increasing size of basis set.

molecules may be regarded as third-order theories in that they neglect, or merely approximate, fourth-order and higher-order terms in the perturbation analysis of the energy. Diagrammatic perturbation theory offers a systematic scheme for extending such calculations. There is considerable interest in the evaluation of fourth-order terms since these terms represent, at least in part, the dominant corrections to most techniques currently being employed in calculating atomic and molecular correlation energies. There is particular interest in the fourth-order components of the correlation energy which involve triple-excitations and quadruple-excitations in intermediate states, since both of these components are omitted in, for example, limited single and double excitation configuration mixing, while the triple excitation component is omitted in coupled pair theories of the type discussed in Section 7.4.

In Table 7.20, the results of full fourth-order diagrammatic many-body perturbation theory calculations on seven molecules are displayed (Wilson and Guest 1981). The fourth-order energy E_4 is divided into a component corresponding to single and double excitations E_{4SD}, triple excitations E_{4T}, and quadruple excitations E_{4Q}. Values are calculated for the Padé approximants $E[3/0]$, $E[2/1]$, $E[4/0]$, and $\mathscr{E}_4$—the latter being

Table 7.20

Full fourth-order calculations using diagrammatic perturbation theory

Molecule	E_2	E_3	E_4	E_{4SD}	E_{4T}	E_{4Q}	$E[3/0]$	$E[2/1]$	$E[4/0]$	$\mathscr{E}_4$
FH	−0.2206	−0.0001	−0.0069	−0.0041	−0.0039	+0.0011	−0.2207	−0.2207	−0.2276	−0.2276
OH_2	−0.2176	−0.0045	−0.0080	−0.0050	−0.0050	+0.0020	−0.2221	−0.2222	−0.2300	−0.2305
NH_3	−0.1989	−0.0118	−0.0026	−0.0003	−0.0050	+0.0027	−0.2107	−0.2114	−0.2133	−0.2137
CH_4	−0.1704	−0.0191	−0.0068	−0.0056	−0.0040	+0.0028	−0.1895	−0.1919	−0.1963	−0.1985
N_2	−0.3291	+0.0098	−0.0293	−0.0178	−0.0180	+0.0065	−0.3193	−0.3196	−0.3486	−0.3461
CO	−0.3076	+0.0048	−0.0284	−0.0172	−0.0161	+0.0049	−0.3028	−0.3029	−0.3312	−0.3299
HCN	−0.3030	+0.0016	−0.0252	−0.0152	−0.0167	+0.0067	−0.3014	−0.3015	−0.3267	−0.3263

Atomic units are used throughout.

the fourth-order invariant defined in (7.6.9). It can be seen in Table 7.20 that the fourth-order energy can often be significantly larger than the third-order energy. This is particularly evident for the multiply bonded molecules N_2, CO, and HCN. However, for these systems, the third-order energy is positive and there is some degree of cancellation between the third-order and the fourth-order energy components. It is perhaps not surprising that the fourth-order energy is larger than the third-order component since singly-excited, triply excited, and quadruply excited configurations first contribute to the correlation energy in fourth-order. In all of the molecules considered in Table 7.20, the fourth-order triple-excitation energy, which is often heuristically neglected, is seen to be chemically significant; that is, it has a magnitude greater than one millihartree, and is certainly as large as, and in the majority of cases larger than, the other fourth-order contributions to the correlation energy.

A comparison of full configuration mixing calculations for the H_2 dimer and for the ground state of the water molecule with full fourth-order diagrammatic perturbation theory calculations is made in Table 7.21. For the H_2 dimer, a minimum basis set has been used; while for the H_2O molecule a double-zeta basis was employed. For the H_2 dimer, two nuclear geometries are considered; one in which quasi-degeneracy effects are small and one in which they are large. It can be seen that full

Table 7.21

Comparison of full configuration mixing and diagrammatic many-body perturbation theory through fourth-order†

System	H_2 dimer Case (a)	H_2 dimer Case (b)	H_2O molecule
E_2	−0.031 053	−0.043 648	−0.139 478
E_3	−0.012 711	−0.018 811	−0.001 391
E_{4S}	−0.000 136	−0.000 002	−0.000 908
E_{4D}	−0.006 023	−0.010 743	−0.003 083
E_{4T}	−0.000 887	−0.000 003	−0.001 364
E_{4Q}	−0.003 051	−0.006 036	−0.000 815
E_4	−0.010 97	−0.016 783	−0.006 170
$E[3/0]$	−0.043 764	−0.062 459	−0.140 869
$E[2/1]$	−0.052 572	−0.076 706	−0.140 883
$E[4/0]$	−0.053 861	−0.079 242	−0.147 039
Double-excitation configuration mixing	−0.052 515	−0.105 552	−0.139 340
Full configuration mixing	−0.053 690	−0.109 196	−0.148 028

† Based on the work of Jankowski and Paldus (1980), Saxe *et al.* (1980), Wilson and Guest (1980), Wilson, Jankowski, and Paldus (1983).
The calculations for the H_2 dimer employed the H4 model defined in Table 7.7. Case (a) involve no quasi-degeneracy effects ($\alpha = 0.500$); case (b) involves significant quasi-degeneracy ($\alpha = 0.005$). Energies are in hartrees.

fourth-order perturbation theory is in excellent agreement with full configuration mixing, within the same basis set, if there is no significant quasi-degeneracy.

Of the components of the correlation energy in fourth-order of the perturbation series for the energy, the triple-excitation component is the most computationally demanding, leading to an algorithm depending on the seventh power of the number of basis functions employed. A simultaneous attack, on both the basis set truncation error and the truncation error resulting from the neglect of fourth-order and higher-order terms in the perturbation expansion, rapidly becomes computationally prohibitive and, furthermore, leads to an increasingly complicated description of electron correlation effects. The method of scaling of the zero-order hamiltonian attempts to overcome this problem.

The rate of convergence of a perturbation series is determined not only by the choice of zero-order wavefunction, but also by the particular reference hamiltonian employed. A more rapidly convergent perturbation expansion can often be obtained by means of a simple modification of the zero-order hamiltonian; namely by introducing a scaling parameter into the zero-order hamiltonian (cf. eqn (7.6.3)). The idea of introducing scaling parameters into the reference spectrum goes back to the work of Feenberg and his collaborators (Goldhammer and Feenberg 1956; Feenberg 1958). More, recently, it has been demonstrated that improved results can often be obtained from second-order perturbation theory calculations of electron correlation energies using large basis sets by introducing a scaling parameter μ into the zero-order hamiltonian $\hat{\mathscr{H}}_0$ giving the scaled reference operator

$$\mu\hat{\mathscr{H}}_0 \tag{7.6.17}$$

(Wilson 1979*a*, 1981*d*). The scaling parameter may be determined from calculations taken to third-order or higher-order in the energy, using a basis set of more moderate size.

If the calculation using the moderate size basis set is taken to third-order in the energy, then the scaling parameter may be chosen so that the scaled second-order energy $\mu^{-1}E_2$ is equal to the [2/1] Padé approximant; the scaling parameter is then given by

$$\mu = \frac{E_2}{E[2/1]} = 1 - \frac{E_3}{E_2}. \tag{7.6.18}$$

This scaling parameter is now used in a calculation using a large basis set which is only taken to second-order in the energy. This procedure has been termed the μ-scaling technique. The application of the μ-scaling technique is illustrated in Table 7.22 for the beryllium atom and for the neon atom. In neon higher-order terms in the perturbation expansion are

Table 7.22

Scaling of second-order correlation energies. Energy in hartree; percentage of empirical correlation energy given in parentheses

	Be	Ne
$E_{\text{empirical}}$	−0.0943	−0.390
E_2 [a]	−0.0757 (80.3%)	−0.3854 (98.8%)
μ^{-1} [b]	1.1972	0.9971
$\mu^{-1}E_2$	−0.0906 (96.1%)	−0.3843 (98.5%)

[a] Eggarter and Eggarter (1978*a*,*b*).
[b] Wilson and Silver (1976), Silver *et al.* (1979).

small and the scaled second-order energy differs little from the second-order energy prior to scaling. On the other hand, for the beryllium atom, higher-order terms in the perturbation series are important and the use of the scaling procedure recovers a further 16 per cent of the correlation energy.

It has been demonstrated that the components of the correlation energy involving triply and quadruply excited states can be significant. Unfortunately, these higher-order terms, and in particular the triply excited terms, are more difficult to evaluate than those involving only double excitations. The scaling procedure can be extended to take account of higher-order terms in the perturbation series. To take account of fourth-order terms the scaling parameter

$$\nu = \frac{E_2}{\mathbb{E}_4} \tag{7.6.19}$$

can be defined, where $\mathbb{E}_4$ is an estimate of the correlation energy through fourth-order. $\mathbb{E}_4$ could, for example, be taken to be $E[4/0]$ or the fourth-order invariant defined in (7.6.9). The fourth-order scaling procedure based on this invariant has been termed (Wilson 1981*d*) ν-scaling. The scaling procedure can easily be extended to take account of, for example, fifth-order terms where the scaling parameter

$$\sigma = \frac{E_2}{E[3/2]}, \tag{7.6.20}$$

in which $E[3/2]$ is the [3/2] Padé approximant to the correlation energy, would prove useful.

The μ-scaling technique should not be regarded as a method for estimating higher-order terms in the perturbation series (Byron and Joachain 1979; Wilson 1979*d*). It should be viewed rather as a method for choosing a scaled zero-order hamiltonian so that the perturbation

Table 7.23

The μ-scaling technique[a]

Molecule	μ(A)	μ(B)	$E[2/1]$(B)	E_2(B)	$\mu(A)^{-1}E_2(B)$
CH_4	0.8882	0.8879	−191.9	−170.4 (11.20%)	−191.9 (0.00%)
NH_3	0.9317	0.9403	−211.5	−198.9 (5.96%)	−213.5 (0.95%)
OH_2	0.9670	0.9793	−222.2	−217.6 (2.07%)	−225.0 (1.26%)
FH	0.9901	0.9995	−220.7	−220.6 (0.05%)	−222.8 (0.95%)
N_2	1.0090	1.0298	−319.6	−329.1 (2.97%)	−326.0 (2.00%)

[a] all energies are in millihartree; values of

$$\left|\frac{E_2(B)-E[2/1](B)}{E[2/1](B)}\right|\times 100\% \quad \text{and} \quad \left|\frac{\mu(A)^{-1}E_2(B)-E[2/1](B)}{E[2/1](B)}\right|\times 100\%$$

are given in parentheses.

series only has to be taken to second-order in the energy. Likewise, the ν-scaling technique should not be regarded as a method for estimating the components of the correlation energy associated with triply and quadruply excited states.

In Tables 7.23 and 7.24, the use of the μ-scaling and of the ν-scaling procedure are, respectively, illustrated for a number of molecular systems. In these tables the smaller of the basis sets is designated A and the larger is designated B. In Table 7.23, the values of $E[2/1]$ obtained using

Table 7.24

The ν-scaling technique[a]

Molecule	ν(A)	ν(B)	$\mathscr{E}_4$(B)	E_2(B)	$\nu(A)^{-1}E_2(B)$
CH_4	0.8632	0.8581	−198.6	−170.4 (14.20%)	−197.4 (0.60%)
NH_3	0.9072	0.9303	−213.8	−198.9 (6.97%)	−219.3 (2.57%)
OH_2	0.9424	0.9436	−230.6	−217.6 (5.64%)	−230.9 (0.13%)
FH	0.9633	0.9692	−227.6	−220.6 (3.08%)	−229.0 (0.62%)
N_2	0.9450	0.9508	−346.2	−329.1 (4.94%)	−348.3 (0.61%)

[a] all energies are in millihartrees; values of

$$\left|\frac{E_2(B)-\mathscr{E}_4(B)}{\mathscr{E}_4(B)}\right|\times 100\% \quad \text{and} \quad \left|\frac{\nu(A)^{-1}E_2(B)-\mathscr{E}_4(B)}{\mathscr{E}_4(B)}\right|\times 100\%$$

are given in parentheses.

basis set B are given; these are the results which are to be approximated by introducing μ determined from a calculation using basis set A into a second-order calculation using basis set B. Values of $E_2(\mathrm{B})$ and $\mu(\mathrm{A})^{-1}E_2(\mathrm{B})$ are given in the final two columns of Table 7.23, together with the percentage difference between each of these values and $E[2/1]$ in parentheses. With the exception of the result for the FH molecule, the introduction of the scaling parameter leads to an improved second-order energy. The average difference between $E_2(\mathrm{B})$ and $E[2/1](\mathrm{B})$ is about 4.5%, whereas the average difference between $\mu(\mathrm{A})^{-1}E_2(\mathrm{B})$ and $E[2/1](\mathrm{B})$ is only about 1.0%. The use of the ν-scaling parameter is illustrated in Table 7.24. $\mathscr{E}_4(\mathrm{B})$ is the fourth-order invariant which is to be approximated by the ν-scaling procedure. The average difference between the $\mathscr{E}_4(\mathrm{B})$ and $E_2(\mathrm{B})$ is about 7.0%, whereas the average difference between $\mathscr{E}_4(\mathrm{B})$ and $\nu(\mathrm{A})^{-1}E_2(\mathrm{B})$ is only about 0.9%. The ν-scaling technique avoids the computation of the triple and quadruple excitation component of the correlation energy with a large basis set.

In employing the scaling procedure, it is clearly desirable that the basis set of moderate size used to determine the scaling parameter and the larger basis set which is used to perform the scaled second-order calculation are related in some well defined fashion. It has been demonstrated that a very powerful approach to the calculation of correlation energies can be obtained by using the scaling technique in conjunction with a universal systematic sequence of even-tempered basis sets (Wilson 1981b). Near-linear dependence in the large basis set can then be avoided and furthermore reliable extrapolation procedures can be employed to obtain estimates of the basis set limit. The use of the μ-scaling technique in conjunction with a universal systematic sequence of even-tempered basis sets is illustrated in Fig. 7.11.

The scaling procedure also forms the basis of a useful method for the analysis of correlation energies. The second-order energy has the form of a sum of pair energies and the scaled second-order energy provides a rigorous basis for a pair analysis of the correlation energy (Wilson and Guest 1981). Furthermore, we note that the scaling technique is not limited to the calculation of correlation energy but can also be employed in the treatment of molecular properties (Wilson 1980*b*; Sadlej and Wilson 1981).

The scaling technique leads to a very efficient computational procedure. Second-order energy calculations with very large basis sets are computationally most tractable, especially on vector-processing computers (Guest and Wilson 1981; Wilson 1981*e*, 1983). One of the most time-consuming parts of most correlation energy calculations is the transformation of integrals over atomic basis functions to integrals over molecular orbitals. (Modern vector-processing computers have proved most highly efficient

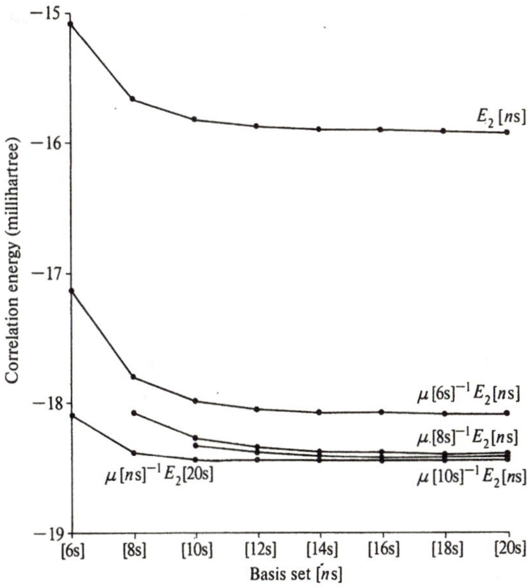

FIG. 7.11. The μ-scaling technique used in conjunction with a universal systematic sequence of even-tempered basis sets.

at performing this transformation (Guest and Wilson 1981).) For second-order energy calculations only, a restricted list of two-electron integrals over molecular orbitals is required. It can be shown (Pendergast and Mayes 1978) that whereas the number of operations required to effect a complete four-index transformation is

$$\tfrac{11}{8}N^5+\tfrac{7}{4}N^4+\tfrac{5}{8}N^3+\tfrac{1}{4}N^2 \tag{7.6.21}$$

where N is the number of basis functions; for the restricted transformation necessary for a second-order calculation the number of operations is

$$\tfrac{1}{2}\{N^4n+3N^3(n^2+n)+N^2(n-3n^3-n^2)+N(n^4-n^2)\} \tag{7.6.22}$$

where n is the number of closed-shell orbitals. The file-handling scheme described by Silver (1978*a*) can be employed to perform second-order calculations in large basis sets. Perturbation calculations including third-order and higher-order terms are only required for basis sets of modest size. It is possible to follow the approach of Handy (1980) for configuration mixing and perform high-order calculations with all of the required integrals stored in the computer central memory.

7.7 Computational aspects of the electron correlation problem

Computers play a central role in the practical implementation of the vast majority of the theoretical apparatus for the accurate determination of the electronic structure of molecular systems which has been considered in this monograph. Indeed, the outlook for performing accurate quantum mechanical calculations for molecules changed radically in the early 1950s with the advent of the digital computer. The problem of effectively utilizing the potentially enormous computing power which will result from rapid and continuous evolution of integrated circuit technology towards smaller, increasingly sophisticated elementary devices, is steadily becoming more and more significant in computational atomic and molecular physics. In this concluding section, it is proposed to say a few words about the future directions which the computational aspects of molecular physics are likely to take.

In a review of molecular quantum mechanics, McWeeny and Pickup (1980) summarize the outlook for the next few years as follows:

> In looking to the future, one thing is clear beyond all doubt: *ab initio* molecular calculations, of 'chemical accuracy', are going to be dominated more and more by the development of computers and highly efficient algorithms. New theories will still be required; . . . , but formal theory will not be enough; the feasibility of the computational implementation will be of paramount importance.

The parallel computer (see, for example, Hockney and Jesshope 1981) will certainly have a most important influence on molecular electronic structure calculations and a central problem of computational molecular physics is to develop efficient techniques for the *ab initio* treatment of electron correlation effects which allow the organization of concurrent computation on a very large scale. Clearly this ability, to organize an algorithm in such a way that calculation is performed in parallel, is not merely a computational problem but is most intricately connected with the theory upon which the computational scheme is based.

Parallel computers have already had a significant impact on molecular electronic structure calculations (see, for example, Guest and Wilson 1981) and have enabled new problems to be investigated, for example, the triple-excitation component of the correlation energy (Wilson and Saunders 1980). Hockney and Jesshope (1981) summarize one important result of experience in using parallel computers so far:

> What has changed with the advent of the parallel computer is the ratio between the performance of a good and a bad computer program. This factor is not likely to exceed a factor of two on a serial machine, whereas factors of ten or more are not uncommon on parallel computers.

In order to adapt theories to take maximum advantage of parallel computers, it is perhaps preferable not to limit the consideration to one particular computer, but to consider the question of parallelism in more general terms. A very useful concept in such discussions is the paracomputer model (see, for example, Wilson 1981*e*, 1983).

A paracomputer (Schwartz 1980) consists of a very large number N of identical processors, each with a conventional order code set, that share a common memory which they can read simultaneously in a single cycle. Although such machines are not physically realizable, since any computing element cannot have more than some fixed number of external connections, they do play a most useful theoretical role in the determination of the limits of parallel computation.

A single processor can clearly simulate an N-processor paracomputer in time $\mathcal{O}(N)$. Parallel algorithms for computational molecular physics can, therefore, be developed using any computing machine and their potential performance on a parallel computer can be investigated by simulation on a serial machine.

An N-processor paracomputer can never be more than N times faster than a serial computer. The approach to this limit will be governed by the degree of parallelism in the algorithm and the theory upon which it is based. It should be noted that there are computations which can never be performed more rapidly on an N-processor paracomputer than on a serial machine. For example, the evaluation of x^{2^k} can be achieved by successive squaring in k cycles on a serial machine; on a paracomputer no power higher than x^{2^m} can be obtained in m cycles of operation and the time required for the whole computation is independent of the number of processors available. An important problem in computational molecular physics is to investigate the extent to which existing theories of electronic structure lead to algorithms which are suited to parallel computation. The paracomputer model (Schwartz 1980) is obviously extremely valuable in such considerations.

Although theoretically useful, the paracomputer model is not physically realizable since no computing element can have more than some fixed number of external connections. A second model, the ultracomputer model (Schwartz 1980), is useful if one wishes to take into account this physical limitation. An ultracomputer consists of a very large number, N, of processors, each with its own memory, but in which each processor can communicate with a fixed number, k, of other processors. An N-processor ultracomputer cannot compute more rapidly than an N-processor paracomputer. The actual performance characteristics of the ultracomputer model will depend on the communication characteristics of the problem. No more than k quantities can be brought together during any cycle and, therefore, the solution of a problem involving N inputs and N outputs will require at least $\mathcal{O}(\log_k N)$ cycles.

There appears to be a close connection between the linked diagram theorem, which was discussed in Chapter 4, and the possibility of devising parallel algorithms for calculations of electron correlation effects in molecules. The linked diagram theorem effectively decouples a system involving a large number of electrons into a series of smaller problems, each of which can be treated concurrently during a calculation (Wilson 1981, 1983). There is a very close connection between the theoretical apparatus employed to describe correlation effects in molecules and the extent to which parallelism can be introduced into algorithms devised to perform calculations.

In conclusion, we point out that the parallel computer will almost certainly have an impact on computational molecular physics second only to that of the first use of digital computers in this field during the 1950s. Calculations of very high accuracy will become routine and numerical studies of larger systems than has hitherto been possible will become tractable. Furthermore, the economics of such calculations will increasingly make the computer a primary source of chemical information for molecules which are difficult to study experimentally.

REFERENCES

Adams, B. G. and Paldus, J. (1979). *Phys. Rev.* **A20,** 1.
Ahlrichs, R. (1974). *Theor. chim. Acta* **33,** 157.
Ahlrichs, R. (1975). *Chem. Phys. Lett.* **34,** 570.
Ahlrichs, R. (1979). *Comput. Phys. Commun.* **17,** 31.
Ahlrichs, R. (1983). In *Methods in computational molecular physics* (ed. G. H. F. Diercksen and S. Wilson). Reidel, Dordrecht.
Ahlrichs, R., Driessler, F., Lischka, H., Staemmler, V., and Kutzelnigg, W. (1975). *J. Chem. Phys.* **62,** 1235.
Ahlrichs, R., Keil, F., Lischka, H., Kutzelnigg, W., and Staemmler, V. (1975). *J. Chem. Phys.* **63,** 455.
Ahlrichs, R. and Kutzelnigg, W. (1968). *J. chem. Phys.* **48,** 1819.
Ahlrichs, R. and Taylor, P. (1981). *J. chim. Phys.* **78,** 315.
Albat, R. and Gruen, N. (1973). *Chem. Phys. Lett.* **18,** 572.
Allen, T. L. and Shull, H. (1961). *J. chem. Phys.* **35,** 1644.
Amos, A. T. (1970). *J. chem. Phys.* **52,** 603.
Amos, A. T. (1972). *Int. J. Quantum Chem.* **6,** 125.
Amos, A. T. (1978). *J. Phys. B: At. Mol. Phys.* **11,** 2053.
Arai, T. (1960). *J. chem. Phys.* **33,** 95.
Atkins, P. W. (1983). *Molecular quantum mechanics* (2nd edn). Oxford University Press.

Bagus, P. S. and Wahlgren, U. I. (1976). *Computers and Chemistry* **1,** 95.
Bagus, P. S., Moser, C. M., Goethals, P., and Verhaegen, G. (1973). *J. Chem. Phys.* **58,** 1886.
Baker, G. A. (1965). *Adv. theor. Phys.* **1,** 1.
Baker, G. A. (1975). *Essentials of Padé approximants.* Academic Press, New York.
Balint-Kurti, G. G. (1975). In *Molecular beam scattering: physical and chemical applications* (ed. K. P. Lawley). Wiley, London.
Balint-Kurti, G. G. and Karplus, M. (1969). *J. chem. Phys.* **50,** 479.
Banerjee, A. and Grein, F. (1976). *Int. J. Quantum Chem.* **10,** 123.
Banerjee, A. and Grein, F. (1977). *J. chem. Phys.* **66,** 1054.
Bardo, R. D. and Ruedenberg, K. (1973). *J. chem. Phys.* **59,** 5956, 5966.
Bardo, R. D. and Ruedenberg, K. (1974). *J. chem. Phys.* **60,** 918.
Barr, T. L. and Davidson, E. R. (1970). *Phys. Rev.* **A1,** 644.
Bartlett, R. J. and Purvis, G. D. (1978*a*). *J. chem. Phys.* **68,** 2114.
Bartlett, R. J. and Purvis, G. D. (1978*b*). *Int. J. Quantum Chem.* **14,** 561.
Bartlett, R. J. and Purvis, G. D. (1980). *Physica Scripta* **21,** 255.
Bartlett, R. J. and Shavitt, I. (1977*a*). *Chem. Phys. Lett.* **50,** 190; erratum (1978) **57,** 157.
Bartlett, R. J. and Shavitt, I. (1977*b*). *Int. J. Quantum Chem. Symp.* **11,** 165, erratum (1978) **12,** 543.
Bartlett, R. J., Shavitt, I. and Purvis, G. D. (1979). *J. chem. Phys.* **71,** 281.
Bartlett, R. J. and Silver, D. M. (1974*a*). *Phys. Rev.* **A10,** 1927.
Bartlett, R. J. and Silver, D. M. (1974*b*). *Chem. phys. Lett.* **29,** 199.

Bartlett, R. J. and Silver, D. M. (1974*c*). *Int. J. Quantum Chem.* **8,** 271.
Bartlett, R. J. and Silver, D. M. (1975*a*). *J. chem. Phys.* **62,** 3258; erratum (1976) **64,** 4578.
Bartlett, R. J. and Silver, D. M. (1975*b*). *Int. J. Quantum Chem. Symp.* **9,** 183.
Bartlett, R. J. and Silver, D. M. (1976*a*). In *Quantum science* (ed. J. L. Calais, O. Goscinski, J. Linderberg, and Y. Ohrn). Plenum, New York.
Bartlett, R. J. and Silver, D. M. (1976*b*). *J. chem. Phys.* **64,** 1260.
Bartlett, R. J., Wilson, S., and Silver, D. M. (1977). *Int. J. Quantum Chem.* **12,** 737.
Bazley, N. W. and Fox, D. W. (1961). *Phys. Rev.* **124,** 483.
Bazley, N. W. and Fox, D. W. (1966). *Phys. Rev.* **148,** 90.
Beckel, C. L. (1976). *J. chem. Phys.* **65,** 4319.
Bender, C. F. and Davidson, E. R. (1966). *J. phys. Chem.* **70,** 2675.
Bender, C. F. and Davidson, E. R. (1969). *Phys. Rev.* **183,** 23.
Bethe, H. A. and Fermi, E. (1932). *Z. Phys.* **77,** 296.
Bingel, W. (1966). *Theor. chim. Acta* **5,** 341.
Binkley, J. S. and Pople, J. A. (1975). *Int. J. Quantum Chem.* **9,** 229.
Bishop, D. M. (1967). *Adv. Quantum Chem.* **3,** 25.
Blau, R., Rau, A. R. P., and Spruch, L. (1973). *Phys. Rev.* **A8,** 119, 131.
Bloch, C. (1958). *Nucl. Phys.* **6,** 329.
Bloch, C. and Horowitz, J. (1958). *Nucl. Phys.* **8,** 91.
Blomberg, M. R. A. and Siegbahn, P. E. M. (1977). *Int. J. Quantum Chem.* **14,** 583.
Blomberg, M. R. A., Siegbahn, P. E. M., and Roos, B. J. (1980). *Int. J. Quantum Chem.* **514,** 229.
Born, M. (1951). *Nach. Akad. Wiss., Göttingen* **1**.
Born, M. and Huang, K. (1954). *Dynamical theory of crystal lattices.* Oxford University Press.
Born, M. and Oppenheimer, J. R. (1927). *Ann. Phys., Leipzig* **84,** 457.
Boys, S. F. (1950). *Proc. R. Soc., Lond.* **A200,** 542.
Boys, S. F. and Bernardi, F. (1970). *Mol. Phys.* **19,** 553.
Boys, S. F., Cook, G. B., Reeves, C. M., and Shavitt, I. (1956). *Nature* **178,** 1207.
Brandow, B. H. (1967). *Rev. mod. Phys.* **39,** 771.
Brandow, B. H. (1975). In *Effective interactions and operators in nuclei* (ed. B. R. Barrett). Springer-Verlag, Berlin.
Brandow, B. H. (1977). *Adv. Quantum Chem.* **10,** 187.
Breit, G. (1929). *Phys. Rev.* **34,** 553.
Breit, G. (1930). *Phys. Rev.* **36,** 382.
Breit, G. (1932). *Phys. Rev.* **39,** 616.
Brenig, W. (1957). *Nucl. Phys.* **4,** 463.
Brillouin, L. (1932). *J. Physique* **7,** 373.
Brink, D. M. and Satchler, G. R. (1968). *Angular momentum.* Clarendon Press, Oxford.
Brooks, B. R. and Schaefer, M. F. (III). (1979). *J. chem. Phys.* **70,** 5092.
Brooks, B. R., Laidig, W. D., Saxe, P., Goddard, J. D., and Schaefer, M. F. (III). (1981). *The unitary group* (ed. J. Hinze). Springer, New York.
Bruecker, K. A. (1954). *Phys. Rev.* **96,** 508.
Bruecker, K. A. (1955*a*). *Phys. Rev.* **97,** 1353.
Brueckner, K. A. (1955*b*). *Phys. Rev.* **100,** 36.
Bruna, P. J., Kramer, W. E., and Vasudevan, K. (1975). *Chem. Phys.* **9,** 91.
Buenker, R. J. and Peyerimhoff, S. (1975). *Theor. chim Acta* **39,** 217.
Buenker, R. J., Peyerimhoff, S., and Butscher, W. (1978). *Mol. Phys.* **35,** 771.
Bunge, C. F. (1976). *Phys. Rev.* **A14,** 1965.

Bunge, C. F. (1980). *Physica Scripta* **21,** 328.
Byron, F. W. and Joachain, C. J. (1979). *J. Phys. B: At. mol. Phys.* **12,** L597.

Cade, P. E. and Huo, W. M. (1967). *J. chem. Phys.* **47,** 614, 649.
Cade, P. E., Sales, K. D., and Wahl, A. C. (1966). *J. chem. Phys.* **44,** 1973.
Cantu, A. A., Klein, D. J., Matsen, F. A., and Seligman, T. H. (1975). *Theor. chim. Acta* **38,** 341.
Carroll, D. P., Silverstone, H., and Metzger, R. M. (1979). *J. chem. Phys.* **71,** 4142.
Čársky, P. and Urban, M. (1980). Ab initio calculations: methods and applications in chemistry. *Lecture Notes in Chemistry* **16.** Springer-Verlag, Berlin.
Cashion, J. K. (1963). *J. chem. Phys.* **39,** 1872.
Cederbaum, L. S. (1973). *Theor. chim. Acta* **31,** 239.
Cederbaum, L. S. and Domcke, W. (1977). *Adv. chem. Phys.* **36,** 205.
Cederbaum, L. S. and Schirmer, J. (1974). *Z. Phys.* **271,** 221.
Chang, E. S., Pu, R. T., and Das, T. P. (1968). *Phys. Rev.* **174,** 1.
Chang, T. C. and Schwarz, W. H. E. (1977). *Theor. chim. Acta* **44,** 45.
Chisholm, C. D. H. (1976). *Group theoretical techniques in quantum chemistry.* Academic Press, London.
Chisholm, C. D. H. and Dalgarno, A. (1966). *Proc. R. Soc., Lond.* **A290,** 264.
Christiansen, P. A. and McCullough, E. A. (1977). *J. chem. Phys.* **67,** 1877, 4142.
Čížek, J. (1966). *J. chem. Phys.* **45,** 4256.
Čížek, J. (1969). *Adv. chem. Phys.* **14,** 36.
Claverie, P., Diner, S., and Malrieu, J. (1967). *Int. J. Quantum Chem.* **1,** 751.
Clementi, E. (1963). *J. chem. Phys.* **38,** 996.
Clementi, E. (1967). *J. chem. Phys.* **46,** 3851.
Clementi, E. and Roetti, C. (1974). *At. Data Nucl. Data Tables* **14,** 177.
Clementi, E., Roothaan, C. C. J., and Yoshimine, M. (1962). *Phys. Rev.* **127,** 1618.
Clementi, E. and Veillard, A. (1966). *J. chem. Phys.* **44,** 3050.
Coester, F. (1958). *Nucl. Phys.* **7,** 421.
Coester, F. and Kummel, H. (1960). *Nucl. Phys.* **17,** 477.
Cohen, E. R. and Taylor, B. N. (1973). *J. phys. Chem. Ref. Data* **2,** 663.
Coleman, A. J. (1963). *Rev. mod. Phys.* **35,** 668.
Coleman, A. J. (1968). *Adv. Quantum Chem.* **4,** 83.
Coleman, A. J. (1977). *Int. J. Quantum Chem.* **11,** 907.
Cook, D. B. (1980). *Mol. Phys.* **42,** 235.
Cooley, J. W. (1961). *Math. Comp.* **15,** 363.
Coolidge, A. S. (1932). *Phys. Rev.* **42,** 189.
Cooper, D. L. and Wilson, S. (1982*a*). *J. Phys. B: At. mol. Phys.* **15,** 493.
Cooper, D. L. and Wilson, S. (1982*b*). *J. chem. Phys.* **76,** 6088.
Cooper, D. L. and Wilson, S. (1982*c*). *J. chem. Phys.* **77,** 5053.
Cooper, D. L. and Wilson, S. (1982*d*). *J. chem. Phys.* **77,** 4551.
Cooper, I. L. and McWeeny, R. (1966). *J. chem. Phys.* **45,** 226.
Coulson, C. A. (1937). *Trans. Faraday Soc.* **33,** 1479.
Coulson, C. A. (1961). *Valence.* Oxford University Press.
Coulson, C. A. (1973). Progress Report, Department of Theoretical Chemistry, University of Oxford, 1972/3.
Coulson, C. A. (1973). Quoted in *Rev. mod. Phys.* **45,** 22.
Coulson, C. A. and Fischer, I. (1949). *Phil. Mag.* **40,** 386.
Csizmadia, I. G., Sutcliffe, B. T., and Barnett, M. (1965). *Can. J. Phys.* **42,** 1645.

Dacre, P. D. (1970). *Chem. Phys. Lett.* **7,** 47.
Dacre, P. D. (1977). *Chem. Phys. Lett.* **50,** 147.
Dalgaard, E. (1979). *Chem. Phys. Lett.* **65,** 559.

Dalgaard, E. and Jorgensen, P. (1978). *J. chem. Phys.* **69,** 3833.
Dalgarno, A. and Stewart, A. L. (1961). *Proc. Phys. Soc., Lond.* **77,** 467.
Das, G. and Wahl, A. C. (1966). *J. chem. Phys.* **44,** 87.
Das, G. and Wahl, A. C. (1967). *J. chem. Phys.* **47,** 2934.
Das, G. and Wahl, A. C. (1972). *J. chem. Phys.* **56,** 3532.
Davidson, E. R. (1972). *Adv. Quantum Chem.* **6,** 235.
Davidson, E. R. (1974). In *The world of quantum chemistry*. Proceedings of the First International Congress on Quantum Chemistry (ed. R. Daudel and B. Pullman). Reidel, Dordrecht.
Davidson, E. R. (1976). *Reduced density matrices in quantum chemistry.* Academic Press, New York.
Davidson, E. R. and Bender, C. F. (1978). *Chem. Phys. Lett.* **59,** 369.
Davidson, E. R. and Silver, D. W. (1977). *Chem. Phys. Lett.* **52,** 403.
Day, B. (1967). *Rev. mod. Phys.* **39,** 719.
Des Cloizeaux, J. (1960). *Nucl. Phys.* **20,** 321.
Diercksen, G. H. F. (1983). In *Methods in computational molecular physics* (ed. G. H. F. Diercksen and S. Wilson). Reidel, Dordrecht.
Dirac, P. A. M. (1929). *Proc. R. Soc., Lond.* **A123,** 714.
Dirac, P. A. M. (1958). *The principles of quantum mechanics* (4th edn). Clarendon Press, Oxford.
Docken, K. K. and Hinze, J. (1972). *J. chem. Phys.* **57,** 4928.
Doggett, G. (1969). *Theor. chim. Acta* **15,** 344.
Doggett, G. (1980). *Ann. Rep.* **C3**. *Chem. Soc.*
Downward, M. J. and Robb, M. A. (1977). *Theor. chim. Acta* **46,** 129.
Drake, G. W. F. and Schlesinger, M. (1977). *Phys. Rev.* **A15,** 1990.
Duch, W. (1980). *Theor. chim. Acta* **57,** 299.
Duch, W. and Karwowski, J. (1979). *Theor. chim. Acta* **51,** 175.
Duch, W. and Karwowski, J. (1981). In *The unitary group for the evaluation of electronic energy matrix elements* (ed. J. Hinze). *Lecture Notes in Chemistry* **22.** Springer-Verlag, Berlin.
Dunham, J. L. (1932). *Phys. Rev.* **41,** 721.
Dunning, T. H., Jr. (1976). *J. chem. Phys.* **65,** 3854.
Dunning, T. H., Jr. and Hay, P. J. (1977). In *Methods of electronic structure theory* (ed. H. F. Schaefer III). Plenum, New York.
Dupuis, M. and King, H. (1977). *Int. J. Quantum Chem.* **11,** 613.
Dutta, C. M., Dutta, N. C., and Das, T. P. (1970). *Phys. Rev. Lett.* **25,** 1695.
Dutta, N. C., Ishihara, I., Matsubara, C., and Das, T. P. (1970). *Int. J. Quantum Chem. Symp.* **3,** 367.
Dutta, N. C., Ishihara, I., and Matsubara, C. (1969). *Phys. Rev. Lett.* **22,** 8.
Dutta, N. C. and Karplus, M. (1975). *Chem. Phys. Lett.* **31,** 455.
Dutta, N. C., Matsubara, C., Pu, R. T., and Das, T. P. (1969). *Phys. Rev.* **177,** 33.
Dykstra, C. E. (1977). *J. chem. Phys.* **67,** 4716.
Dykstra, C. E. (1978*a*). *J. chem. Phys.* **68,** 1829.
Dykstra, C. E. (1978*b*). *J. chem. Phys.* **68,** 4244.
Dykstra, C. E., Schaefer, H. F., and Meyer, W. (1976). *J. chem. Phys.* **65,** 2740, 5141.

Ebbing, D. D. (1963). *J. chem. Phys.* **36,** 1361.
Ebbing, D. D. and Henderson, R. C. (1965). *J. chem. Phys.* **45,** 2225.
Eckart, C. (1935). *Phys. Rev.* **47,** 552.
Eden, R. J. and Francis, N. C. (1955). *Phys. Rev.* **97,** 1366.
Edmiston, C. and Ruedenberg, K. (1963). *Rev. mod. Phys.* **35,** 457.

Edmonds, A. R. (1957). *Angular momentum in quantum mechanics*. Princeton University Press.
Eggarter, E. and Eggarter, T. P. (1978*a*). *J. Phys. B: At. mol. Phys.* **11,** 1157.
Eggarter, E. and Eggarter, T. P. (1978*b*). *J. Phys. B: At. mol. Phys.* **11,** 2069.
Eggarter, E. and Eggarter, T. P. (1978*c*). *J. Phys. B: At. mol. Phys.* **11,** 2969.
Eggarter, T. P. and Eggarter, E. (1978*d*). *J. Phys. B: At. mol. Phys.* **11,** 3635.
El Baz, E. and Castel, B. (1972). *Graphical methods of spin algebras*. Dekker, New York.
Elder, M. (1973). *Int. J. Quantum Chem.* **7,** 75.
England, W. B. (1980). *J. chem. Phys.* **72,** 2108.
Epstein, P. S. (1926). *Phys. Rev.* **28,** 695.

Feenberg, E. (1958). *Ann. Phys., N.Y.* **3,** 292.
Feller, D. F. and Ruedenberg, K. (1979). *Theor. chim. Acta* **52,** 231.
Feshbach, H. (1958). *Ann. Phys., N.Y.* **5,** 357.
Feshbach, H. (1962). *Ann. Phys., N.Y.* **19,** 287.
Fetter, A. L. and Walecka, J. D. (1971). *Quantum theory of many-particle systems*. McGraw-Hill, New York.
Feynman, R. P. (1939). *Phys. Rev.* **59,** 340.
Feynman, R. P. (1948). *Rev. mod. Phys.* **20,** 367.
Fischer, C. F. (1977). *The Hartree–Fock methods for atoms*. Wiley, New York.
Fletcher, R. (1970), *Mol. Phys.* **19,** 55.
Flores, J. and Moshinsky, M. (1967). *Nucl. Phys.* **A93,** 81.
Fock, V. (1930). *Z. Phys.* **61,** 126; **62,** 795.
Fock, V. (1932). *Z. Phys,* **75,** 622.
Fock, V. (1950). *Dokl. Akad. Nauk. SSSR* **73,** 735.
Fougere, P. F. and Nesbet, R. K. (1966). *J. chem. Phys.* **44,** 285.
Frantz, L. M. and Mills, R. L. (1960). *Nucl. Phys.* **15,** 16.
Freed, K. (1971). *Phys. Rev.* **A3,** 578.
Freed, K. (1971). *Ann. Rev. phys. Chem.* **22,** 313.
Freeman, D. F. and Karplus, M. (1976). *J. chem. Phys.* **64,** 2641.
Frisch, M. J., Krishnan, R. and Pople, J. A. (1980). *Chem. Phys. Lett.* **75,** 66.

Gabriel, J. R. (1961). *Proc. Cambridge philos. Soc.* **57,** 330.
Gallup, G. A. (1968). *J. chem. Phys.* **48,** 1752.
Gallup, G. A. (1969*a*). *J. chem. Phys.* **50,** 1206.
Gallup, G. A. (1969*b*). *J. chem. Phys.* **50,** 1215.
Gallup, G. A. (1973). *Adv. Quantum Chem.* **7,** 113.
Gallup, G. A. and Norbeck, J. M. (1976). *J. chem. Phys.* **64,** 2179.
Gammel, J. L. and McDonald, F. A. (1966). *Phys. Rev.* **142,** 1242.
Garpman, S., Lindgren, I., Lindgren, J., and Morrison, J. (1975). *Phys. Rev.* **A11,** 758.
Gaunt, J. A. (1929). *Proc. R. Soc., Lond.* **122,** 513.
Gelfand, I. M. and Tsetlin, M. L. (1950*a*). *Dokl. Akad. Nauk. SSSR* **71,** 825.
Gelfand, I. M. and Tsetlin, M. L. (1950*b*). *Dokl. Akad. Nauk. SSSR* **71,** 1070.
Gellman, M. and Low, F. (1951). *Phys. Rev.* **84,** 350.
Gerratt, J. (1968). *Ann. Rep. Chem. Soc., Lond.* **65,** 3.
Gerratt, J. (1971). *Adv. at. mol. Phys.* **7,** 141.
Gerratt, J. (1974). *Specialist periodical reports: theoretical chemistry* **1,** 60. The Chemical Society, London.
Gerratt, J. and Lipscomb, W. (1968). *Proc. natn. Acad. Sci.* **59,** 339.
Gerratt, J. and Raimondi, M. (1980). *Proc. R. Soc., Lond.* **A371,** 525.

Gilbert, T. L. (1964). In *Molecular orbitals in chemistry, physics and biology* (ed. B. Pullman and P.-O. Löwdin). Academic Press, London.
Gillespie, R. J. (1973). *Molecular geometry*. Van Nostrand Rheinhold, New York.
Gillespie, R. J. and Nyholm, R. S. (1957). *Q. Rev.* **11,** 339.
Goddard, W. A. (1967*a*). *Phys. Rev.* **157,** 73.
Goddard, W. A. (1967*b*). *Phys. Rev.* **157,** 81.
Goddard, W. A. (1970). *Int. J. Quantum Chem. Symp.* **3,** 593.
Goddard, W. A., Dunning, T. H., Hunt, W. J., and Hay, P. J. (1973). *Accts chem. Res.* **6,** 368.
Goldhammer, P. and Feenberg, E. (1956). *Phys. Rev.* **101,** 1233.
Goldstone, J. (1957). *Proc. R. Soc. Lond.* **A239,** 267.
Golebiewski, A., Hinze, J., and Yurtsever, E. (1979). *J. chem. Phys.* **70,** 1101.
Goodisman, J. (1968). *J. chem. Phys.* **48,** 2981.
Goodisman, J. (1969). *J. chem. Phys.* **50,** 903.
Goodisman, J. (1973). *Diatomic interaction potential theory*, Vols 1 and 2 Academic Press, New York.
Gordon, R. (1968). *J. chem. Phys.* **48,** 4984.
Goscinski, O. (1967). *Int. J. Quantum Chem.* **1,** 769.
Goscinski, O. and Brandas, E. (1968). *Chem. Phys. Lett.* **2,** 299.
Goscinski, O. and Brandas, E. (1970). *Phys. Rev.* **A1,** 552.
Gouyet, J. F. (1973). *J. chem. Phys.* **59,** 4637.
Gouyet, J. F., Schranner, R., and Seligman, T. H. (1975). *J. Phys A: Math. gen. Phys.* **8,** 285.
Grant, I. P. (1970). *Adv. Phys.* **19,** 747.
Green, S. (1971). *J. chem. Phys.* **54,** 827.
Green, S. (1974). *Adv. chem. Phys.* **25,** 179.
Green, S., Bagus, P. S., Liu, B., McLean, A. D. and Yoshimine, M. (1972). *Phys. Rev.* **A5,** 1614.
Grein, F. and Banerjee, A. (1975). *Int. J. Quantum Chem. Symp.* **9,** 147.
Grein, F. and Chang, T. C. (1971). *Chem. Phys. Lett.* **12,** 44.
Guest, M. F. and Saunders, V. R. (1974). *Mol. Phys.* **28,** 819.
Guest, M. F. and Wilson, S. (1980). *Chem. Phys. Lett.* **72,** 49.
Guest, M. F. and Wilson, S. (1981). In *Supercomputers in chemistry*, Symposium Proceedings (ed. P. Lykos and I. Shavitt). American Chemical Society, Washington DC.

Hall, G. G. (1951). *Proc. R. Soc., Lond.* **A205,** 541; **A208,** 328.
Hall, G. G. (1959). *Rep. Prog. Phys.* **22,** 1.
Hall, G. G. and Lennard-Jones, J. E. (1950). *Proc. R. Soc., Lond.* **A202,** 155.
Hamermesh, M. (1962). *Group theory*. Addison-Wesley, Reading, Mass., USA.
Handy, N. C. (1980). *Chem. Phys. Lett.* **74,** 280.
Harris, F. E. (1967*a*). *Adv. Quantum Chem.* **3,** 61.
Harris, F. E. (1967*b*). *J. chem. Phys.* **46,** 2769.
Harris, F. E. (1967*c*). *J. chem. Phys.* **47,** 1047.
Hartree, D. R. (1927). *Proc. Cambridge philos. Soc.* **24,** 89.
Hartree, D. R. (1948). *Proc. Cambridge philos. Soc.* **45,** 230.
Hartree, D. R. (1957). *The calculation of atomic structures*. Wiley, New York.
Hartree, D. R. and Black, M. M. (1933). *Proc. R. Soc., Lond.* **A139,** 311.
Hartree, D. R., Hartree, W., and Swirles, B. (1939). *Phil. Trans. R. Soc.* **238,** 229.
Harter, W. G. (1973). *Phys. Rev.* **A8,** 2819.
Harter, W. G. and Patterson, C. W. (1976*a*). *Phys. Rev.* **A13,** 1067.

Harter, W. G. and Patterson, C. W. (1976*b*). *A unitary calculus for electronic orbitals*. Springer-Verlag, Berlin.
Hata, J. (1975). *J. chem. Phys.* **62,** 1221.
Hayes, E. F. and Parr, R. G. (1967). *J. chem. Phys.* **47,** 3961.
Hegarty, D. and Robb, M. A. (1979*a*). *Mol. Phys.* **37,** 1455.
Hegarty, D. and Robb, M. A. (1979*b*). *Mol. Phys.* **38,** 1795.
Hegarty, D. and Robb, M. A. (1980). In *Electron correlation*, Proc. Daresbury study weekend (ed. M. F. Guest and S. Wilson). Science Research Council, London.
Heine, V. (1960). *Group theory in quantum mechanics*. Macmillan, London.
Heitler, W. and London, F. (1927). *Z. Phys.* **44,** 455.
Heitler, W. and Rumer, G. (1931). *Z. Phys.* **68,** 12.
Hellman, H. (1937). *Einfuhrung in die Quantenchemie*. Franz Deutiche, Leipzig.
Herring, C. (1966). In *Magnetism* Vol. IIb (ed. G. T. Rado and H. Shul). Academic Press, London.
Herzberg, G. (1950). *Spectra of diatomic molecules*. Van Nostrand, Princeton, New Jersey.
Hibbert, A. (1975). *Rep. Prog. Phys.* **38,** 1217.
Hillier, I. H. and Saunders, V. R. (1970). *Int. J. Quantum Chem.* **4,** 503.
Hinchliffe, A. and Bounds, D. G. (1978). *Specialist periodical reports: theoretical chemistry* **3,** 70. The Chemical Society, London.
Hinze, J. (1973). *J. chem. Phys.* **59,** 6424.
Hinze, J. (1981) (ed.). *The unitary group for the evaluation of electronic energy matrix elements, Lecture Notes in Chemistry* **22**. Springer-Verlag, Berlin.
Hinze, J. and Roothaan, C. C. J. (1967). *Prog. theor. Phys. Suppl.* **40,** 37.
Hirao, K. (1974). *J. chem. Phys.* **60,** 3247.
Hirschfelder, J. O. and Linnett, J. W. (1950). *J. chem. Phys.* **18,** 130.
Hirschfelder, J. O., Byers Brown, W., and Epstein, S. (1964). *Adv. Quantum Chem.* **1,** 255.
Hockney, R. W. (1979). *Contemp. Phys.* **20,** 149.
Hockney, R. W. and Jesshope, C. R. (1981). *Parallel computers: architecture, programming and algorithms*. Adam Hilger, Bristol.
Hoffmann, H. M., Lee, S. Y., Richert, J., and Weidenmuller, H. (1974). *Ann. Phys., N.Y.* **85,** 410.
Hohenberg, P. and Kohn, W. (1964). *Phys. Rev.* **136,** B864.
Horak, Z. (1965). In *Modern quantum chemistry* (ed. O. Sinanoğlu). Academic Press, New York.
Hose, G. and Kaldor, U. (1979). *J. Phys. B: At. mol. Phys.* **12,** 3827.
Hose, G. and Kaldor, U. (1980). *Phys. Script.* **21,** 357.
Howat, G., Trsic, M., and Goscinski, O. (1977). *Int. J. Quantum Chem.* **11,** 283.
Hubač, I. (1980). *Int. J. Quantum Chem.* **17,** 195.
Hubač, I. and Čársky, P. (1978). *Top. Curr. Chem.* **75,** 97.
Hubač, I. and Čársky, P. (1980). *Phys. Rev.* **A22,** 2392.
Hubač, I. and Kvasnicka, V. (1977). *Croatica Chemica Acta* **49,** 677.
Hubač, I., Kvasnicka, V., and Holubac, A. (1973). *Chem. Phys. Lett.* **23,** 381.
Hubač, I. and Urban, M. (1977). *Theor. chim. Acta* **45,** 185.
Hubač, I., Urban, M., and Kellö, V. (1979). *Chem. Phys. Lett.* **62,** 584.
Hubbard, J. (1957). *Proc. R. Soc., Lond.* **A240,** 539.
Hubbard, J. (1958*a*). *Proc. R. Soc., Lond.* **A243,** 336.
Hubbard, J. (1958*b*). *Proc. R. Soc., Lond.* **A244,** 199.
Hugenholtz, N. H. (1957). *Physica* **23,** 481.
Hund, F. (1928). *Z. Phys.* **51,** 759.

Hund, F. (1931). *Z. Phys.* **73,** 1.
Hurley, A. C. (1976*a*). *Electron theory of small molecules.* Academic Press, New York.
Hurley, A. C. (1976*b*). *Electron correlation in small molecules.* Academic Press, New York.
Hurley, A. C., Lennard-Jones, J. E., and Pople, J. A. (1953). *Proc. R. Soc., Lond.* **A220,** 446.
Hylleraas, E. A. (1929). *Z. Phys.* **54,** 347.
Hylleraas, E. A. and Undheim, B. (1930). *Z. Phys.* **65,** 759.

Ihrig, E., Rosensteel, G., Chow, H., and Trainor, L. E. H. (1976). *Proc. R. Soc., Lond.* **A348,** 339.
Isihara, E. and Poe, R. T. (1972). *Phys. Rev.* **A6,** 111.
Itagaki, T. and Saika, A. (1977). *Chem. Phys. Lett.* **52,** 530.
Itagaki, T. and Saika, A. (1979). *J. chem. Phys.* **70,** 2378.
Itoh, T. (1965). *Rev. mod. Phys.* **37,** 157.

James, H. M. and Coolidge, A. S. (1933). *J. chem. Phys.* **1,** 825.
James, H. M. and Coolidge, A. S. (1935). *J. chem. Phys.* **3,** 129.
James, H. M. and Coolidge, A. S. (1938). *J. chem. Phys.* **6,** 730.
Jankowski, K. and Malinowski, P. (1978). *Chem. Phys. Lett.* **54,** 68.
Jankowski, K. and Malinowski, P. (1980*a*). *Phys. Rev.* **A22,** 51.
Jankowski, K. and Paldus, J. (1980*b*). *Int. J. Quantum Chem.* **18,** 214.
Johansson, A., Kollman, P., and Rothenberg, S. (1973). *Theor. chim. Acta* **29,** 167.
Johnson, M. B. and Baranger, M. (1971). *Ann. Phys., N.Y.* **62,** 172.
Jonsson, B., Roos, B. O., Taylor, P. R., and Siegbahn, P. E. M. (1981). *J. chem. Phys.* **74,** 4566.
Jordan, K. D. (1935). *J. mol. Spectrosc.* **56,** 329.
Jordan, K. D., Kinsey, J. L., and Silbey, R. (1974). *J. chem. Phys.* **61,** 911.
Judd, B. R. (1967). *Second quantization and atomic spectroscopy.* The Johns Hopkins University Press, Baltimore.

Kaldor, U. (1973*a*). *Phys. Rev.* **A7,** 427.
Kaldor, U. (1973*b*). *Phys. Rev. Lett.* **31,** 1338.
Kaldor, U. (1975*a*). *J. chem. Phys.* **62,** 4634.
Kaldor, U. (1975*b*). *J. chem. Phys.* **63,** 2199.
Kaldor, U. (1976). *J. comp. Phys.* **20,** 432.
Kaplan, I. G. (1975). *Symmetry of many-electron systems* (translated from the Russian by J. Gerratt). Academic Press, New York.
Kato, T. (1966). *Perturbation theory for linear operations.* Springer-Verlag, Berlin.
Karwowski, J. (1973*a*). *Theor. chim. Acta* **29,** 151.
Karwowski, J. (1973*b*). *Chem. Phys. Lett.* **19,** 279.
Kellö, V., Urban, M., Hubač, I., and Čársky, P. (1978). *Chem. Phys. Lett.* **58,** 83.
Kelly, H. P. (1963). *Phys. Rev.* **131,** 684.
Kelly, H. P. (1964*a*). *Phys. Rev.* **A134,** 1450.
Kelly, H. P. (1964*b*). *Phys. Rev.* **B136,** 896.
Kelly, H. P. (1966). *Phys. Rev.* **144,** 39.
Kelly, H. P. (1968). *Adv. theor. Phys.* **2,** 75.
Kelly, H. P. (1969*a*). *Adv. chem. Phys.* **14,** 129.
Kelly, H. P. (1969*b*). *Phys. Rev. Lett.* **23,** 255.
Kelly, H. P. (1970*a*). *Int. J. Quantum Chem. Symp.* **3,** 349.

Kelly, H. P. (1970*b*). *Phys. Rev.* **A1,** 274.
Kelly, H. P. (1979). *Comput. Phys. Commun.* **17,** 99.
Kelly, H. P. and Ron, A. (1971). *Phys. Rev.* **A4,** 11.
Kelly, H. P. and Sessler, A. (1963). *Phys. Rev.* **132,** 2091.
Kendrick, J. and Hillier, I. H. (1976). *Chem. Phys. Lett.* **41,** 483.
Killingbeck, J. (1977). *Rep. Prog. Phys.* **40,** 963.
Kim, H. and Hirschfelder, J. O. (1967). *J. chem. Phys.* **47,** 1005.
Kirtman, B. and Cole, S. (1978). *J. chem. Phys.* **69,** 5055.
Kistenmacher, H., Popkie, H., and Clementi, E. (1972). *J. chem. Phys.* **58,** 5627.
Klahn, B. and Bingel, W. A. (1977*a*). *Theor. chim. Acta* **44,** 9.
Klahn, B. and Bingel, W. A. (1977*b*). *Theor. chim. Acta* **44,** 27.
Klein, D. J. (1974). *J. chem. Phys.* **61,** 785.
Klessinger, M. and McWeeny, R. (1965). *J. chem. Phys.* **42,** 3348.
Klonover, A. and Kaldor, U. (1978). *J. Phys. B: At. mol. Phys.* **11,** 1623.
Klonover, A. and Kaldor, U. (1979). *J. Phys. B: At. mol. Phys.* **12,** 323.
Kolos, W. (1970). *Adv. Quantum Chem.* **5,** 99.
Kolos, W. and Roothaan, C. C. J. (1960). *Rev. mod. Phys.* **32,** 169.
Kolos, W. and Wolniewicz, L. (1963). *Rev. mod. Phys.* **35,** 169.
Kolos, W. and Wolniewicz, L. (1964). *J. chem. Phys.* **41,** 3663.
Koopmans, T. A. (1933). *Physica* **1,** 104.
Kotani, M., Amemiya, A., Ishiguro, E., and Kimuro, T. (1963). *Tables of molecular integrals* (2nd edn). Maruzen, Tokyo.
Krenciglowa, E. M. and Kuo, T. T. S. (1974). *Nucl. Phys.* **A235,** 171.
Krishnan, R. and Pople, J. A. (1978). *Int. J. Quantum Chem.* **14,** 91.
Krishnan, R., Binkley, J. S., Seeger, R., and Pople, J. A. (1980). *J. chem. Phys.* **72,** 650.
Krishnan, R., Frisch, M. J., and Pople, J. A. (1980). *J. chem. Phys.* **72,** 1980.
Kronig, R. de L. (1930). *Band spectra and molecular structure.* Cambridge University Press.
Kryachko, E. S. (1980). *Int. J. Quantum Chem.* **18,** 1029.
Kumar, K. (1962). *Perturbation theory and the nuclear many-body problem.* North Holland, Amsterdam.
Kuo, T. T. S., Lee, S. Y., and Radcliff, K. F. (1971). *Nucl. Phys.* **A176,** 65.
Kuprievich, V. A. and Schramko, O. V. (1975). *Int. J. Quantum Chem.* **9,** 1009.
Kutzelnigg, W. (1964). *J. chem. Phys.* **40,** 3640.
Kutzelnigg, W. (1973). *Top. curr. Chem.* **41,** 31.
Kutzelnigg, W. (1977). In *Methods of electronic structure theory* (ed. H. F. Schafer III). Plenum Press, New York.
Kutzelnigg, W., Del Re, G., and Berthier, G. (1968). *Phys. Rev.* **172,** 49.
Kutzelnigg, W., Meunier, A., Levy, B., and Berthier, G. (1977). *Int. J. Quantum Chem.* **12,** 77.
Kvasnicka, V. (1974). *Czech. J. Phys.* **B24,** 605.
Kvasnicka, V. (1975*a*). *Czech. J. Phys.* **B25,** 371.
Kvasnicka, V. (1975*b*). *Theor. chim Acta* **36,** 297.
Kvasnicka, V. (1975*c*). *Phys. Rev.* **A12,** 1159.
Kvasnicka, V. (1977*a*). *Adv. chem. Phys.* **36,** 345.
Kvasnicka, V. (1977*b*). *Czech. J. Phys.* **B27,** 599.
Kvasnicka, V. (1977*c*). *Chem. Phys. Lett.* **51,** 165.
Kvasnicka, V. and Holubec, A. (1975). *Chem. Phys. Lett.* **32,** 489.
Kvasnicka, V., Holubec, A., and Hubač, I. (1974). *Chem. Phys. Lett.* **24,** 361.
Kvasnicka, V. and Hubač, I. (1974). *J. chem. Phys.* **60,** 4483.
Kvasnicka, V., Laurinc, V., and Hubač, I. (1974). *Phys. Rev.* **A10,** 2016.

Kvasnicka, V., Laurinc, V., and Biskupic, S. (1979). *Chem. Phys. Lett.* **67,** 81.
Kvasnicka, V., Laurinc, V., and Biskupic, S. (1980). *Mol. Phys.* **39,** 143.
Kvasnicka, V., Laurinc, V., and Biskupic, S. (1981). *Czech. J. Phys.* **B31,** 41.

Ladik, J. and Čížek, J. (1980). *J. chem. Phys.* **73,** 2357.
Landau, L. D. and Lifshitz, L. M. (1958). *Quantum mechanics.* Pergamon Press, Oxford.
Landau, L. D. and Lifshitz, L. M. (1965). *Quantum mechanics* (2nd edn). Pergamon Press, Oxford.
Langhoff, P. W. and Hernandez, A. J. (1976). *Int. J. Quantum Chem. Symp.* **10,** 337.
Langhoff, S. R. and Davidson, E. R. (1974). *Int. J. Quantum Chem.* **8,** 61.
Langmuir, I. (1919). *J. Am. chem. Soc.* **41,** 868, 1543.
Le Clerque, J. (1976). *Int. J. Quantum Chem.* **10,** 439.
Lee, T. and Das, T. P. (1972). *Phys. Rev.* **A6,** 968.
Lee, T., Dutta, N. C., and Das, T. P. (1970*a*). *Phys. Rev. Lett.* **25,** 204.
Lee, T., Dutta, N. C., and Das, T. P. (1970*b*). *Phys. Rev. Lett.* **25,** 1695.
Lee, T., Dutta, N. C., and Das, T. P. (1970*c*). *Phys. Rev.* **A1,** 995.
Lee, T., Dutta, N. C., and Das, T. P. (1971). *Phys. Rev.* **A4,** 1410.
Leinaas, J. M. and Kuo, T. T. S. (1978). *Ann. Phys.* **111,** 19.
Lengsfield, B. H. (1980). *J. chem. Phys.* **73,** 382.
Lennard-Jones, J. E. (1930). *Proc. R. Soc., Lond.* **A129,** 598.
Lennard-Jones, J. E. (1949). *Proc. R. Soc., Lond.* **A197,** 14.
Levy, B. (1969). *Chem. Phys. Lett.* **4,** 17.
Levy, B. (1970). *Int. J. Quantum Chem.* **6,** 297.
Levy, B. (1973). *Chem. Phys. Lett.* **18,** 59.
Levy, B. (1978). In *Proc. Fourth Semin. Comput. Methods Quantum Chem.* (ed. B. Roos and G. H. F. Diercksen). Max-Planck-Institut für Physik und Astrophysik, Munchen.
Levy, M. and Perdew, J. P. (1982). *Int. J. Quantum Chem.* **21,** 511.
Lewis, G. N. (1916). *J. Am. chem. Soc.* **38,** 762.
Linderberg, J. and Ohrn, Y. (1973). *Propagators in quantum chemistry.* Academic Press, New York.
Linnett, J. W. and Ricra, A. (1969). *Theor. chim. Acta* **15,** 196.
Lindgren, I. (1974). *J. Phys. B: At. mol. Phys.* **7,** 2441.
Lindgren, I. (1978). *Int. J. Quantum Chem. Symp.* **12,** 33.
Lindgren, I. and Morrison, J. (1982). *Atomic many-body theory.* Springer-Verlag, Berlin.
Littlewood, D. E. (1950). *The theory of group characters and matrix representations of groups* (2nd edn). Clarendon Press, Oxford.
Liu, B. and McLean, A. D. (1973). *J. chem. Phys.* **59,** 4557.
Louck, J. D. (1970). *Am. J. Phys.* **38,** 3.
Louck, J. D. and Galbraith, H. W. (1972). *Rev. mod. Phys.* **44,** 540.
Louck, J. D. and Galbraith, H. W. (1976). *Rev. mod. Phys.* **48,** 69.
Longuet-Higgins, H. C. (1948). *Proc. Phys. Soc.* **60,** 270.
Löwdin, P.-O. (1950). *J. chem. Phys.* **18,** 365.
Löwdin, P.-O. (1951). *J. chem. Phys.* **19,** 1396.
Löwdin, P.-O. (1955*a*). *Phys. Rev.* **97,** 1474.
Löwdin, P.-O. (1955*b*). *Phys. Rev.* **97,** 1490.
Löwdin, P.-O. (1955*c*). *Phys. Rev.* **97,** 1509.
Löwdin, P.-O. (1956). *Adv. Phys.* **5,** 1.
Löwdin, P.-O. (1959). *Adv. chem. Phys.* **2,** 207.

Löwdin, P.-O. (1960). *Rev. mod. Phys.* **32,** 328.
Löwdin, P.-O. (1961). *J. chem. Phys.* **35,** 78.
Löwdin, P.-O. (1962*a*). *J. math. Phys.* **3,** 969.
Löwdin, P.-O. (1962*b*). *J. math. Phys.* **3,** 1171.
Löwdin, P.-O. (1962*c*). *Rev. mod. Phys.* **34,** 520.
Löwdin, P.-O. (1964). *Rev. mod. Phys.* **36,** 966.
Löwdin, P.-O. (1965). *J. chem. Phys.* **43,** S1, S175.
Löwdin, P.-O. (1970). *Adv. Quantum Chem.* **5,** 185.

Ma, S. and Brueckner, K. A. (1968). *Phys. Rev.* **165,** 18.
Manne, R. (1977). *Int. J. Quantum Chem. Symp.* **11,** 195.
MacDonald, J. K. L. (1933). *Phys. Rev.* **43,** 830.
McLean, A. D. and Yoshimine, M. (1967). Tables of linear molecule wave functions. *Suppl. IBM J. Res. Develop.* (1968) **12,** 206.
McWeeny, R. (1950). *Nature* **166,** 21.
McWeeny, R. (1954). *Proc. R. Soc., Lond.* **A223,** 63.
McWeeny, R. (1959). *Proc. R. Soc., Lond.* **A253,** 242.
McWeeny, R. (1960). *Rev. mod. Phys.* **32,** 335.
McWeeny, R. (1970). *Spins in chemistry*. Academic Press, New York.
McWeeny, R. (1974). *Mol. Phys.* **28,** 1273.
McWeeny, R. (1975). *Chem. Phys. Lett.* **35,** 13.
McWeeny, R. (1982). *The shape and structure of molecules*. Oxford University Press.
McWeeny, R. and Cooper, I. L. (1966). *J. chem. Phys.* **45,** 226.
McWeeny, R. and Mizuno, Y. (1961). *Proc. R. Soc., Lond.* **A259,** 554.
McWeeny, R. and Pickup, B. T. (1980). *Rep. Prog. Phys.* **43,** 1065.
McWeeny, R. and Ohno, K. (1960). *Proc. R. Soc., Lond.* **A255,** 367.
McWeeny, R. and Sutcliffe, B. T. (1963). *Proc. R. Soc., Lond.,* **A273,** 103.
McWeeny, R. and Sutcliffe, B. T. (1976). *Methods of molecular quantum mechanics* (2nd edn). Academic Press, New York.
March, N. H. (1957). *Adv. Phys.* **6,** 1.
March, N. H. (1975). *Self-consistent fields in atoms*. Pergamon, Oxford.
March, N. H. (1981). *Spec. Period. Rep.: Theor. Chem.* **4,** 92. The Royal Society of Chemistry, London.
March, N. H., Young, W. H., and Sampanthar, S. (1967). *The many-body problem in quantum mechanics*. Cambridge University Press.
Massey, H. S. W. (1976). *Negative ions*. Cambridge University Press.
Massey, H. S. W. (1979). *Adv. at. molec. Phys.* **15,** 2.
Matsen, F. A. (1964). *Adv. Quantum Chem.* **1,** 59.
Matsen, F. A. (1970). *J. Am. Chem. Soc.* **92,** 3525.
Matsen, F. A. (1974). *Int. J. Quantum Chem. Symp.* **8,** 379.
Matsen, F. A. (1976). *Int. J. Quantum Chem. Symp.* **10,** 525.
Matsen, F. A. (1978). *Adv. Quantum Chem.* **12,** 223.
Matsen, F. A. and Klein, D. J. (1971). *J. phys. Chem.* **75,** 1860.
Mattuck, R. D. (1976). *A guide to Feynman diagrams* (2nd edn). McGraw-Hill, New York.
Messiah, A. (1967). *Quantum mechanics* (Vols 1 and 2). North Holland Pub. Co., Amsterdam.
Mehler, E. L., Ruedenberg, K., and Silver, D. M. (1970). *J. chem. Phys.* **52,** 1181.
Meunier, A., Levy, B., and Berthier, G. (1976). *Int. J. Quantum Chem.* **10,** 1061.
Meyer, W. (1971). *Int. J. Quantum Chem. Symp.* **5,** 341.
Meyer, W. (1973). *J. chem. Phys.* **58,** 1017.

Meyer, W. (1974). *Theor. chim. Acta* **35,** 227.
Meyer, W. (1976). *J. chem. Phys.* **64,** 2901.
Meyer, W. (1977). In *Methods of electronic structure theory. Mod. theor. Chem.* **3,** 413.
Meyer, W. and Rosmus, P. (1975). *J. chem. Phys.* **63,** 2356.
Mezey, P. (1979). *Theor. chim. Acta* **53,** 187.
Micha, D. (1970). *Phys. Rev.* **A1,** 755.
Miller, J. H. and Kelly, H. P. (1971). *Phys. Rev. Lett.* **26,** 679.
Miller, K. and Ruedenberg, K. (1968*a*). *J. chem. Phys.* **48,** 3414.
Miller, K. and Ruedenberg, K. (1968*b*). *J. chem. Phys.* **48,** 3464.
Mills, I. M. (1974). *Spec. Period. Rep.: Theor. Chem.* **1,** 110. The Chemical Society, London.
Moeller, Chr. and Plesset, M. S. (1934). *Phys. Rev.* **46,** 618.
Moffitt, W. E. (1954). *Rep. Prog. Phys.* **17,** 173.
Morita, T. (1963). *Prog. Theor. Phys.* **29,** 351.
Morrison, R. C. and Gallup, G. A. (1969). *J. chem. Phys.* **50,** 1214.
Morrison, H. (ed.) (1968). *The quantum theory of many-particle systems.* Gordon and Breach, New York.
Moshinsky, M. (1966). In *Many-body problems and other selected topics in theoretical physics* (ed. M. Moshinsky, T. Brody, and G. Jacobs). Gordon and Breach, New York.
Moshinsky, M. (1968). *Group theory and the many-body problem.* Gordon and Breach, New York.
Moshinsky, M. and Seligman, T. H. (1971). *Ann. Phys., N.Y.* **66,** 311.
Moss, R. E. (1973). *Advanced molecular quantum mechanics.* Chapman and Hall, London.
Muftakhova, F. A. (1976). *Int. J. Quant. Chem.* **10,** 1077.
Mulder, J. J. C. (1966). *Mol. Phys.* **10,** 479.
Mulliken, R. S. (1928). *Phys. Rev.* **32,** 186, 761.
Mulliken, R. S. (1932). *Phys. Rev.* **41,** 49.
Mulliken, R. S. and Ermler, W. C. (1977). *Diatomic molecules: results of ab initio calculations.* Academic Press, New York.
Musher, J. I. and Schulman, J. M. (1968). *Phys. Rev.* **173,** 93.

Nesbet, R. K. (1955). *Proc. R. Soc., Lond.* **A230,** 312, 322.
Nesbet, R. K. (1965). *Adv. chem. Phys.* **9,** 321.
Nesbet, R. K. (1969). *Adv. chem. Phys.* **14,** 1.
Newman, D. J. (1969). *J. phys. Chem. Solids* **30,** 1709.
Newman, D. J. (1970). *J. phys. Chem. Solids* **31,** 1143.
Norbeck, J. M. and McWeeny, R. (1975). *Chem. Phys. Lett.* **34,** 206.

Oberlechner, G., Owono–N'–Guema, F., and Richert, J. (1970). *Nuovo Cim.* **B68,** 23.
Offerman, R. (1976). *Nucl. Phys.* **A223,** 368.
Offerman, R., Eg, W., and Kummel, H. (1976). *Nucl. Phys.* **A273,** 349.
Ostlund, N. S. and Merrifield, D. L. (1976). *Chem. Phys. Lett.* **39,** 612.

Padé, H. (1892). *Ann. sci. Ecole norm. sup. Paris (Suppl.)* **9,** 3.
Paldus, J. (1974). *J. chem. Phys.* **61,** 5321.
Paldus, J. (1975). *Int. J. Quantum Chem. Symp.* **9,** 165.
Paldus, J. (1976*a*). *Phys. Rev.* **A14,** 1620.
Paldus, J. (1976*b*). *Theor. Chem.: Adv. Perspect.* **2,** 131.

Paldus, J. (1977). *J. chem. Phys.* **67,** 303.
Paldus, J. (1977). In *Electrons in finite and infinite structures* (ed. P. Phariseau and L. Scheire). Plenum, New York.
Paldus, J. (1979). In *Group theoretical methods in physics, Proc. Seventh Int. Colloq. Integr. Conf. Group Theory Math. Phys.* (ed. W. Beiglboeck, A. Bohm, and E. Takasugi). Springer, New York.
Paldus, J. (1980). In *Electron correlation,* Proc. Daresbury study weekend (ed. M. F. Guest and S. Wilson). Science Research Council, London.
Paldus, J. (1981). In *The unitary group for the evaluation of electronic energy matrix elements* (ed. J. Hinze) *Lecture Notes in Chemistry* **22,** 1. Springer, Berlin.
Paldus, J., Adams, B. G., and Čížek, J. (1977). *Int. J. Quantum Chem.* **11,** 813.
Paldus, J. and Boyle, M. J. (1980*a*). *Phys. Scripta* **21,** 295.
Paldus, J. and Boyle, M. J. (1980*b*). *Phys. Rev.* **A22,** 2299.
Paldus, J. and Čížek, J. (1973). In *Energy, structure and reactivity,* Proc. 1972 Boulder Summer Res. Conf. Theor. Chem. (ed. D. W. Smith and W. B. McRae). Wiley, New York.
Paldus, J. and Čížek, J. (1975). *Adv. Quantum Chem.* **9,** 105.
Paldus, J., Čížek, J., and Shavitt, I. (1972). *Phys. Rev.* **A5,** 50.
Paldus, J. and Wong, H. C. (1973). *Comput. Phys. Commun.* **6,** 1.
Paldus, J. and Wormer, P. E. S. (1978). *Phys. Rev.* **A18,** 827.
Paldus, J. and Wormer, P. E. S. (1979). *Int. J. Quantum Chem.* **16,** 1321.
Palke, W. E. and Goddard, W. A. (1969). *J. chem. Phys.* **50,** 4524.
Parks, J. M. and Parr, R. G. (1958). *J. chem. Phys.* **28,** 335.
Parks, J. M. and Parr, R. G. (1960). *J. chem. Phys.* **32,** 1657.
Parr, R. G. (1963). *The quantum theory of molecular electronic structure.* Benjamin, New York.
Parr, R. G., Ellison, F. O., and Lykos, P. (1956). *J. chem. Phys.* **24,** 1106.
Patterson, C. W. and Harter, W. G. (1977). *Phys. Rev.* **A15,** 2372.
Pauncz, R. (1979). *Spin eigenfunctions: construction and use.* Plenum, New York.
Pendergast, P. and Hayes, E. F. (1978). *J. comp. Phys.* **26,** 236.
Pitzer, K. (1979). *Acc. chem. Res.* **12,** 271.
Polezzo, S. (1975). *Theor. chim Acta* **38,** 211.
Pople, J. A. (1973). In *Energy, structure and reactivity,* Proc. 1972 Boulder Res. Conf. Theor. Chem. Wiley, New York.
Pople, J. A., Binkley, J. S., and Seeger, R. (1976). *Int. J. Quantum Chem. Symp.* **10,** 1.
Pople, J. A., Krishnan, R., Schlegel, H. B., and Binkley, J. S. (1978). *Int. J. Quantum Chem.* **14,** 545.
Pople, J. A., Seeger, R., and Krishnan, R. (1977). *Int. J. Quantum Chem. Symp.* **11,** 149.
Price, S. L. and Stone, A. J. S. (1979). *Chem. Phys. Lett.* **65,** 127.
Primas, H. (1961). *Helv. phys. Acta* **34,** 1961.
Primas, H. (1963). *Rev. mod. Phys.* **35,** 710.
Primas, H. (1965). In *Modern quantum chemistry* (ed. O. Sinanoğlu) Vol. 2. Academic Press, New York.
Prime, S. and Robb, M. A. (1976). *Theor. chim. Acta* **42,** 181.
Prime, S. and Robb, M. A. (1977). *Chem. Phys. Lett.* **47,** 527.
Purvis, G. D. and Bartlett, R. J. (1978). *J. chem. Phys.* **68,** 2114.
Pu, R. T. and Chang, E. S. (1966). *Phys. Rev.* **151,** 31.
Pyper, N. C. and Gerratt, J. (1977). *Proc. R. Soc., Lond.* **A355,** 407.
Pyykkö, P. (1978). *Adv. Quantum Chem.* **11,** 353.

Raffenetti, R. C. (1973*a*). *J. chem. Phys.* **58,** 4452.
Raffenetti, R. C. (1973*b*). *J. chem. Phys.* **59,** 5936.
Raffenetti, R. C. (1975). *Int. J. Quantum Chem. Symp.* **9,** 289.
Raffenetti, R. C. and Ruedenberg, K. (1973). *J. chem. Phys.* **59,** 5978.
Raffenetti, R. C., Hsu, K., and Shavitt, I. (1977). *Theor. chim. Acta* **45,** 33.
Raimondi, M., Tantardini, G. F., and Simonetta, M. (1975). *Mol. Phys.* **30,** 703, 797.
Raimondi, M., Campion, W., and Karplus, M. (1977). *Mol. Phys.* **34,** 1483.
Redmon, L. T., Purvis, G. D., and Bartlett, R. J. (1978). *J. chem. Phys.* **69,** 5386.
Redmon, L. T., Purvis, G. D., and Bartlett, R. J. (1980). *J. chem. Phys.* **72,** 986.
Reeves, C. M. (1963). *J. chem. Phys.* **39,** 1.
Reeves, C. M. and Harrison, J. (1963). *J. chem. Phys.* **39,** 11.
Rettrup, S. (1977). *Chem. Phys. Lett.* **47,** 59.
Rettrup, S. and Sharma, C. (1977). *Theor. chim. Acta* **46,** 63, 73.
Richards, W. G. (1967). *Trans. Faraday Soc.* **63,** 257.
Richards, W. G. (1983). *Quantum pharmacology* (2nd edn). Butterworths, London.
Richards, W. G., Trivedi, H., and Cooper, D. L. (1981). *Spin–orbit coupling in molecules.* Clarendon Press, Oxford.
Richert, J., Schucan, T. H., Scrubel, M. H., and Weidenmuller, H. A. (1976). *Ann. Phys., N.Y.* **96,** 139.
Riley, M. and Dalgarno, A. (1971). *Chem. Phys. Lett.* **9,** 382.
Robb, M. A. (1973). *Chem. Phys. Lett.* **20,** 274.
Robb, M. A. (1974). In *Computational techniques in quantum chemistry and molecular physics* (ed. G. H. F. Diercksen, B. T. Sutcliffe, and A. Veillard). D. Reidel.
Robb, M. A. (1981). In *The unitary group for the evaluation of electronic matrix elements* (ed. J. Hinze). *Lecture Notes in Chemistry* **22**. Springer, Berlin.
Robb, M. A. and Hegarty, D. (1978). In *Correlated wavefunctions.* Proc. Daresbury study weekend (ed. V. R. Saunders). Science Research Council, London.
Robb, M. A. and Wilson, S. (1980). *Mol. Phys.* **40,** 1333.
Robbe, J. M. and Schamps, J. (1976). *J. chem. Phys.* **65,** 5420.
Robinson, G. de B. (1961). *Representation theory of the symmetric group.* University of Toronto Press.
Roman, P. (1965). *Advanced quantum theory.* Addison-Wesley, Reading, Mass., USA.
Roos, B. (1972). *Chem. Phys. Lett.* **15,** 153.
Roos, B. O. (1983). In *Methods in computational molecular physics* (ed. G. H. F. Diercksen and S. Wilson). D. Reidel.
Roos, B. O. and Siegbahn, P. E. M. (1977). In *Methods of electronic structure theory: Mod. theor. Chem.* **3**. (ed. H. F. Schafer III). Plenum, New York.
Roos, B. O., Taylor, P. R., and Siegbahn, P. E. M. (1980). *Chem. Phys.* **48,** 157.
Roothaan, C. C. J. (1951). *Rev. mod. Phys.* **23,** 69.
Roothaan, C. C. J. (1960). *Rev. mod. Phys.* **32,** 179.
Roothaan, C. C. J. and Bagus, P. S. (1963). *Methods Comput. Phys.* **2,** 47.
Roothaan, C. C. J., Detrich, J., and Hopper, D. G. (1979). *Int. J. Quantum Chem. Symp.* **13,** 93.
Rosenberg, B. J. and Shavitt, I. (1975). *J. chem. Phys.* **63,** 2163.
Rosensteel, G., Ihrig, E., and Trainor, L. E. H. (1975). *Proc. R. Soc., Lond.* **A344,** 387.
Rosmus, P. and Meyer, W. (1977). *J. chem. Phys.* **66,** 13.
Rosmus, P. and Meyer, W. (1978). *J. chem. Phys.* **69,** 2745.

Rossky, P. and Karplus, M. (1976). *J. chem. Phys.* **67,** 5419.
Rossky, P. and Karplus, M. (1977). *J. chem. Phys.* **67,** 5419.
Ruedenberg, K. (1971). *Phys. Rev. Lett.* **27,** 1105.
Ruedenberg, K., Cheung, L. M., and Elbert, S. T. (1979). *Int. J. Quantum Chem.* **16,** 1069.
Ruedenberg, K. and Poshuta, R. D. (1972). *Adv. Quantum Chem.* **6,** 267.
Ruedenberg, K. and Sundbarg, K. R. (1976). In *Quantum science* (ed. J. L. Calais, O. Goscinski, J. Linderberg, and Y. Ohrn). Plenum, New York.
Ruedenberg, K., Raffenetti, R. C., and Bardo, R. (1973). In *Energy structure and reactivity*. Proc. 1972 Boulder Res. Conf. Theor. Chem. Wiley, New York.
Rumer, G. (1932). *Nachr. Ges. Wiss., Gottingen* p. 337.
Rutherford, D. E. (1948). *Substitutional analysis.* Edinburgh University Press.
Ruttink, P. J. A. and van Lenthe, J. H. (1977). *Theor. chim. Acta* **44,** 97.

Sadlej, A. J. (1977*a*). *Mol. Phys.* **34,** 731.
Sadlej, A. J. (1977*b*). *Chem. Phys. Lett.* **47,** 50.
Sadlej, A. J. (1978). *Theor. chim. Acta* **47,** 205.
Sadlej, A. J. (1979). *J. phys. Chem.* **83,** 1653.
Sadlej, A. J. and Wilson, S. (1981). *Mol. Phys.* **44,** 299.
Salmon, W. I. (1974). *Adv. Quantum Chem.* **8,** 37.
Salmon, W. I. and Ruedenberg, K. (1972*a*). *J. chem. Phys.* **57,** 2776.
Salmon, W. I. and Ruedenberg, K. (1972*b*). *J. chem. Phys.* **57,** 2792.
Salmon, W. I., Ruedenberg, K., and Cheung, L. M. (1972). *J. chem. Phys.* **57,** 2787.
Salzman, W. (1968). *J. chem. Phys.* **49,** 3035.
Sandeman, I. (1940). *Proc. R. Soc., Edinb.* **60,** 210.
Sanders, P. G. H. (1969). *Adv. chem. Phys.* **14,** 365.
Sanders, P. G. H. (1971). In *Atomic physics and astrophysics* (ed. M. Chrestien and E. Lipworth). Gordon and Breach, London.
Sarma, C. R. and Dinesha, K. V. (1978). *J. math. Phys.* **19,** 1662.
Sarma, C. R. and Rettrup, S. (1977). *Theor. chim. Acta* **46,** 63.
Sasaki, F. (1974). *Int. J. Quantum Chem.* **8,** 605.
Sasaki, F. (1977). *Int. J. Quantum Chem. Symp.* **11,** 125.
Saunders, V. R. (1983). In *Methods in computational molecular physics* (ed. G. H. F. Diercksen and S. Wilson). D. Reidel.
Saunders, V. R. and Guest, M. F. (1974). *Quantum chemistry—the state of the art* (ed. V. R. Saunders and J. Brown). Science Research Council, London.
Saxe, P., Schaefer, H. F., and Handy, N. C. (1981). *Chem. Phys. Lett.* **79,** 202.
Schaefer, H. F. (1972). *The electronic structure of atoms and molecules.* Addison-Wesley, Reading, Mass., USA.
Schmidt, M. W. and Ruedenberg, K. (1979). *J. chem. Phys.* **71,** 3951.
Schucan, T. H. and Weidenmuller, H. A. (1972). *Ann. Phys., N.Y.* **73,** 108.
Schucan, T. H. and Weidenmuller, H. A. (1973). *Ann. Phys., N.Y.* **76,** 483.
Schulman, J. M. and Kaufman, D. N. (1970). *J. chem. Phys.* **53,** 477.
Schulman, J. M. and Kaufman, D. N. (1972). *J. chem. Phys.* **57,** 2328.
Schwartz, C. M. (1962). *Phys. Rev.* **126,** 1015.
Schwartz, C. M. (1963). *Methods Comput. Phys.* **2,** 241.
Schwartz, J. T. (1980). *ACM Trans. Prog. Lang. Syst.* **2,** 484.
Schweber, S. S. (1961). *An introduction to relativistic quantum field theory.* Row-Peterson, Evanston, Illinois.
Schwinger, J. (1965). In *Quantum theory of angular momentum* (ed. L. C. Biedenharn and H. Van Dam). Academic Press, New York.

Seeger, R. and Pople, J. A. (1977). *J. chem. Phys.* **66,** 3045.
Seeger, R., Krishnan, R., and Pople, J. A. (1978). *J. chem. Phys.* **68,** 2519.
Serber, R. (1934*a*). *Phys. Rev.* **45,** 461.
Serber, R. (1934*b*). *J. chem. Phys.* **2,** 697.
Shavitt, I. (1977*a*). *Int. J. Quantum Chem. Symp.* **11,** 131.
Shavitt, I. (1977*b*). In *Methods of electronic structure theory. Mod. Theor. Chem.* **3** (ed. H. F. Schaefer III). Plenum, New York.
Shavitt, I. (1978). *Int. J. Quantum Chem. Symp.* **12,** 5.
Shavitt, I. (1979*a*). *Chem. Phys. Lett.* **63,** 421.
Shavitt, I. (1979*b*). *New methods in computational quantum chemistry and their application in modern supercomputers.* Battelle Columbus Laboratories.
Shavitt, I. (1980). In *Electron correlation.* Proc. Daresbury study weekend (ed. M. F. Guest and S. Wilson). Science Research Council, London.
Shavitt, I. (1981). In *The unitary group for the evaluation of electronic energy matrix elements* (ed. J. Hinze). *Lecture Notes in Chemistry* **22**. Springer, Berlin.
Shull, H. (1959). *J. chem. Phys.* **30,** 1405.
Shull, H. and Ebbing, D. D. (1957). *J. chem. Phys.* **28,** 812.
Shull, H. and Löwdin, P.-O. (1959). *J. chem. Phys.* **30,** 617.
Sidgwick, N. V. and Powell, H. M. (1940). *Proc. R. Soc., Lond.* **A176,** 153.
Siegbahn, P. E. M. (1978). *Chem. Phys. Lett.* **55,** 386.
Siegbahn, P. E. M. (1979). *J. chem. Phys.* **70,** 5391.
Siegbahn, P. E. M. (1980). *J. chem. Phys.* **72,** 1647.
Siegbahn, P. E. M. (1983). In *Methods in computational molecular physics* (ed. G. H. F. Diercksen and S. Wilson). D. Reidel.
Silver, B. L. (1976). *Irreducible tensor methods. An introduction for chemists.* Academic Press, New York.
Silver, D. M. (1969). *J. chem. Phys.* **50,** 5108.
Silver, D. M. (1971). *J. chem. Phys.* **55,** 1461.
Silver, D. M. (1978*a*). *Comput. Phys. Commun.* **14,** 71.
Silver, D. M. (1978*b*). *Comput. Phys. Commun.* **14,** 81.
Silver, D. M. (1980). *Phys. Rev.* **A21,** 1106.
Silver, D. M. and Bartlett, R. J. (1976). *Phys. Rev.* **A13,** 1.
Silver, D. M., Mehler, E. L., and Ruedenberg, K. (1970). *J. chem. Phys.* **52,** 1174, 1206.
Silver, D. M. and Nieuwpoort, W. C. (1978). *Chem. Phys. Lett.* **57,** 421.
Silver, D. M. and Wilson, S. (1977). *J. chem. Phys.* **67,** 5552.
Silver, D. M. and Wilson, S. (1978). *J. chem. Phys.* **69,** 3787.
Silver, D. M. and Wilson, S. (1982). *Proc. R. Soc., Lond.* **A383,** 477.
Silver, D. M., Wilson, S., and Bartlett, R. J. (1977). *Phys. Rev.* **A16,** 477.
Silver, D. M., Wilson, S., and Bunge, C. F. (1979). *Phys. Rev.* **A19,** 1375.
Silver, D. M., Wilson, S., and Nieuwpoort, W. C. (1978). *Int. J. Quantum Chem.* **14,** 635.
Silverstone, H. J., Carroll, D. P., and Silver, D. M. (1978). *J. chem. Phys.* **68,** 616.
Simons, G., Parr, R. J., and Finlan, J. M. (1973). *J. chem. Phys.* **59,** 3229.
Simons, J. P. (1977). *Ann. Rev. phys. Chem.* **28,** 15.
Sims, J. and Hagstrom, S. (1971*a*). *J. chem. Phys.* **55,** 4699.
Sims, J. and Hagstrom, S. (1971*b*). *J. chem. Phys.* **4,** 908.
Sinanoğlu, O. (1961*a*). *Phys. Rev.* **122,** 493.
Sinanoğlu, O. (1961*b*). *Proc. R. Soc., Lond.* **A260,** 379.
Sinanoğlu, O. (1962). *J. chem. Phys.* **36,** 706.
Sinanoğlu, O. (1964). *Adv. chem. Phys.* **6,** 315.
Sinanoğlu, O. and Skutnik, B. (1968). *Chem. Phys. Lett.* **1,** 699.

Sinanoğlu, O. and Brueckner, K. A. (1970). *Three approaches to electron correlation in atoms*. Yale University Press, New Haven, USA.
Siu, A. K. Q. and Davidson, E. R. (1970). *Int. J. Quant. Chem.* **4,** 223.
Slater, J. C. (1929). *Phys. Rev.* **34,** 1293.
Slater, J. C. (1930). *Phys. Rev.* **35,** 210.
Slater, J. C. (1931). *Phys. Rev.* **38,** 1109.
Slater, J. C. (1951*a*). *J. chem. Phys.* **19,** 220.
Slater, J. C. (1951*b*). *Quantum theory of matter*. McGraw-Hill, New York.
Slater, J. C. (1953). *Rev. mod. Phys.* **25,** 199.
Slater, J. C. (1960). *Quantum theory of atomic structure* (Vols 1 and 2). McGraw-Hill, New York.
Slater, J. C. (1965). *Quantum theory of molecules and solids*. McGraw-Hill, New York.
Steinborn, O. (1983). In *Methods in computational molecular physics* (ed. G. H. F. Diercksen and S. Wilson). D. Reidel.
Stern, P. and Kaldor, U. (1976). *J. chem. Phys.* **64,** 2002.
Stevenson, A. F. and Crawford, M. F. (1938). *Phys. Rev.* **54,** 314, 375.
Sutcliffe, B. T. (1966). *J. chem. Phys.* **45,** 235.
Sutcliffe, B. T. (1980). In *Electron correlation*. Proc. Daresbury study weekend (ed. M. F. Guest and S. Wilson). Science Research Council, London.
Sutcliffe, B. T. (1980). In *Quantum dynamics of molecules* (ed. R. G. Woolley). Plenum, New York.
Sutcliffe, B. T. (1983). In *Methods of computational molecular physics* (ed. G. H. F. Diercksen and S. Wilson). D. Reidel.

Taylor, B. N., Parker, W. H., and Langenberg, D. N. (1969). *Rev. mod. Phys.* **41,** 375.
Taylor, P. R. (1981). *J. chem. Phys.* **74,** 1256.
Taylor, P. R., Bacskay, G. B., Hush, N. S., and Hurley, A. C. (1976). *Chem. Phys. Lett.* **41,** 444.
Taylor, P. R., Bacskay, G. B., Hush, N. S., and Hurley, A. C. (1978). *J. chem. Phys.* **69,** 1971, 4607.
Thouless, D. J. (1961). *The quantum mechanics of many-body systems*. Academic Press, New York.
Tolmachev, V. V. (1969). *Adv. chem. Phys.* **14,** 421, 471.
Tuan, D. T. (1970). *Chem. Phys.* **7,** 115.

Uhlenbeck, G. E. and Goudsmit, S. (1925). *Naturwiss.* **13,** 953.
Uhlenbeck, G. E. and Goudsmit, S. (1929). *Nature* **117,** 264.
Urban, M. and Hobza, P. (1975). *Theor. chim. Acta* **36,** 215.
Urban, M., Hubač, I., Kellö, V., and Noga, J. (1980). *J. chem. Phys.* **72,** 3378.
Urban, M., Kellö, V., and Hubač, I. (1977). *Chem. Phys. Lett.* **51,** 170.
Urban, M., Kellö, V., and Čársky, P. (1977). *Theor. chim. Acta* **45,** 205.
Urban, M. and Kellö, V. (1979). *Mol. Phys.* **38,** 1621.
Urban, M. and Lavicky, T. (1972). *Chem. Phys. Lett.* **16,** 563.

van Duijneveldt, F. (1971). *IBM Tech. Res. Rept.* RJ948.
Van Lenthe, J. H. and Balint-Kurti, G. G. (1980). *Chem. Phys. Lett.* **76,** 138.
Van Vleck, J. H. and Sherman, A. (1935). *Rev. mod. Phys.* **7,** 167.
Veillard, A. (1966). *Theor. chim. Acta* **4,** 22.
Veillard, A. and Clementi, E. (1968). *J. chem. Phys.* **49,** 2415.

Wahl, A. C. and Das, G. (1970). *Adv. Quantum Chem.* **5,** 261.
Wahl, A. C. and Das, G. (1977). In *Methods of electronic structure theory* (ed. H. F. Schaefer). Plenum, New York.
Walch, S., Dunning, T. H., Raffenetti, R. C., and Bobrowitz, F. (1980). *J. chem. Phys.* **72,** 406.
Weidenmuller, H. A. (1975). In *Effective interactions and operators in nuclei* (ed. B. R. Barrett). Springer, Berlin.
Weinbaum, S. (1933). *J. chem. Phys.* **1,** 593.
Weinhold, F. (1972). *Adv. Quantum Chem.* **6,** 299.
Weinstein, D. H. (1934). *Proc. natn. Acad. Sci. US* **20,** 529.
Werner, H. J. and Meyer, W. (1980). *J. chem. Phys.* **73,** 2342.
Weyl, H. (1928). *Gruppentheorie und Quantenmechanik.* Hirzel, Leipzig.
Weyl, H. (1936). *Classical groups.* Princeton University Press.
Wehl, H. (1964). *The theory of groups and quantum mechanics.* Dover, New York.
Whiffen, D. H. (1978). *Pure appl. Chem.* **50,** 75.
Wick, G. C. (1950). *Phys. Rev.* **80,** 268.
Wigner, E. P. (1931). *Gruppentheorie und ihre Anwendung auf die Quantenmechanik der Atomspektren.* Vieweg, Braunschweig.
Wigner, E. P. (1935). *Math. naturw. Anz. ungar. Akad. Wiss.* **53,** 475.
Wigner, E. P. (1959). *Group theory and its application to the quantum mechanics of atomic spectra* (transl. J. J. Griffin). Academic Press, New York.
Wigner, E. P. and Seitz, F. (1933). *Phys. Rev.* **43,** 804.
Wilson, S. (1976). *J. chem. Phys.* **64,** 1692.
Wilson, S. (1977*a*). *Int. J. Quantum Chem.* **12,** 604.
Wilson, S. (1977*b*). *J. chem. Phys.* **67,** 4491.
Wilson, S. (1977*c*). *J. chem. Phys.* **67,** 5088.
Wilson, S. (1977*d*). *Chem. Phys. Lett.* **49,** 168.
Wilson, S. (1978*a*). *Mol. Phys.* **35,** 1, 1381.
Wilson, S. (1978*b*). *Comput. Phys. Commun.* **14,** 91.
Wilson, S. (1978*c*). In *Correlated wavefunctions.* Proc. Daresbury study weekend (ed. V. R. Saunders). Science Research Council, London.
Wilson, S. (1979*a*). *J. Phys. B: At. mol. Phys.* **12,** L657,; *ibid* **13,** 1505.
Wilson, S. (1979*b*). *Chem. Phys. Lett.* **66,** 255.
Wilson, S. (1979*c*). *J. Phys. B: At. mol. Phys.* **12,** 1623.
Wilson, S. (1979*d*). *J. Phys. B: At. mol. Phys.* **12,** L599.
Wilson, S. (1980*a*). In *Electron correlation.* Proc. Daresbury study weekend (ed. M. F. Guest and S. Wilson). Science Research Council, London.
Wilson, S. (1980*b*). *Mol. Phys.* **39,** 525.
Wilson, S. (1980*c*). *Int. J. Quantum Chem.* **18,** 905.
Wilson, S. (1980*d*). *Chem. Rev.* **80,** 263.
Wilson, S. (1980*e*). *Theor. chim. Acta* **57,** 53.
Wilson, S. (1980*f*). *Theor. chim. Acta* **58,** 31.
Wilson, S. (1981*a*). *Chem. Phys. Lett.* **81,** 467.
Wilson, S. (1981*b*). *Theor. chim. Acta* **59,** 71.
Wilson, S. (1981*c*). *Specialist periodical report: theoretical chemistry* **4,** 1. The Royal Society of Chemistry, London.
Wilson, S. (1981*d*). *J. Phys.* **B14,** L31.
Wilson, S. (1981*e*). *Proc. 5th Seminar on Computational Methods in Quantum Chemistry, Groningen.*
Wilson, S. (1982*a*). *J. Phys.* **B15,** L191.
Wilson, S. (1982*b*). *Theor. chim. Acta* **61,** 343.
Wilson, S. (1983). In *Methods in computational molecular physics* (ed. G. H. F. Diercksen and S. Wilson). Reidel, Dordrecht.

Wilson, S. and Gerratt, J. (1974). *Proc. SRC Atlas Symp.: Quantum chemistry, the state of the art* (ed. V. R. Saunders and J. Brown). Science Research Council, London.
Wilson, S. and Gerratt, J. (1975). *Mol. Phys.* **30,** 777.
Wilson, S. and Gerratt, J. (1979). *J. Phys. B: At. mol. Phys.* **12,** 339.
Wilson, S. and Guest, M. F. (1980). *Chem. Phys. Lett.* **73,** 607.
Wilson, S. and Guest, M. F. (1981). *J. Phys. B: At. mol. Phys.* **14,** 1709.
Wilson, S. and Sadlej, A. J. (1981). *Theor. chim. Acta* **60,** 19.
Wilson, S. and Saunders, V. R. (1979). *J. Phys. B: At. mol. Phys.* **12,** L403; *ibid* **13,** 1505.
Wilson, S. and Saunders, V. R. (1980). *Comput. Phys. Commun.* **19,** 293.
Wilson, S. and Silver, D. M. (1976). *Phys. Rev.* **A14,** 1949.
Wilson, S. and Silver, D. M. (1977*a*). *J. chem. Phys.* **67,** 1649.
Wilson, S. and Silver, D. M. (1977*b*). *J. chem. Phys.* **66,** 5400.
Wilson, S. and Silver, D. M. (1978*a*). *Proc. Fourth Semin. Comput. Problems Quantum Chem.* (ed. B. Roos and G. H. F. Diercksen). Max-Planck-Institut für Physik und Astrophysik, Munich.
Wilson, S. and Silver, D. M. (1978*b*). *Mol. Phys.* **36,** 1539.
Wilson, S. and Silver, D. M. (1979*a*). *Int. J. Quantum Chem.* **15,** 683.
Wilson, S. and Silver, D. M. (1979*b*). *Comput. Phys. Commun.* **17,** 47.
Wilson, S. and Silver, D. M. (1979*c*). *Chem. Phys. Lett.* **63,** 367.
Wilson, S. and Silver, D. M. (1979*d*). *Chem. Phys. Lett.* **63,** 367.
Wilson, S. and Silver, D. M. (1979*e*). *Theor. chim. Acta* **54,** 83.
Wilson, S. and Silver, D. M. (1980). *J. chem. Phys.* **72,** 2159.
Wilson, S. and Silver, D. M. (1982). *J. chem. Phys.* **77,** 3674.
Wilson, S., Jankowski, K., and Paldus, J. (1983). *Int. J. Quantum Chem.* **23,** 1781.
Wilson, S., Silver, D. M., and Bartlett, R. J. (1977). *Mol. Phys.* **33,** 1177.
Wilson, S., Silver, D. M., and Farrell, R. A. (1977). *Proc. R. Soc., Lond.* **A356,** 363.
Winter, N. W., Ermler, W., and Pitzer, R. M. (1973). *Chem. Phys. Lett.* **19,** 179.
Wong, H. C. and Paldus, J. (1973). *Comput. Phys. Commun.* **6,** 9.
Wormer, P. E. S. (1981). In *The unitary group for the evaluation of electronic energy matrix elements* (ed. J. Hinze). *Lecture Notes in Chemistry* **22**. Springer, Berlin.
Wormer, P. E. S. and Paldus, J. (1979). *Int. J. Quantum Chem.* **16,** 1307.
Wormer, P. E. S. and Paldus, J. (1980). *Int. J. Quantum Chem.* **18,** 841.

Yamanouchi, T. (1937). *Proc. phys.-math. Soc. Jap.* **19,** 436.
Yamanouchi, T. (1938). *Proc. phys.-math. Soc. Jap.* **20,** 245.
Yamanouchi, T. (1948). *J. phys. Soc. Jap.* **3,** 245.
Yeager, D. L. and Jorgenson, P. (1979). *J. chem. Phys.* **71,** 755.
Yoshimine, M. (1973). *J. comput. Phys.* **11,** 449.
Young, W. H. and March, N. H. (1958). *Phys. Rev.* **109,** 1854.
Yutsis, A. P. and Bandzaitis, A. A. (1977). *Theory of angular momentum in quantum mechanics* (2nd edn). Mokslas, Vilnius.
Yutsis, A. P., Levinson, I. B., and Vanagas, V. (1962). *Mathematical apparatus of the theory of angular momenta.* Israel Program for Scientific Translations, Jerusalem.

Ziman, J. (1969). *Elements of advanced quantum theory.* Cambridge University Press.

INDEX